The Latin American Studies Book Series

The Latin American Studies Book Series promotes quality scientific research focusing on Latin American countries. The series accepts disciplinary and interdisciplinary titles related to geographical, environmental, cultural, economic, political and urban research dedicated to Latin America. The series publishes comprehensive monographs, edited volumes and textbooks refereed by a region or country expert specialized in Latin American studies.
The series aims to raise the profile of Latin American studies, showcasing important works developed focusing on the region. It is aimed at researchers, students, and everyone interested in Latin American topics.
Submit a proposal: Proposals for the series will be considered by the Series Advisory Board. A book proposal form can be obtained from the Publisher, Juliana Pitanguy (juliana.pitanguy@springer.com).

Santiago López
Editors

Socio-Environmental Research in Latin America

Interdisciplinary Approaches Using GIS and Remote Sensing Frameworks

Springer

Editor
Santiago López
University of Washington Bothell
Bothell, WA, USA

ISSN 2366-3421 ISSN 2366-343X (electronic)
The Latin American Studies Book Series
ISBN 978-3-031-22679-3 ISBN 978-3-031-22680-9 (eBook)
https://doi.org/10.1007/978-3-031-22680-9

This Springer imprint is published by the registered company Springer Nature Switzerland AG
The registered company address is: Gewerbestrasse 11, 6330 Cham, Switzerland

Preface

Socio-environmental (SE) research is one of the most vibrant areas of inquiry and knowledge production ascribed to current scientific agendas across the globe. It is also well understood that SE problems, by definition, reach across a diversity of disciplines and require different modes of analysis that could enable not only digging deeper into the unique features of place, societies, and individuals but also identifying general spatio-temporal patterns and trends that allow studying human-environment interactions over time.

GIScience (the science behind geotechnologies like geographic information systems and remote sensing and the processes of generating geographic knowledge) has allowed bridging the divide between the conceptual work of scholars and the empirical work of those who attempt to apply geographical principles and theories to changing and ever-evolving SE conditions. Thus, GIScience needs and continues to engage with a variety of actors and research practices, both remote and participatory, while expanding and diversifying its epistemological frameworks to respond to legitimate citizens' concerns about knowledge generation, data ownership, and information access.

This book presents relevant examples of SE research that highlight the challenges and opportunities of using geotechnologies in interdisciplinary settings across the vast, culturally and environmentally mega-diverse region known as Latin America. This is not a complete compilation, and it does not try to be. Each chapter prods deeply into relevant SE issues identified by researchers from Latin America and elsewhere doing applied empirical work in the region. In highlighting depth over breadth, some thematic content, applications, and locales were sacrificed. However, the intention is that students, academics, professionals, policy makers, and general audiences will not only learn about Latin American SE issues and GIScience applications but, in doing so, will also develop an interest in geographic exploration either through geotechnologies and *in situ* field work or through a combination of approaches that will continue beyond the themes, locations, and methods presented in this book. Geography is indeed an open discipline and GIScience an inclusive interdisciplinary framework. We hope that this book will inspire not only those who have profound knowledge of the themes, technologies, methods, and geographic locations covered in this compilation but also those who are just starting their career paths and geographic exploration of Latin American landscapes.

Bothell, WA, USA Santiago López

Acknowledgments

I would like to thank all the authors and working groups who contributed their time and effort to make this volume a reality. This acknowledgment is particularly pertinent in the context of the current global crisis caused by the COVID-19 pandemic, which has affected a variety of students, researchers, and academics in different and profound ways, from disruptions to field work activities and academic productivity to changes in life priorities. I am deeply grateful to all of those who were able to contribute despite the challenging times, and for bearing with my questions and comments during the review and copyediting phases of the chapters. I would also like to take this opportunity to thank the editorial team at Springer, especially Zachary Romano for his encouragement to develop this book, the several times he met with me to go over the publication process, and for answering all my questions. I would also like to thank Ragavendar Mohan for all the editorial support, guidance, and prompt reply to all my emails. Finally, thank you to my wife, Becky. From reading early drafts of my chapters and reviewing additional ones to keeping the kids busy so I could edit and write, she was as important to this volume getting completed as I was.

Contents

About the Contributors

Aurelia Ahua is an Indigenous Technician from the *Nacionalidad Waorani del Ecuador* (NAWE) with extensive experience in mapping land use in the Ecuadorian Amazon using remotely sensed and geodetic data. Email: N/A.

Stephen Aldrich is a Professor of Geography in the Department of Earth and Environmental Systems at Indiana State University. Terre Haute, IN. He earned his PhD in Geography with a Specialization in Environmental Science and Public Policy at Michigan State University, an MA in Geography from Michigan State University, and a BA in Geography at Clark University. His research focuses on social drivers of land change (particularly in the Brazilian Amazon), how human activity causes changes to the environment more generally, and how 3D data capture and analysis methods can enhance our understanding of environmental and cultural phenomena. Email: steve.aldrich@indstate.edu

Fernando Alfaro is an Associate Professor at GEMA Center for Genomics, Ecology & Environment of the Universidad Mayor and Researcher at the Patagonia Interdisciplinary Research Station of the Pontificia Universidad Católica de Chile (UC) and the Observatory Homme-Milieux International Patagonia-Bahia Exploradores of the LabEx DRIIHM, INEE-CNRS. He holds a PhD in Ecology from *Universidad de Chile* (UC) and a BA in Biology from the *Universidad Mayor de San Simón* (Bolivia). His research interests include terrestrial ecosystem ecology, ecosystem processes, biogeochemistry, functional synchrony between contrasting environments, and ecological stoichiometry and Global Change. Email: fernando.alfaro@mayor.cl

Desiree Estilita Alvarado is the Lieutenant EP, and head of the Department of Geographical Information Systems at the *Instituto Geográfico Naciona*l (IGN) in Lima, Peru. She works on Peru's *Cartas Nacionales*. She holds a master's in environmental science from the UNSNM, and a bachelor's in Geographic Engineering from the *Universidad Nacional Federico Villareal*. She is also Professor of Technical

Higher Education at the *Escuela Nacional Geomática* and has participated in two GTASO workshops. Email: dalvaradoc @ign.gob.pe

Jorge Washington Vela Alvarado is a Principal Professor in the Agronomic Science Department at *Universidad Nacional de Ucayali* (UNU) in Pucallpa, Ucayali, Peru. He is a founding member of the Amazon Borderlands Research Center/*Centro de Investigación de Fronteras Amazónicas* (CIFA). He is a specialist in biodiversity and animal production and management, and a founding member of the Transboundary Geographic Group of the Southwestern Amazon (GTASO). He holds a PhD in Environment and Sustainable Development from the Inca Garcilaso de la Vega University, a master's degree in Animal Production from the *Universidad Nacional Agraria La Molina* (UNALM), and a bachelor's degree in Zootechnology from the *Universidad Agraria de la Selva* (UNAS). Email: jvelaunu@gmail.com

Cesar I. Alvarez Mendoza is an Assistant Professor at the Universidad Politécnica Salesiana in Quito, Ecuador. He obtained a PhD in Remote Sensing from the University of Porto, Portugal, an MA in Environmental Management and a BEng in Geographical Engineering from The Armed Forces University (ESPE) in Sangolquí, Ecuador. His research focuses on the use of satellite and aircraft remote sensing (including drone systems) on environmental and public health applications and geospatial data science.

Abraham Boyotai is an Indigenous Technician from the *Nacionalidad Waorani del Ecuador* (NAWE) with extensive experience in mapping indigenous land use and territories on the Ecuadorian Amazon using remotely sensed and geodetic data. Email: N/A

Vera Reis Brown is the former Executive Director of the *Secretaria de Estado do Meio Ambiente e das Políticas Indígenas do Acre* (SEMAPI) and former coordinator of the *Centro Integrado de Geoprocessamento e Monitoramento Ambiental* (CIGMA). She has experience in the management of natural resources, transboundary fluvial systems, environmental risk, and environmental reserves with additional expertise in ecology and environmental education. She holds a doctorate in environmental engineering from *Universidade de São Paulo/São Carlos*, a master's degree in environmental engineering from *Universidade de São Paulo/São Carlos*, and a bachelor's degree in Biology from *Universidade Santo Amaro*. Email: diretoriaexecutiva.sema@gmail.com

Oscar Calva is a Geographic Engineer from the Armed Forces University in Sangolquí, Ecuador at GeoIS, who has extensive experience in mapping land use in the Ecuadorian amazonía using remotely sensed, geodetic data and field surveys. Email: N/A

Piero Enmanuel Rengifo Cardenas is a GIS, Remote Sensing, and Biodiversity Conservation Specialist at the *Sociedad Peruana de Derecho Ambiental* (SPDA) in

Puerto Maldonado, Madre de Dios, Peru. He formerly fulfilled the same role at *Conservación Amazónica* (ACCA). He holds a master's degree in Ecology and Environmental Management at the *Universidad Nacional San Antonio Abad del Cusco* (UNSAAC) and a bachelor's degree in Forestry Engineering from the *Universidad Nacional de Ucayali* (UNU). Email: prengifo@conservacionam-azoica.org

Vitor Pessoa Colombo is a PhD candidate in the School of Architecture, Civil and Environmental Engineering, at École Polytechnique Fédérale de Lausanne (Switzerland). He holds a Master of Science in Architecture from Accademia di Architettura (Mendrisio, Switzerland), and a Bachelor of Science in Architecture from *École Polytechnique Fédérale de Lausanne*. His research interests include habitat and health issues related to rapid, largely "informal" urban growth in the Global South, reflecting on the role of geographic information as a tool for both spatial transformation and social development. Email: vitor.pessoacolombo@gmail.com

Sonaira Souza da Silva is Professor of Environmental Science at *Universidade Federal do Acre – Campus Floresta* (UFAC) in Cruzeiro do Sul, Acre, Brazil, and director of the *Laboratório de Geoprocessamento Aplicado ao Meio Ambiente* (LabGAMA). She holds a Doctor of Tropical Forestry Science from the *Instituto Nacional de Pesquisas da Amazonia* (INPA); a Master of Acroecology, AgroSustainability, and Agricultural Production from UFAC; and a Bachelor of Agronomic Engineering from UFAC. Email: sonaira.silva@ufac.br

José Frank de Melo Silva is a Geoprocessing Specialist at the *Comissão Pró-Índio do Acre* in Rio Branco, Acre, Brazil. He is currently working on a master's degree at the Universidade Federal do Rondonia. His work supports Indigenous rights and environmental sustainability in the Amazon. Email: frank@cpiacre.org.br

Antonio Willian Flores de Melo is a Professor at *Universidade Federal do Acre-Campus Floresta* (UFAC) in Cruzeiro do Sul, Acre, Brazil, and former executive director of the *Instituto de Mudanças Climáticas do Acre*. He holds a Doctor of Tropical Forest Science from the *Instituto Nacional de Pesquisas da Amazonia* (INPA), a Master of Ecology of Agroecosystems from the University of São Paulo (USP), and a BEng in Agronomic Engineering from UFAC. An expert in remote sensing, GIS, and soils, he is a lead facilitator and founding member of the Transboundary Geographic Group of the Southwestern Amazon (GTASO). Email: willian.flores@ufac.br

Enrique R. Flores is a Professor of Range Management and Principal Investigator at the Rangeland Ecology and Utilization Laboratory at *Universidad Nacional Agraria La Molina*. He has an MS and PhD in Range Science from Utah State University, USA. Professor Flores has published scientific articles and book chapters on topics related to rangeland ecology, grazing management, and adaptation and vulnerability of Puna herders to climate change. Email: efm@lamolina.edu.pe

Francisca Flores-Galaz is the coordinator of both the Patagonia Interdisciplinary Research Station of the *Pontificia Universidad Católica de Chile* (PUC) and the Observatory Homme-Milieux International Patagonia-Bahia Exploradores of the LabEx DRIIHM, INEE-CNRS. She holds a BA in Geography and a BA in Anthropology from UC. Her research interests are geographic-environmental interactions, landscape geography, socio-territorial impacts of tourism, and migration and interculturality. Email: ffflores@uc.cl

Matthew Fry is an Associate Professor of Geography at the University of North Texas. He holds a PhD in Geography from the University of Texas at Austin. His examines human-environment interactions with a focus on natural resources, energy, landscape change, and urbanization. Governance, property, and justice are central themes of his research in Latin America, Mexico, and Texas. Email: mfry@unt.edu

Didier Galop is the research director at CNRS, director of the Observatory Homme-Milieux Pyrénées Haut Vicdessos, and co-director of the Observatory Homme-Milieux International Patagonia-Bahia Exploradores of the Labex DRIIHM, INEE-CNRS. He also leads the GEODE UMR 5602 Laboratory at the *Université Toulouse Jean Jaurès*. He holds a PhD, a MA, and a BA in Geography from the same university. His research focuses on the study of the relationships between environment and society and long-term co-evolution processes. His research is based on an interdisciplinary retro-observation of socioecological processes that combines the multiproxy study of the paleobiological content of sedimentary archives with eco-historical approaches. Email: didier.galop@univ-tlse2.fr

David Orlando González Gamarra is a Principal Professor at the National University of San Antonio Abad del Cusco where he teaches and mentors in the Science, Ecology and Environmental Management Master's Program as postgraduate and undergraduate thesis advisor. He holds a doctorate degree in Education from UNMSM and a master's degree in economics from the Universidad Nacional San Agustín de Arequipa. Email: ogonzalezgamarra@gmail.com

Andrés Gerique Zipfel is a Postdoctoral Researcher at the Institute of Geography of the Friedrich-Alexander-University Erlangen-Nürnberg (FAU), Germany, and associate consultant at the Bullinger-Institute for Negotiation, Zurich. He holds a Habilitation and a PhD in Geography from the FAU and a MA in agricultural engineering, specialty of environmental protection and rural development, from the Justus-Liebig-University Giessen, Germany. His research interests include human-environment interactions, especially in protected areas, biodiversity conservation and climate change, land use and tourism, and ethnobotany with an emphasis on Latin America and Europe. Email: andres.gerique-zipfel@fau.de

Katharina Heider is a Research Associate at the Institute of Geography at the University of Hamburg, Germany. She holds a doctorate degree in Earth System Science as well as a Master and Bachelor of Science in Geography from the University of Hamburg, Germany. Her research interests are human-environment interactions, sustainability, smallholder agriculture, and climate change with a focus on GIS and mixed methods. Email: katharina.heider@uni-hamburg.de

Andrew Hilburn is an Associate Professor of Geography in the Department of Social Sciences at Texas A&M International University in Laredo, Texas, USA. He holds a PhD in Geography from the University of Kansas, a MS in Geography from the University of Southern Mississippi, and a BS in Geography from the University of South Alabama. He conducts research on a broad range of human-environmental issues in Mexico and the US-Mexico Borderlands using both field methods and geospatial techniques. Email: ahilburn@tamiu.edu

Miguel Alva Huayaney is an Associate Professor at the *Universidad Nacional Mayor de San Marcos* (UNMSM) in Lima, Peru, and responsible for the Geomatics Laboratory there. He holds a master's degree in Geographic Information Systems (GIS) from the Universidad Nacional Federico Villareal, a Master of Geography from UNMSM, and a Bachelor of Geography from UNMSM. Email: malvah@unmsm.edu.pe

Alina Álvarez Larrain is an Assistant Researcher in the National Council for Scientific and Technical Research (CONICET) in the Institute of Dating and Archaeometry (InDyA), Argentina. She was a Post-doctoral Fellow in P-GIS at the Center for Research in Environmental Geography (CIGA) from the National Autonomous University of Mexico (UNAM), and a PhD in Archaeology from University of Buenos Aires (UBA). Her research focuses on the archeology of the Southern Andes and participatory mapping for landscapes archaeology studies. Email: alarrain@indya.unju.edu.ar

Elaine Lopes is an Environmental Scientist working on a Doctorate in Forestry Engineering at the *Universidade Federal de Lavras*. Currently her work as a GIS Analyst for PROGEN focuses on the closure of mines and mining activities. Her GeoScientific data analysis improves decision making and mitigation of mining systems. She holds a master's degree in Environmental Science from the *Universidade Federal de Minas Gerais* (UFMG) and a bachelor's degree in Forestry Engineering from the *Universidade Federal do Acre* (UFAC). Email: lopes.elaine7@gmail.com

Juan Miguel Rodriguez-Lopez is a Postdoctoral Research Associate and Principal Investigator of the COVIDGI project at the Institute of Geography, University of

Hamburg, Germany. His research interests are human-environment interactions, computing social science, cities, and climate change with emphasis on quantitative methods. Email: miguel.rodriguez@uni-hamburg.de

Santiago López is an Associate Professor in the School of Interdisciplinary Arts and Sciences and director of the Environmental Education and Research Center at the University of Washington–Bothell, Bothell, WA. He holds a PhD in Geography and the Environment from University of Texas at Austin, a MA in Geography from Arizona State University, and a BEng in Geographical Engineering from the Armed Forces University (ESPE) in Ecuador. His research interests include human–environment dynamics, land use and land cover transformations, and climate change with an emphasis on GIScience applications. Email: cslopez@uw.edu

Adolfo Maldonado has a PhD in Collective Health, Environment, and Society, from the Simón Bolívar Andean University of Ecuador, a master's degree in Agent Strategies and Development Cooperation Policies from the University of the Basque Country, a diploma in tropical medicine from the University of Barcelona, and a medical degree from the University of Granada. He is co-founder of the non-profit organization *Clínica Ambiental*. His interests include public health in areas impacted by the oil industry. He also provides support to social organizations with the conviction that public health-disease processes depend on the quality of the soil that is stepped on. Email: salud@accionecologica.org

Javier A. Ñaupari is a Professor of Rangeland Ecology and Management, and chair of the Animal Production Department at La Molina University, Lima. Dr. Ñaupari, a Fulbright alumnus, received his PhD in Natural Resources from the University of Idaho and MS degree in Range Animal Production at La Molina University. His research area is remote sensing and geographic information systems applied to ecological processes in mountain range ecosystems. Email: jnaupariv@lamolina.edu.pe

Gilberto Nenkimo is the President of the *Nacionalidad Waorani del Ecuador* (NAWE). Son of Waorani leaders, with extensive knowledge of the Waorani way of life and the governance context and rules of use of their territory. Email: N/A

Jason Nesbitt is an Associate Professor in the Department of Anthropology at Tulane University, USA. He holds a PhD in Anthropology from Yale University, an MA in Anthropology from Trent University, and a BA in Archaeology from Simon Fraser University in Canada. He specializes in the archaeology of the Central Andes, with a focus on early complex societies of Peru and human responses to climate changes. Email: jnesbitt@tulane.edu

Maria Luiza Pinedo Ochoa is an *Indigenista* and an expert within the Public Policy and Regional Action Program at the *Comissão Pró-Índio do Acre* in Rio Branco, Acre, Brazil. Her studies were conducted at Estácio de Sá and UniNorte.

His work supports Indigenous rights and environmental sustainability in the Amazon. Email: malu@cpiacre.org.br

Jorge Olea-Peñaloza is an Assistant Professor in the Department of Environmental Sciences at the Universidad Católica of Temuco (UCT) and researcher at the Patagonia Interdisciplinary Research Station of the *Pontificia Universidad Católica de Chile* (PUC) and the Observatory Homme-Milieux International Patagonia-Bahia Exploradores of the LabEx DRIIHM, INEE-CNRS. He holds a PhD in Geography from the UC and a MA and BA in History from the *Universidad de Chile* (UC). His main research interests are the society-nature system through the lens of rural and historical geography, along with environmental history. Email: jolea@uct.cl

Bertha Balbín Ordaya deceased, is the former Geography Commission Representative for the Peruvian National Section of the PanAmerican Institute of Geography and History (PAIGH). She is also the former Professor of Geography and Chair of the *Facultad de Geografía* at the *Universidad Nacional Mayor de San Marcos* (UNMSM). She held a master's degree in Geography from the *Universidad de Chile* (UC), and bachelor's degree in Geography from UNMSM. A commemorative photograph of her is displayed in the museum at Peru's Instituto Geográfico Nacional (IGN), where she is the first woman to be so honored. Email: N/A

Julio C. Postigo is an Assistant Professor in the Department of Geography at Indiana University. He holds a PhD in Geography and the Environment and a MA in Latin American Studies from The University of Texas at Austin. His current research investigates the synergistic impacts of climate change and economic development on pastoral social-ecological systems. His work also examines the generation of local knowledge among small farmers, the building of institutional resilience to global change. Email: jpostigo@iu.edu

Claire Victoria Powell is a Senior Marketing Associate at Homethrive in Baltimore, MD, USA, and a former student researcher at the University of Richmond where she was a member of the Amazon Mapping Team and obtained a bachelor's degree in Environmental Studies and Latin American Latino and Iberian Studies (LALIS). Email: claire.powell@richmond.edu

Ana Puyol is the former Executive Director of EcoCiencia, is a biologist with a master's degree on Environmental Education and Sustainable Development – Cátedra Unesco, Madrid, Spain, with a long trajectory in the planning and execution of conservation and development projects. Email: N/A.

Jorge Qüense is an Associate Professor at the Institute of Geography, and head of the Diploma Program in Geographic Information Systems (GIS) at the *Pontificia Universidad Católica de Chile* (PUC). He is also a researcher at the Patagonia Interdisciplinary Research Station UC and the Observatory Homme-Milieux

International Patagonia-Bahia Exploradores of the LabEx DRIIHM, INEE-CNRS. He holds a PhD in Environment from the Joseph Fourier University (UJF) in Grenoble, France. He completed a Diploma of Advanced Studies (DEA) in Management of Mountain Areas at UJF, France. He holds a BA in Geography from the PUC. He specializes in GIS, cartography, and remote sensing, with an interest in multitemporal studies. Email: jquense@uc.cl

Alejandro Salazar-Burrows is an Associate Professor at the Institute of Geography at the *Pontificia Universidad Católica de Chile* (PUC) and director of the Patagonia Interdisciplinary Research Station UC. He is also the director of the Observatory Homme-Milieux International Patagonia-Bahia Exploradores of the LabEx DRIIHM, INEE-CNR (France). He holds a PhD in Social Sciences from the AgroParisTech and a master's degree in integrated Territories Management (France). His research interests are social and territorial re-composition and peri-urbanization phenomena, urban-rural relations associated with new ruralities, human-environment interactions, and environmental geography of the Western Patagonia. Email: asalazab@uc.cl

David Seward Salisbury is an Associate Professor in the Department of Geography, Environment, and Sustainability at the University of Richmond, VA, USA. He is Chair of the Conference of Latin American Geography (CLAG) and the principal representative for the Geography Commission of the United States National Section for the PanAmerican Institute of Geography and History (PAIGH). He holds degrees from the University of Texas, the University of Florida, and Middlebury College, and served as a Peace Corps Volunteer (Guatemala). Dr. Salisbury is the Co-I of a NASA SERVIR Amazonia grant with PI Dr. Spera. His applied approach focuses on resource conflict, infrastructure impacts, transboundary political ecology, Indigenous rights, conservation, and climate change in the Amazon borderlands shared by Peru and Brazil. Email: dsalisbu@richmond.edu

Alexandre Pereira Santos is a PhD researcher at the Center for Earth Systems Research and Sustainability (CEN) at the University of Hamburg. He holds an MSc in Contemporary Urbanism from the Federal University of Pelotas and a BArch from the Federal University of Rio Grande do Sul, Brazil. His research interests include urban socio-environmental interaction, vulnerability, climate change, and health with mixed and interdisciplinary methods and a focus on GIScience, agent-based modelling, and participatory techniques. Email: alexandre.pereira.santos@uni-hamburg.de

Sidney Novoa Sheppard is Director of GIS and technology for conservation at Conservación Amazónica (ACCA) in Lima, Peru. He holds a master's degree in Conservation Biology from the *Escuela Superior de Conservación Ambiental e Sustentabilidad*-ESCAS and a bachelor's degree in Ecology from the *Universidad Nacional Agraria La Molina* (UNALM). He is the lead member of Peru's SERVIR Amazonia team from ACCA. Email: snovoa@conservacionamazoica.org

Rodrigo Sierra is a Principal Consultant at GeoIS with a PhD in Geography from Ohio State University. He has worked and taught on issues related to deforestation policy analysis and modeling and indigenous territorialities for almost three decades. Email: rsierra@geo-is.com

Stephanie A. Spera is an Assistant Professor of Geography, Environment, and Sustainability at the University of Richmond, and principal director of a NASA SERVIR Amazonia grant and founding researcher of the Amazon Borderlands Spatial Analysis Team (ABSAT) in Virginia, USA. She holds a PhD in Earth, Environmental, and Planetary Sciences from Brown University, and a Bachelor's in Earth and Planetary Sciences and Film and Media Studies from Washington University in St. Louis. Email: sspera@richmond.edu

Pedro Tipula Tipula is the Coordinator of the *Sistema de Información sobre Comunidades Campesinas del Perú* (SICCAM) of the *Instituto para el Bien Común* (IBC) in Lima, Peru, and is a leading Peruvian representative for the *Red Amazónica* (RAISG). He holds a master's degree in Land Use Policy and Management and a bachelor's degree in Geography from the *Universidad Nacional Mayor de San Marcos* (UNMSM). An expert in GIS, cartography, and Indigenous territoriality, he is a lead facilitator and founding member of the Transboundary Geographic Group of the Southwestern Amazon (GTASO). Email: ptipula@ibcperu.org

Kim André Vanselow is a Research Associate and Lecturer at the Institute of Geography at Friedrich-Alexander-University Erlangen-Nürnberg, Germany. His research focus is on biogeography, vegetation ecology, and human-environmental interaction, in particular in high mountains. His regional interest is mainly in Central Asia, northern and southern Africa, and Latin America. Email: kim.vanselow@fau.de

Javier Vargas is a Biologist with a master's degree in Biological Sciences from Universidad de Chile, and a project coordinator at EcoCiencia with extensive experience in community relation with indigenous communities to support conservation and development efforts. Email: N/A

Sylvia Villacís is a Geographer with a master's degree in Environmental Management from Massey University, New Zealand. She has extensive experience in mapping land use in the Ecuadorian Amazon using remotely sensed and geodetic data. Email: N/A

Chapter 1
Introduction: Interdisciplinarity, GIScience, and Socio-Environmental Research in Latin America

Santiago López

Abstract Accelerated environmental change worldwide threatens the sustainability of the Earth system. Calls for research that involves community outreach and solutions to raising environmental concerns have triggered integrative research approaches to assess the myriad of factors affecting human-environment interactions. Some of the proposed solutions have required a re-thinking of human-environment relationships and the roles that science, policy makers, governmental and non-governmental organizations, communities, and individuals play in achieving local or regional sustainability. Interdisciplinary applied research in Latin America, the main regional focus of this book, has advanced in the past three decades to address issues of land degradation, forest loss, conservation of biodiversity, economic development, and more. Some of the solutions have been techno-scientific and embedded in geographic information science, which has allowed the identification of not only locations threatened by environmental degradation, but the factors, geographic patterns, and spatio-temporal trends of socio-environmental change. The contribution of geographic information science toward the advancement of socio-environmental knowledge in Latin America is undeniable. Drawing on research from various countries and biogeographical regions across Latin America, the chapters in this book bring new interdisciplinary insights, deeply rooted in geographic information science and technologies, on the complex socio-environmental dynamics that characterizes this diverse region, thus contributing to a vibrant research area within human-environment geography.

Keywords Interdisciplinary · Socio-environmental interactions · GIScience · Latin America

---

S. López (✉)
School of Interdisciplinary Arts and Sciences, University of Washington Bothell, Bothell, WA, USA
e-mail: cslopez@uw.edu

S. López (ed.), *Socio-Environmental Research in Latin America*, The Latin American Studies Book Series, https://doi.org/10.1007/978-3-031-22680-9_1

1 Interdisciplinary Socio-Environmental Research

The past two decades have been characterized by increasing calls for socio-environmental (SE) research that involve outreach and pragmatic solutions to raising environmental issues and concerns (Turner II et al. 2016). These calls have been triggered by accelerated SE change worldwide that threatens the sustainability of the Earth system and its human and non-human elements. Further, SE systems (SES) (i.e., closely connected social and biophysical subsystems that affect each other) have experienced accelerated transformations that test their capacity to respond and adapt to endogenous and exogenous pressures, ultimately affecting their resilience and long-term survival (Postigo 2014). Threats on climate dynamics, land cover, and nutrient cycling are of particular concern since they are intimately entwined with the health and functioning of the biosphere (Mahli et al. 2020). These requests have also challenged researchers to be more responsive to societal needs and set research priorities that are directly aligned with such demands. In this way, research approaches have become more integrative than in the past, implementing SES frameworks that have required a re-thinking of human-environment relationships and the roles that science, policy makers, governmental and non-governmental organizations, communities, and individuals play in reaching local or regional sustainability goals.

Since SE change is systemic and cumulative, its analysis requires a suite of approaches and frameworks to unravel the connections between its human and physical drivers and their consequences for local populations and the environments where they interact (Brown 2017; Turner II et al. 1990). Thus, the requests have also been for 'interdisciplinary' research with emphasis on long term, place-based monitoring, mapping, and analysis of SES. Although there is no widely accepted consensus about the meaning of interdisciplinarity, in this book we use Hicks et al.'s (2010) conceptualization, who defined it as the 'production of research which crosses disciplinary boundaries.' Interdisciplinarity is not new; academic disciplines have been frequently combined to form new disciplines to provide better answers to emerging questions. We could distinguish, for example, between narrow or small interdisciplinarity, which implies collaboration between similar disciplines (e.g., between the biological and ecological sciences), and wide or big interdisciplinarity which involves collaboration between distant disciplines (e.g., between the natural/physical sciences and social sciences/humanities) (Morillo et al. 2003). This latter type does not only involve academics, but also policy makers and other stakeholders (e.g., people who can not only help co-define research questions and develop models, but also actively participate in data collections, scientific experiments, or in-depth conversations about specific research components). Interdisciplinary approaches have helped develop not only top-bottom characterizations of SES dynamics, but also bottom-up solutions to specific public concerns (Nakashima et al. 2018; Nielsen and D'haen 2014; Valdivia et al. 2010) and better understandings of SE change patterns than single-disciplinary responses.

The emphasis on both humans and their environments recognizes the complexity of past and present conditions affecting nature and society at various spatio-temporal scales. Technoscientific responses to address these complexities are commonly embedded in wide interdisciplinarity and constructed around common theoretical frameworks and research questions (Pulver et al. 2018). SES research frameworks that have emerged in the past couple of decades include coupled human-environment systems (Liu et al. 2007a, b) resilience (Folke 2006; Carpenter et al. 2001; Gunderson and Holling 2001), human-ecosystem relationships (Machlis et al. 1997), vulnerability (Turner II et al. 2003), social-ecological systems (Ostrom 2009), and integrated assessment of ecosystem functions and services (de Groot et al. 2002). This body of research has helped to progressively contextualize SE interactions and embed them in historical, political, and economic contexts to inform vulnerability, hazards (including land degradation and deforestation), adaptation, and ecosystem management (Fig. 1.1). Similarly, all these frameworks draw some attention to the importance of scale and acknowledge multi-scalar connections among SE features and processes (Pulver et al. 2018). Thus, geospatial approaches have been critical not only to delimit the geographic scope of SES research, but also to ease the integration of general scientific knowledge (*nomothetic geography*) and the unique characteristics of places (*idiographic geography*). These kinds of approaches, for instance, have allowed the 'telecoupling' (or remote connection) of spatially distant ecosystems and local processes via a range of social (e.g., socio-economic or health outcomes) and biophysical processes (e.g., greenhouse emissions), or a range of

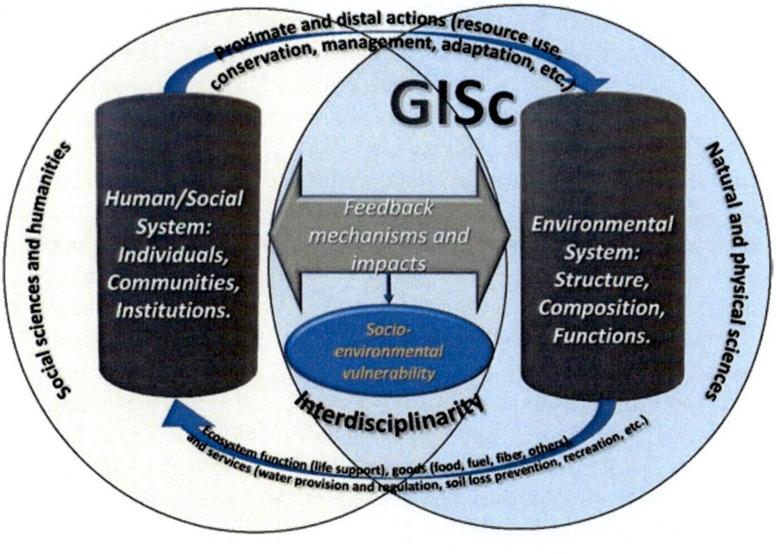

Fig. 1.1 Conceptualization of a social and environmental system and the traditional study domains of the social sciences/humanities and the natural/physical sciences. Geographic information science (GIScience) may serve as a bridging scientific framework that allows the integration of diverse disciplines

distal (e.g., institutional conditions) and proximate factors (e.g., agricultural expansion), deeming them critical to climate change adaptation, land use land cover (LULC) change, or spatial epidemiological concerns.

2 Geographic Information Science Epistemology and Interdisciplinarity

The integration of geospatial technologies and *in-situ* or field-based SE research has opened up opportunities for advancing our understanding of the causes, consequences, and processes of environmental degradation and social change. Nevertheless, this type of integration comes with its own difficulties. For instance, one of the challenges that remote sensing (RS) has faced in the past is the difficulty of its integration with the social sciences and humanities, although this situation has been slowly changing since the end of the twentieth century. The integration of these scientific approaches requires the synthesis not only of data but also of quite different epistemologies. While geographic information science (GIScience) has been traditionally embedded in logical positivism and structuralism that connect knowledge production with formal universal laws based on verifiable, valid, and replicable observations (Leszczynski 2017), the social sciences (including the geohumanities) on the other hand, have been shifting towards a post-structuralist epistemology, moving away from pre-established, socially-constructed structures and top-down models in the search for more nuanced, bottom-up, understandings of human and environmental dynamics (Warf and Sui 2010). These different views of knowledge production undeniably affect those who do interdisciplinary research since they commonly trigger a sense of insecurity and marginalization as the questions and issues central to their research are somewhat tangential to the specialized disciplines (Rindfuss and Stern 1998).

How can different views of knowledge production be integrated in both theory and praxis in current SE research? Some alternatives include collaborative research design approaches that include mixed methodologies in which data and methods are combined in different ways to put emphasis on the qualitative or quantitative data and/or analysis techniques. These approaches necessarily need a measure of sensitivity and open communication among researchers from different fields, which implies active listening, curiosity, and understanding of each other's perspectives and potential contributions to joint efforts (Bridle et al. 2013). This communicative attitude has been referred to as "appreciative inquiry" (Graybill et al. 2006) and has been key to successful interdisciplinary programs. Recent advances in participatory and qualitative geographic information systems (GIS) have enabled some of these communication lines for a more refined exploration of social and spatio/environmental phenomena with the goal of generating more robust and appropriate policy responses (Elwood 2006). Consequently, there is a continuous need to develop novel forms of inquiry and methodological approaches, capable of accounting for the hybrid character of SE change and across geographic scales. In this context,

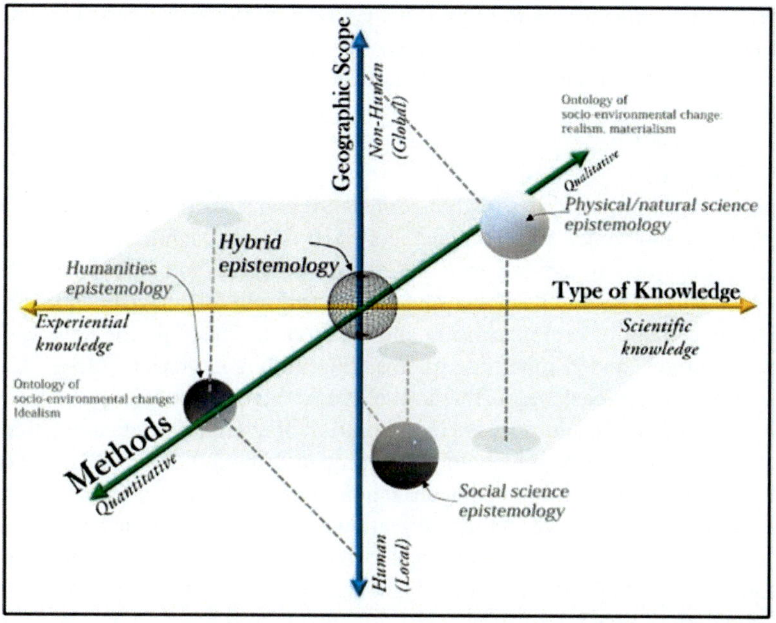

Fig. 1.2 A hybrid epistemological conceptual framework depicting the connection between geographic scope, type of knowledge, and methods. (Source: Modified from Lópezet al. (2017)

hybrid epistemological frameworks could allow the integration of different types of knowledge, methods, and geographic scopes (Fig. 1.2). Such a framework allows us to create a middle ground space that delimits the boundary, process, and scope of SE knowledge by integrating different analytical approaches that are often not compatible (Watson-Verran and Turnbull 1995). GIScience has been directly incorporating such hybridities since the late 1990s, especially through its engagement with mixed method practices and the evolution of a more inclusive and flexible epistemology (Elwood et al. 2012).

3 Geographic Information Science and Socio-Environmental Research in Latin-America

Latin American landscapes are diverse, complex, and undergoing significant transformations with implications for the sustainability of SES in the region, currently threatened by accelerated forest cover reduction (Manners and Varela-Ortega 2017), the replacement of native forests with exotic woody species (Schwartz et al. 2020), and land degradation (Metternicht et al. 2010). Although there are intricate connections between proximate causes and underlying forces of such changes and their operation at different spatio-temporal scales, in Latin America, these entanglements have been often associated with global change dynamics engrained in national neoliberal economic reforms and neo-extractivist agendas (Bebbington et al. 2018). These processes are affecting the ways humans relate to their local and regional environments, with major consequences for climate dynamics (Kalnay and Cai 2003), biotic diversity and ecosystem services (Daily 2000), and the sustainability of SES (Turner II et al. 2003).

Many of the solutions developed to address SE concerns in Latin America have been technical and rooted in GIScience, comprising both quantitative and qualitative GIS approaches (including participatory mapping), RS, and global navigational satellite systems (GNSS). GIScience approaches to study SES have been facilitated by the enhancement of remote sensors and their resolutions, and the advancement of geotechnologies and computing systems from mobile devices to super computers, cloud-based services, and artificial intelligence algorithms to capture and process geolocated data. In addition, the incorporation of alternative data collection and analytical methods (e.g., volunteered geographic information or VGI) have allowed not only more comprehensive characterizations of SES than in the past, but more meaningful ones. At the core of this integration is the recognition that SE change is multifaceted and driven by social, political, economic, cultural, and biophysical factors that affect human decision making in unique and intricate ways. Thus, the implementation of new interdisciplinary GIScience-based practices has been key to environmental justice concerns, conservation, development, adaptation, and mitigation strategies.

These past few decades have also been characterized not only by the evolution of geotechnologies (software and hardware), geospatial approaches, and novel concepts oriented toward the advancement of GIScience, but by increased human capacity through rigorous undergraduate and post-graduate training at higher education institutions worldwide. In Latin America, bachelor's in science and engineering degrees together with master's and doctoral programs with a focus on GIScience have proliferated. Different forums have also emerged since the earlier applications of GIS, RS, and GNSS in the 1980s to serve as knowledge exchange spaces for interaction among Latin American and Latin Americanist professionals, academics, and students. Some of these forums include international partnerships such as the Pan-American Institute of Geography and History (IPGH), the Ibero-American Conference of Geographic Information Systems (CONFIBSIG), the Latin-American chapters of Geoscience and Remote Sensing Society (GRSS), the Society of Latin-American Specialists on Remote Sensing (SELPER), or nationally sponsored forums like the Brazilian Symposium on Remote Sensing (SBSR) or the French Institute of Research for Development (IRD, or ex-ORSTOM) to name a few. Space research institutes like National Institute of Space Research (INPE) in Brazil or the Mario Gulich Institute for Advanced Space Studies in Argentina are examples of long-term programs, deeply rooted in RS and GIS technologies since their creation in the 1970s and early 2000s respectively, designed to advance SE research in the region.

4 Applied Geospatial Research in Latin America

Remote sensing technologies and products are not new. Aerial photographs, for instance, have been used for more almost a century and satellite images for more than four decades. Since their introduction of analog remote sensing products (e.g., photographs) in Latin America in the 1900s, remote sensing applications have focused on natural resource management, urban planning, land development, risk-hazards assessments, and monitoring of surface-level lithosphere processes including volcanism, geomorphology, air and water quality changes, climate change, habitat fragmentation, floods, hurricanes, landslides, droughts, and forest fires (Chuvieco et al. 2008). In addition, since the early 1990s, there has been an increased emphasis on remote sensing applications in conservation (Aide et al. 2019; Boyle et al. 2014; Leisher et al. 2013), invasion ecology processes (Hoyos et al. 2010), ecosystem function monitoring (Abelleira-Martínez et al. 2016), forest assessments and agricultural expansion (Graesser et al. 2015), soil erosion and land degradation (Metternicht et al. 2010), and carbon loss and storage (Chadwick and Asner 2016, Harris et al. 2012). Socio-geological RS applications in mining and natural resource extraction assessments (Mazabanda et al. 2020; Alvarez-Berríos and Aide 2015; Swenson et al. 2011) have also increased in recent years. Most of this research comes from regions of Latin America where remote sensing has been critical for the development of environmental policy such as Brazil and Argentina (Villalón-Turrubiates et al. 2016) or areas that have been identified as hot spots or fronts of environmental degradation as in the case of areas in Mexico, Ecuador, Bolivia, or Peru.

GIS technology has experienced a sustained growth in research since its introduction in Latin America in the late 1980s (Buzai and Robinson 2010). GIS approaches have been more widespread and interdisciplinary than RS applications and incorporate RS and field-collected socio-economic data to some extent. In recent years, citizen science approaches have helped incorporate local voices into the research process in different areas of Latin America with successful results (Mena et al. 2019; Cunha et al. 2017), especially when adapted to specific cultural settings (Ceccaroni and Piera 2017). The successful integration of multiple sources of geographic information, including VGI, offers SE researchers an unprecedented opportunity to fine tune models, theory, and policy (Elwood et al. 2012). In general, while remote sensing has been mostly used for mapping, monitoring, and analyses of physical processes and geo-ecological features, the data fusion power of GIS that allows linking socio-economic surveys, *in-situ* measurements, and participatory research approaches that guarantee the incorporation of situated experiential knowledge, offer additional opportunities to address important geo-political and socio-economic processes that affect people's livelihoods (Salisbury et al. 2012). Processes such as territorial organization, land rights adjudication, or social resistance against extractive industries are currently widespread and considerably shaping Latin American landscapes. GIScience is at the forefront of many regional initiatives that are helping address such actions.

In summary, the contribution of GIScience toward the advancement of SE knowledge in Latin America is undeniable. Work produced by several research teams by the end of the twentieth century and beginning of the twenty-first century established some key methodological and theoretical foundations when pursuing the integration of social and environmental data at scales consistent with regional and subregional SE analysis (cf. Caldas et al. 2007; Fox et al. 2003; Soares-Filho et al. 2002; Liverman et al. 1998; Sierra 1996, among others). Currently, the theoretical and practical approaches that lead to the coupling of SE features and processes have probably reached a mature state and sophisticated geospatial modeling frameworks such as spatially-explicit regression analysis and agent-based or individual-based modeling approaches have evolved to deal with the spatio-temporal complexities of SE interactions in Latin American urban and rural areas (cf. Santos et al. 2019; España et al. 2018; Arima 2016; López 2014; Barros 2012, among others). Further, recent continental-level analyses of land use and land cover change have also contributed to understand non-linear patterns of SES dynamics in a mega-diverse region threatened by continuous human pressures (Aide et al. 2013, 2019; Grau and Aide 2008).

5 Scope and Purpose of This Book

The purpose of this book is to present current application of GIScience research that aims to advance our interdisciplinary understanding of SES and SE interactions, past and present, in a region that has experienced accelerated environmental degradation in recent decades. Specifically, this book aims to address the following overarching questions:

1. *How are interdisciplinary approaches using RS and GIS uniquely facilitating a dialog between the geophysical sciences, natural sciences, social sciences, and the humanities as conceived by researchers doing work in Latin America?*
2. *How can multiple narratives be introduced to change the ways in which RS and GIS are practiced in this region?*

We attempt to answer these questions through a sample of current applications of geospatial technologies and analyses of literature that emerge from Latin American and Latin Americanist scholarship. The different chapters in this volume showcase the challenges and opportunities of connecting disciplinary expertise rooted in GIScience and interdisciplinary approaches to the practice of SE research. Many of the case studies in this volume are embedded in hybrid epistemological frameworks implemented through a range of mixed methods that span over multiple disciplines such as land change science, watershed science, political ecology, environmental history and archaeology, physical geography, ecology, health sciences, human geography, and others. The chapters presented here present ways of overcoming linear, deterministic, and representational explanations of SE change in favor of non-linear but historically and politically situated SE interactions. We are cognizant that this coverage is not complete, but it offers a robust sample of current SE research practice in terms of themes, theory and methodology, and regional coverage.

Drawing on research from various countries and biogeographical regions across Latin America, the chapters in this book bring new insights on SE dynamics from an interdisciplinary perspective rooted in GIScience, thus contributing to a vibrant research area in human-environment geography.

References

Abelleira-Martínez OJ, Fremier AK, Gunter S et al (2016) Scaling up functional traits for ecosystem services with remote sensing: concepts and methods. Ecol Evol 6(13):4359–4371

Aide TM, Clark ML, Grau HR et al (2013) Deforestation and reforestation of Latin America and the Caribbean (2001–2010). Biotropica 45(2):262–271

Aide TM, Grau HR, Graesser J et al (2019) Woody vegetation dynamics in the tropical and subtropical Andes from 2001 to 2014: satellite image interpretation and expert validation. Glob Change Biol 25:2112–2126

Alvarez-Berríos NL, Aide TM (2015) Global demand for gold is another threat for tropical. Environ Res Lett 10:1–11

Arima EY (2016) A spatial probit econometric model of land change: the case of infrastructure development in Western Amazonia, Peru. PLoS ONE 11(3):1–22

Barros J (2012) Exploring urban dynamics in Latin American cities using and agent-based simulation approach. In: Heppenstall A, Crooks A, Linda MS et al (eds) Agent-based models of geographical systems. Springer, Heidelberg, pp 571–589

Bebbington AJ, Sauls LA, Rosa H et al (2018) Conflicts over extractivist policy and the forest frontier in Central America. Rev Lat Am Caribb Stud 106:103–132

Boyle SA, Kennedy CM, Torres J et al (2014) High-resolution satellite imagery is an important yet underutilized resource in conservation biology. PLoS ONE 9(1):1–11

Bridle H, Vrieling A, Cardillo M et al (2013) Preparing for an interdisciplinary future: a perspective from early-career researchers. Futures 53:22–32

Brown K (2017) Global environmental change II: planetary boundaries – a safe operating space for human geographers? Prog Hum Geogr 41(1):118–130

Buzai GD, Robinson DJ (2010) Geographic information systems (GIS) in Latin America, 1987-2010: a preliminary overview. J Lat Am Geogr 9(3):9–31

Caldas M, Walker R, Arima E et al (2007) Theorizing land cover and land use change: the peasant economy of Amazonian deforestation. Ann Am Assoc Geogr 97(1):86–110

Carpenter SR, Walker B, Anderies JM et al (2001) From metaphor to measurement: resilience of what to what? Ecosystems 4:765–781

Ceccaroni L, Piera J (2017) Analyzing the role of citizen science in modern research. IGI Global, Hershey

Chadwick KD, Asner GP (2016) Organismic-scale remote sensing of canopy foliar traits in lowland tropical forests. Remote Sens 8(87):1–16

Chuvieco E, Opazo S, Sione W et al (2008) Global burned-land estimation in Latin America using MODIS composite data. Ecol Appl 18(1):64–79

Cunha D, Marques JF, de Resende JC et al (2017) Citizen science participation in research in the environmental sciences: key factors related to projects' success and longevity. An Acad Bras Ciênc 89(3):2229–2245

Daily GC (2000) Management objectives for the protection of ecosystem services. Environ Sci Pol 3:333–339

de Groot RS, Wilson MA, Boumans RM (2002) A typology for the classification, description and valuation of ecosystem functions, goods and services. Ecol Econ 41(3):393–408

Elwood S (2006) Critical issues in participatory GIS: deconstructions, reconstructions, and new research directions. Trans GIS 10(5):693–708

Elwood S, Goodchild MF, Sui DZ (2012) Researching volunteered geographic information: spatial data, geographic research, and new social practice. Ann Am Assoc Geogr 102(3):571–590
España G, Grefenstette J, Perkins A et al (2018) Exploring scenarios of chikungunya mitigation with a data-driven agent-based model of the 2014–2016 outbreak in Colombia. Sci Rep 8:12201
Folke C (2006) Resilience: the emergence of a perspective for social-ecological systems analyses. Glob Environ Change 16:253–267
Fox J, Rindfuss RR, Walsh SJ et al (2003) People and the environment: approaches for linking household and commnity surveys to remote sensing and GIS. Kluwer Academic Publishers, Boston
Graesser J, Aide TM, Grau HR et al (2015) Cropland/pastureland dynamics and the slowdown of deforestation in Latin America. Environ Res Lett 10:1–10
Grau HR, Aide TM (2008) Globalization and land use transitions in Latin America. Ecol Soc 13(2):16
Graybill JK, Dooling S, Shandas V et al (2006) A rough guide to interdisciplinarity: graduate student perspectives. Bioscience 56(9):757–763
Gunderson LH, Holling CS (2001) Panarchy: understanding transformations in human and natural systems. Island, Washington
Harris NL, Brown S, Hagen SC et al (2012) Baseline map of carbon emissions from deforestation in tropical regions. Science 336(6088):1573–1576
Hicks CC, Fitzsimmons C, Polunin NV (2010) Interdisciplinarity in the environmental sciences: barriers and frontiers. Environ Conserv 337(4):464–477
Hoyos LE, Gavier-Pizarro GI, Kuemmerle T et al (2010) Invasion of glossy privet (Ligustrum lucidum) and native forest loss in the sierras Chicas of Córdoba, Argentina. Biol Invasions 12:3261–3275
Kalnay E, Cai M (2003) Impact of urbanization and land-use change on climate. Nature 423:528–531
Leisher C, Touval J, Hess SM et al (2013) Land and forest degradation inside protected areas in Latin America. Diversity 5:779–795
Leszczynski A (2017) Epistemological critiques. In: Wilson JP (ed) The geographic information science & technology body of knowledge. DOI: https://doi.org/10.22224/gistbok/2017.4.1
Liu J, Dietz T, Carpernter SR et al (2007a) Complexity of coupled human and natural systems. Science 317(5844):1513–1516
Liu J, Dietz T, Carpenter SR et al (2007b) Coupled human and natural systems. Ambio 36(8):639–649
Liverman D, Moran EF, Rindfuss RR et al (1998) People and pixels: linking remote sensing and social science. National Academy Press, Washington
López S (2014) Modeling agriculatural change through logistic regression and cellular automata: a case study on shifting cultivation. J Geogr Inf Syst 6:220–235
López S, Jung J-K, López-Sandoval MF (2017) A hybrid-epistemological approach to climate change research: linking scientific and local knowledge sysems in the Ecuadorian Andes. Anthropocene 17:30–45
Machlis GE, Force JE, Burch WR (1997) The human ecosystem part I: the human ecosystem as an organizing concept in ecosystem management. Soc Nat Resour 10(4):347–367
Mahli Y, Franklin J, Seddon N et al (2020) Climate change and ecosystems: threats, opportunities and solutions. Philos Trans R Soc Lond B 375
Manners R, Varela-Ortega C (2017) Analysing Latin America and Caribbean forest vulnerability from socio-economic factors. Environ Sci 14(1):109–130
Mazabanda C, Kemper R, Thieme A et al. (2020, September 9). https://maaproject.org/mirador-ecuador/. Accessed 23 Jan 2021
Mena CF, Arsel M, Pellegrini L et al (2019) Community-based monitoring of oil extraction: lesson learned in the Ecuadorian Amazon. Soc Nat Resour 33(3):406–417
Metternicht G, Zinck JA, Blanco PD et al (2010) Remote sensing of land degradation: experiences from Latin America and the Caribbean. J Environ Qual 39:42–61

Morillo F, Bordons M, Isabel G (2003) Interdisciplinarity in science: a tentative typology of disciplines and research areas. J Am Soc Inf Sci Technol 54(13):1237–1249
Nakashima D, Krupnik I, Rubis J (2018) Indigenous knowledge for climate change assessment and adaptation. Cambridge University Press, Cambridge
Nielsen J, D'haen S (2014) Asking about climate change: reflections on methodology in qualitative climate change research published in global environmental change since 2000. Glob Environ Change 24:402–409
Ostrom E (2009) A general framework for analzing sustainability of social-ecological systems. Science 325(5939):419–422
Postigo JC (2014) Perception and resilience of Andean populations facing climate change. J Ethnobiol 34(3):383–400
Pulver S, Ulibarri N, Sobocinski KL et al (2018) Frontiers in socio-environmental research: components, connections, scale, and context. Ecol Soc 23(3):23
Rindfuss R, Stern PC (1998) Linking remote sensing and social science: the need and the challenges. In: Liverman D, Moran EF, Rindfuss R et al (eds) People and pixels: linking remote sensing and social science. National Academy Press, Washington, pp 1–27
Salisbury DS, Flores de Melo AW, Tipula-Tipula P (2012) Transboundary political ecology in the Peru-Brazil borderlands: mapping workshops, geographic information, anp d socio-environmental impacts. Rev Geogr 152:105–115
Santos F, Graw V, Bonilla S (2019) A geographically weighted random forest approach for evaluate forest change drivers in the northern Ecuadorian Amazon. PLoS ONE 14(12):1–37
Schwartz NB, Aide TM, Graesser J et al (2020) Reversals of reforestation across Latin America limit climate mitigation potential of tropical forests. Front For Glob Change 3(85):1–10
Sierra R (1996) Dynamics and patterns of deforestation in the western Amazon: the Napo deforestation front, 1986–1996. Appl Geogr 20(1):1–16
Soares-Filho BS, Cerqueira GC, Pennachin CL (2002) DINAMICA—a stochastic cellular automata model designed to simulate the landscape dynamics in an Amazonian colonization frontier. Ecol Model 154:217–235
Swenson JJ, Carter CE, Domec JC et al (2011) Gold mining in the Peruvian Amazon: global prices, deforestation, and mercury imports. PLoS ONE 6(4):1–7
Turner BL II, Kasperson RE, Meyer WB et al (1990) Two types of global environmental change: definitional and spatial-scale issues in their human dimensions. Glob Environ Change 1(1):14–22
Turner BL II, Kasperson RE, Matson PA et al (2003) A framework for vulnerability analysis in sustainability science. Proc Natl Acad Sci 100(14):8074–8079
Turner BL II, Esler KJ, Bridgewater P et al (2016) Socio-environmental systems (SES) research: what have we learned and how can we use this information in future research programs. Curr Opin Environ Sustain 19:160–168
Valdivia C, Gilles SJ, García M et al (2010) Adapting to climate change in Andean ecosystems: landscapes, capitals, and perceptions shaping rural livelihood strategies and linking knowledge systems. Ann Am Assoc Geogr 100(4):818–834
Villalón-Turrubiates IE, Scavuzzo CM, Feitosa RQ et al (2016) Special issue on applied earth observation and remote sensing in Latin America. IEEE J Sel Top Appl Earth Obs Remote Sens 9(12):5287–5293
Warf B, Sui D (2010) From GIS to neogeography: ontological implications and theories of truth. Ann GIS 16(4):197–209
Watson-Verran H, Turnbull D (1995) Science and other Indigenous knowledge systems. In: Jasanoff S, Markle GE, Peterson JC et al (eds) Handbook of science and technology studies. SAGE Publications, Thousand Oaks, pp 115–139

Chapter 2
Using Spatial Time-Series and Field Data to Understand Cultural Drivers of Land Change: Connecting Land Conflict and Land Change in Eastern Amazonia

Stephen Aldrich

Abstract Since the initial alarm raised by Latin American researchers in the early 1970s, the Amazon has been the focus of many geographic information science (GIScience) approaches to understand the drivers of deforestation. Research in this region has helped drive land cover change research forward, including the theme of cultural drivers (sometimes called social-process drivers) – causal chains through which social interactions between unequal environmental decision-makers lead to land change. This chapter focuses on how social-process drivers (those involving chains of interactions between people, institutions, laws, and regulations) can be uncovered through the combination of historical research, field interviews, remote sensing, geographic information systems, and statistical modeling (spatial, pooled, fixed-, and random-effects regression), in a specific region where deforestation has been exacerbated by social process drivers of deforestation. In Southeastern Pará, Brazil a historical legacy of land conflict, driven by social policy, inequality, superimposed landholdings, bureaucratic confusion, and the legacy of a military government, gave rise to rapid deforestation undertaken by an unplanned and shifting assemblage of actors who looked to benefit from the chaos generated by an ongoing war over land rights. Understanding who is responsible for deforestation in complicated social and ecological environments requires a convergence of research methods that span a wide variety of disciplines and involve data covering long time periods. This study makes it clear that social process drivers are an important component of land change, but they work together with more traditional economic and policy-related drivers to hasten deforestation.

S. Aldrich (✉)
Department of Earth & Environmental Systems, Indiana State University, Terre Haute, IN, USA
e-mail: steve.aldrich@indstate.edu

S. López (ed.), *Socio-Environmental Research in Latin America*, The Latin American Studies Book Series, https://doi.org/10.1007/978-3-031-22680-9_2

Keywords Deforestation · Amazon · Time-series · Human-environment relationships · GIS · Remote sensing

1 Introduction

Alarm regarding the pace and effects of deforestation in the Amazon began to grow in the late 1970s and early 1980s (e.g., Fearnside 1982; Hecht 1981; Goodland and Irwin 1975), constituting one of the most significant environmental concerns in Latin America. At the same time, earth-observing satellite imagery became more accessible for spatial scientists as the Landsat satellite series, and other similar earth-observing orbital platforms, began producing accumulated data which could be used to show change over time (Wulder et al. 2019; Williams et al. 2006, 2018; Lauer et al. 1997). While other sensitive environments were also undergoing rapid change, such as the Pantanal, Central America's lowland and upland forests, and Mexico's Yucatan and Chiapan forests to name just a few, remote sensing had one of the strongest impacts on the science of land change in Amazonia.

There have been nearly innumerable studies of land change, and the processes generating land changes (both proximate and underlying), in the Amazon that have made use of remote sensing and field-collected geospatial data (Caldas et al. 2007; Pan et al. 2004; McCracken et al. 2002; Messina and Walsh 2001; Pfaff 1999; Lambin 1997; Moran et al. 1994; Dale et al. 1993; Mausel et al. 1993 are just some examples). Furthermore, geospatial time-series, particularly the combination of remotely sensed and field data, have been very useful at uncovering trends and tendencies regarding deforestation and the processes that drive it in Amazonia (Aldrich et al. 2006, 2020; Simmons et al. 2007, 2019; Godar et al. 2012; Messina and Walsh 2005; Moran et al. 2003; Walker 1996, 2003) and elsewhere in Latin America.

With the insights gained through land use and land cover analyses across Latin America, among many other world regions, the dominant frameworks for understanding the drivers of land use and land cover change continue to be refined. While some suites of land change drivers have been relatively well explored, such as economic factors, policy and institutional factors, and demographic factors (for a classic example see Table 2 of Geist and Lambin 2001), others are less understood. Of these suites of drivers of land use and land cover change, so-called "cultural factors" (and associated social-process drivers) are particularly thorny problems to address given the lack of systematic data, the diversity of decision-making approaches that actors on the ground employ, and the difficulties of elaborating unspoken norms and tendencies.

Cultural drivers of land use and land cover change derive from values, attitudes, and beliefs, or from individual behaviors (Geist and Lambin 2001). Some examples of cultural drivers that fall in the category of values, attitudes, or beliefs include a frontier/cowboy mentality which sees the environment as something that must be tamed (Kröger 2020), the fear of political uprising or communism (Simmons et al. 2019), or "green hell" perspectives (Hecht and Rajão 2020). Other cultural factors of land use and land cover change include individual behaviors, often occurring in overlapping physical and institutional spaces (Simmons et al. 2007; Schmink and Wood 1992; Bunker 1985).

This chapter focuses on one particular cultural driver of deforestation: land conflict. Land conflict can be a significant driver of land cover change in the Amazon (Aldrich et al. 2020; Simmons et al. 2007, 2019; Brown et al. 2016a; Araujo et al. 2009; Hall 1989; Foweraker 1981a) and elsewhere (Messina et al. 2006; Lucas and Warren 2013; O'Loughlin and Witmer 2010; UNEP 2009; Unruh 2009; Wulan et al. 2004; Kull 2002; Percival and Homer-Dixon 1998; Putzel 1992). Contention has been suggested as a driver of land change for more than 50 years (e.g., Schmink 1982; Foweraker 1981b), and continues to be a concern (e.g., Rodríguez-de-Francisco et al. 2021; Woods et al. 2021). Conflict can influence forests in a variety of ways. Forests can serve, for example, as a place for property occupiers to hide when facing conflict with law enforcement or other landholders. They can also be seen as an indicator that a property is not being used productively (productive use is required by Brazil's constitution), which could promote violence against property owners or their estate (e.g., theft, vandalism, forced squatting). Violence can trigger spillover-effects which leads to deforestation (e.g., fire is used as a weapon in conflict to remove forests, forests are removed in acts of spite). Forests also have economic value, meaning there may be a "race" in contentious situations to liquidate natural capital before one's rivals. While the results of the statistical analyses that address these themes are reported in detail elsewhere (Aldrich 2015; Aldrich et al. 2020, 2012a), it is my intent in this chapter to argue that more detailed work is necessary to improve our understanding of dynamic processes by using the approach we have developed in addressing land conflict and deforestation. My specific broad-level hypotheses regarding land conflict and deforestation were: (1) social movements have different ideas about what makes a property attractive for occupation than other stakeholders, (2) largeholders actively deforest their land to bolster their land claims in the face of conflict, (3) deforestation does not stop when landholdings are expropriated for agrarian reform settlements, and (4) contentious properties have greater overall deforestation than uncontentious ones. To test these hypotheses, I relied on a mixed-methods approach that combines times-series analysis using pooled ordinary least squares (OLS), and fixed -and random- effects econometric models with field interviews, site visits, archival research, and remote sensing and geographic information systems (GIS) analyses. In this chapter, I focus on the insights these mixed-methods methods enabled using a spatial time-series of conflict on largeholdings (and the settlements some of them became) over multiple decades in the Southeastern region of Pará, Brazil.

Disentangling land conflict from other drivers of land cover change in a contentious place can be very difficult given that a variety of drivers of change may be taking place simultaneously, and contention is often a difficult topic on which to gather high quality information. Given the heterogeneous suite of land cover drivers active in a contentious place, any research in this domain will employ a mix of methods to unravel the connection between change and conflict.

1.1 Southeastern Pará

Southeastern Pará encompasses about 20,000 km^2 and is home to the Kayapó, Apyterewa, Xikrin do Cateté, Badjônkôre, Menkragnoti, Sororó, and Mãe Maria Indigenous reserves. Early European colonization of this region focused primarily on finding Indigenous slave laborers (Hemming 1978), and then pivoted to non-timber forest product extraction, particularly caucho rubber and Brazil Nuts, which a portion of the region, known as the Brazil Nut Polygon, is actually named after (Jadão 1984).

Later, starting with the construction of the Transamazon highway (BR-230) and PA-150 (the road linking the region with the capital of the State of Pará, Belém) in the 1960s and 1970s, the region was opened to more systematic colonization and development, creating significant problems regarding land access rights, leading to conflict. The initial period of conflict began with the arrival of large ranchers, pursuing incentives (Serrão 1990; Hecht 1985; Browder 1988; Fearnside 1979), and when the region's extractivist oligarchs figured out that ranching could be more profitable (Aldrich et al. 2020). These newly minted ranchers quickly realized that the "empty" lands they had been granted were being worked by smaller farmers, many of whom worked in the extractive economy. What ensued has been called the "*luta posseira*" (fight for ownership), and many hundreds of people, mostly smallholders, lost their lives (Simmons 2004). At the same time, the region experienced rapid and violent militarization as the Brazilian Military worked to exterminate Maoist revolutionaries (Morais and Silva 2012; Nascimento 2000).

During the slow collapse of Brazil's military dictatorship in the late 1970s and early 1980s social movements remained active, albeit in a hidden manner, advocating for the rights of small farmers, the landless, and the rural poor through rural workers unions and other groups (Simmons et al. 2002; Welch 1999; Wolford 2004; Wright and Wolford 2003). As Brazil redemocratized these social movements, such as the *Federação dos Trabalhadores e Trabalhadoras na Agricultura do Pará* (FETAGRI, the Federation of Agricultural Workers of Pará) and the *Movimento dos Trabaladores Sem Terra* (MST, the Landless Rural Worker's Movement) came forward and acted more publicly, even undertaking direct action for land reform through overt occupation of land (Wolford 2004; Simmons et al. 2002; de Almeida 1990). Since 1980, conflict has taken the lives of more than 500 people in Southeastern Pará alone (Aldrich et al. 2020), while forest cover in the Brazil Nut Polygon (see Fig. 2.1) dropped from around 88.5% in 1984 to 7.8% by 2010. Deforestation and land conflict have been shown to be connected to each other, and to exhibit some spatial concentration (Brown et al. 2016b; Aldrich et al. 2012b).

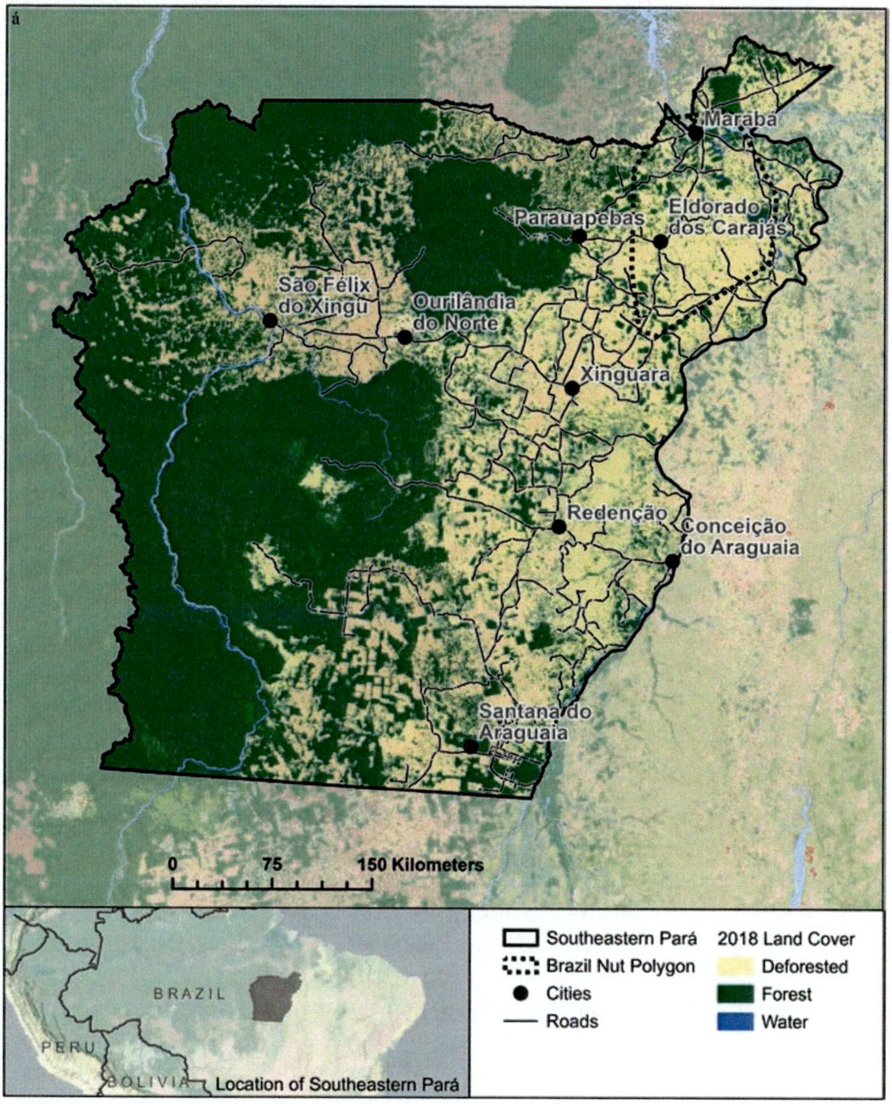

Fig. 2.1 Southeastern Pará, Brazil. The analysis discussed in this chapter focuses on the "Brazil Nut Polygon"

2 Building a Spatial Time-Series of Conflict and Deforestation

A spatial time-series has significant benefits over more traditional cross-sectional remote sensing and field-based data collection techniques. Cross-sectional analyses, such as regression or t-tests, provide snapshots of the status of important variables at a given point in time, and are often used to compare averages of variables like forest cover in different years. However, this practice can run afoul of the assumptions of inferential statistics. For example, statistical analyses assume that independent variable observations are independent from each other, but that is simply not true when comparing years (e.g., the amount of forest on a property in 2014 influences how much forest there is on it today) or locations (e.g., a certain municipality may have more deforestation than another). One solution to this issue is the application of spatially-aware methods like spatial regression, spatial sorting, or cluster analysis, but these approaches only resolve spatial autocorrelation issues; most interesting problems articulate over time, so methods requiring panel-style/longitudinal data may be required to address temporal and spatial autocorrelation concerns. For this reason, I built a spatial time-series on conflict, agronomic characteristics, and deforestation with my colleagues (Aldrich et al. 2020) and spent a significant amount of time in the field verifying our conclusions.

Social-process drivers of land change (a subset of "cultural" drivers that have less to do with cultural concepts like "belief" or a "sense of belonging" and more to do with social-interactions among economic actors) may be among the most difficult in terms of data collection. There are few officially collected datasets, and those that do exist have shortcomings. By way of example, in the case of land conflict in Southeastern Pará, records on conflict collected by the Pastoral Land Commission (e.g., CPT 2010, which is one of a series of annual reports) appear to undercount conflict-related deaths (Aldrich et al. 2020). Similarly, while municipal-level deforestation data for the Amazon is available annually through the PRODES project (PRODES 2018), many land cover changes, and land conflicts, happen at a much finer scale. This means that adequately addressing the conflict-cover change nexus cannot rely solely on official data.

Data assembly began with creating a land cover timeseries. The study area, a subset of Southeastern Pará known as the Brazil Nut Polygon, covers two Landsat 5 scenes. The classification process involved geometric and radiometric correction (the data were processed before Level 2 Landsat data were available), with geometric correction being accomplished by 30+ correspondence points between raw imagery and a Landsat orthophoto dataset created by the Tropical Rain Forest Information Center (TRFIC) at Michigan State University. Radiometric correction for the 54 Landsat 5 scenes (2 per year for a 27-year period, 1984–2010) was performed following the recommendations of Chander and Markham (2003). Landsat 5 data provided annual coverage that was relatively cloud-free, which I classified to a level-1 land cover classification (mature forest, forest regrowth, deforested, water, and cloud classes) using an unsupervised cluster classifier to generate at least 255 separate spectral classes. I then collapsed and labeled each class into the classification scheme mentioned above using a combination of field-based observation and higher-resolution ancillary data to develop class labels. The procedure employed was similar to Aldrich et al. (2012b) and Simmons et al. (2007). Finally, cloud and water pixels were standardized across all years, and an accuracy assessment was completed. Accuracy assessment was performed using field-collected ground verification points. These were randomly located using GIS functions, then revised given transportation and access constraints (i.e., it is hard/impossible to get permission to enter some private properties, and some properties are so remote it would be difficult, dangerous, and time-consuming to access). A total of 94,911.87 hectares in the two-scene study area were verified in 2007 using the Global Positioning System (GPS) and coordinate geometry (compass and laser rangefinder) to create verification polygons. The minimum accuracy across all classes (except clouds, which were impossible to verify) was 83% (Aldrich 2009).

Data on land conflict and related violence is readily available in Brazil from a few different data sources (e.g., CPT 2018; DataLuta 2017), and these data have been used in a variety of high-quality analyses (Brown et al. 2016a, Simmons 2004, 2005; Costa 1999). However, these reports often focus on recording the most overt forms of land conflict-related violence rather than the social and public violence that can come from denunciations, legal rulings, and public statements made by prosecutors, law enforcement, or politicians. Because more subtle forms of violence, or even fairly neutral public discourse, can affect landholder decisions regarding land use, a more complete cataloging of conflict was necessary to understand conflict-driven deforestation. In order to assemble such a catalog, I read and encoded more than 6000 pages, photographed from the two consistently published newspapers in the region, *Opinião!* and *O Correio do Tocantins*. A further 2572 pages were encoded by two Brazilian graduate assistants from the Federal University of Pará. Information encoded included locations or properties referenced in a conflict, the response of police and private gunmen, number of deaths, etc. (see Aldrich et al. 2020 supplemental information SI-1 for more information on the process). Encoding focused on the hard-news items from each newspaper article, focusing on the where, when, and what of each conflict event, and developed organically over the first few hundred articles. If a new theme developed (e.g., labor violations being discussed as a reason for land occupation by social movements) then it was added, and previous articles were reevaluated for the new theme. Once these conflict events were identified, I tied them spatially to the largeholding that they were associated with using a cadastral map created by the Superintendency for Amazonian Development, a now-defunct agency charged with encouraging economic development in the region (SUDAM 1990). This process yielded a highly-detailed, spatially explicit, dataset detailing every conflict event reported between 1984 and 2010, along with facts about the conflict such as whether gunmen or the police were involved, if environmental violations by largeholders were reported, if labor violations on the property were reported, whether the largeholder was known as an "enemy" of land reform, or whether there were other public denunciations of wrongdoing on the land in question.

In addition to land cover and conflict, any systematic analysis of social-process drivers of deforestation should control for more traditional infrastructure and economic conditions (among others). Because market price, production, and diversity of production data were not available over the entire study period (1984–2010, ending with the complete privatization of largeholdings in the region, which were previously mostly exploited under long-term lease agreements with the State of Pará, Brazil), I proxied for market access with cost-distance variables (distance to roads, distance to urban centers, with distance incorporating road transport and surface roughness). Agricultural suitability/productive capacity were controlled for by variables regarding soil suitability (EMBRAPA and IBGE 1999) and distance to water. Given the multidecadal timescale of the study, I also controlled for salient climate influences on production by including annual precipitation from NOAA/OAR/ESRL PSD (Xie and Arkin 1997, updated regularly, see: https://psl.noaa.gov/). Inflation rates and Brazil's Gross Domestic Product were also considered as they have changed over time. All of these features were added to the spatiotemporal dataset as well, meaning that each of the 180 largeholdings in the region had a full accounting of land cover, conflict, infrastructure, economic, and physical aspects for each year from 1984 to 2010.

Building such a comprehensive dataset with limited extra support (e.g., research assistants) took years. Data assembly spanned a decade itself (approximately late 2003 through 2014, with some updates in the years since). The intensive nature of assembling such a dataset counterbalances the power and utility of understanding a complex process; while a comprehensive understanding of conflict-driven deforestation in a region provides insights that serve beyond the time period and study area, the invested labor makes it understandable that many researchers cannot commit the time and resources to deconstruct complex processes in this way. In addition, other researchers in this project put in significant time assisting with data collection, analysis, and writing up results, so fully preparing such a comprehensive dataset will almost always be a team effort.

One complicating factor in research like this is that history is poorly documented, even though historical factors may explain contemporary behaviors. For example, in the case of land conflict, the particularly noteworthy brutality of one wealthy largeholder family can change the trajectory of land occupation in an entire region. Understanding this history takes much more than looking at a database on conflict events or deforestation; speaking with all types of stakeholders is a very important step. My work in Southeastern Pará involved dozens of interviews with ranchers, landless movement members, successfully settled small farmers, rural worker's union leaders, government officials, and academic researchers living and working in the region. The picture that they paint of the evolution of land conflict, and the responses to it, is much more nuanced than the presentations of previous (very good) monographs and articles discussing subsidies (Hecht 1985), Indigenous and smallholder conflict (Schmink 1982), and conflict related to wildcat mining (Schmink and Wood 1992). It also contradicts, in some ways, the general sense that the military dictatorship eliminated and coopted political organization around landless and rural worker's rights.

Key informant interviews flesh out the skeleton that is provided by systematic databases, placing real people into the story, and can, when coded into standardized data, even drive statistical analyses (Aldrich 2015). They can also be fraught with risk ranging from physical danger (threats of violence, including brandishing of weapons), or from having potential future interviews cut off by an informant's reaction to a question (i.e., one informant may warn-off others from talking to a researcher). Nevertheless, these interviews are invaluable to understanding the context for any action. In one interview in 2013, undertaken with Eugenio Arima and Cynthia Simmons in Marabá, we interviewed a man who had lived in Southeastern Pará since the heyday of the Brazil Nut trade. His account of the ongoing semi-covert operation of the rural worker's unions and landless movements beginning in the early 1970s made it clear that the poor had some power even during the dictatorship. Interestingly, he also described the process by which Brazil Nut oligarchs (see Emmi 1999) gradually transitioned to cattle ranchers as their children, sent to elite domestic and/or foreign Universities for education, returned home with new ideas about agricultural production and methods to tap into the wealth of the state (i.e., subsidies for ranching). While many new ranchers were certainly from outside the region, including international corporations such as Volkswagen (Acker 2017), many families with long histories in the region also converted their holdings to ranches during the heyday of subsidies. More contemporaneously, these ranchers also have very well-developed views of why some private landholdings have problems with conflict and why others do not (Aldrich 2015). For example, ranchers believe absentee landlords are likely to be targeted for property occupation by social movements, and that social movements prefer properties with less forest on them because they are easier to plant.

Other informants vastly improved our knowledge of the activity of social movements which spun out of the long simmering rebellion that lived on in the rural worker's unions (*Syndicatos dos Trabalhadores Rurais*, or STRs), including locally important movements such as FETAGRI, CONTAG (*Confederação Nacional dos Trabalhadores da Agricultura*), *Federação dos Agricultores/as Familiares do Brasil* (FETAFRI), Via Campesina, and the ubiquitous (at least in academic literature) MST, to name only a few. The interactions between these groups are certainly fertile ground for further research as they certainly worked with, and against, each other during the history of conflict in Southeastern Pará.

The role of the state in the land reform process is also very important, and government officials have interesting opinions about why land conflict has been a problem in Southeastern Pará. Officials at the Institute for Colonization and Agrarian Reform (INCRA) in Marabá were happy to tell me what types of land characteristics make a property prime for conflict and land occupation, although their opinions were quite different from other stakeholders, and mostly not supported by actual action of the social movements themselves (Aldrich 2015). Indeed, INCRA's statements focused on the agronomic potential of land rather than their strategic value to social movements trying to affect change. For example, INCRA officials indicated that access to water and proximity to market centers would increase the odds of land conflict, offering the explanation that settlers need easy access to markets and need to live near water. Social movement leaders saw their needs differently, often looking toward other factors such as a largeholder reputation for being vicious during conflicts, which might help them publicize the struggle of the landless, or the presence of labor or environmental violations that could improve the odds of a victory in court (Aldrich 2015). Without extensive key informant interviews we would know little about this problematic mismatch between actions on the ground by social movements and their largeholder adversaries and the ideas about agricultural suitability espoused by the officials tasked with addressing these conflicts.

In situ visits with informants on outlier properties, like the ones shown in Fig. 2.2, were the only way to understand what was happening on largeholdings (or former largeholdings) which ran counter to the trend of near-total deforestation or exceeded even the deforestation that would be expected according to regression models (Aldrich et al. 2012b). The landholding (polygon) on the right in Fig. 2.2 was one of a handful that appeared to remain highly forested in the initial imagery from 1984 and remained as recently as December 2020. This would constitute a highly

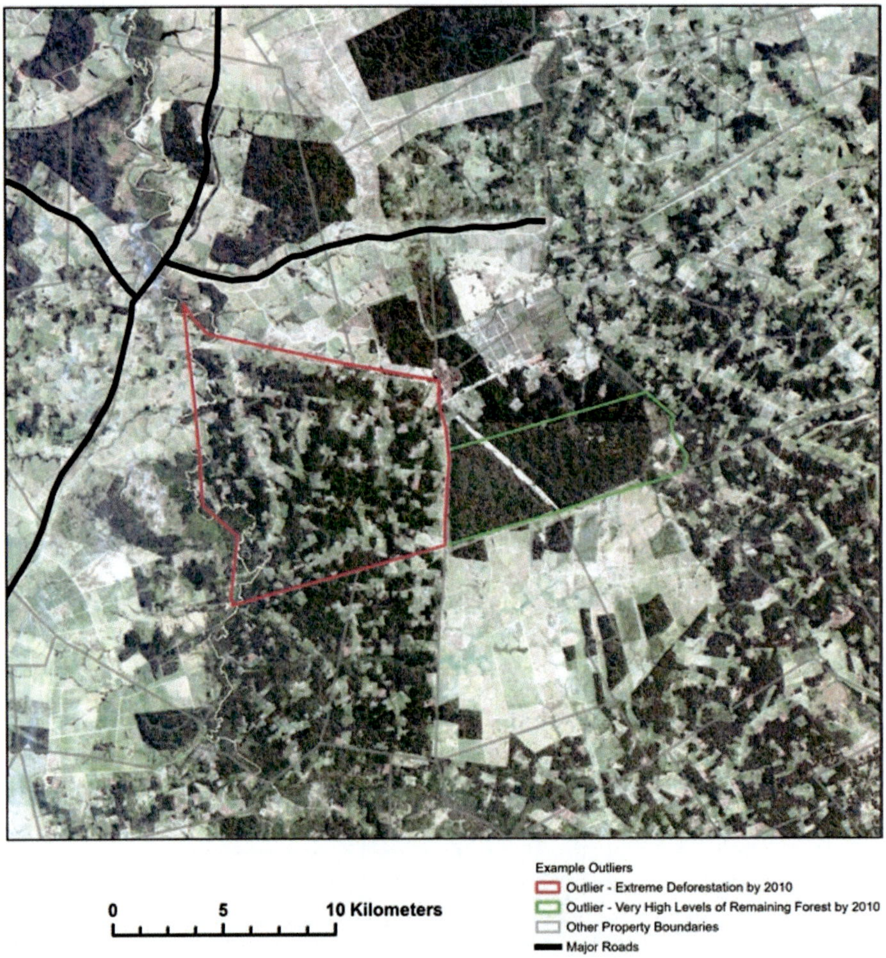

Fig. 2.2 Two outliers that were substantially more forested (polygon on the right) and substantially less forested (polygon on the left) in 2010 than regression models would suggest. In situ interviews showed that the forests of the right polygon (and some adjacent areas) had been replaced with planted Teak, and adjacent areas (left polygon) had been partially deforested to create charcoal

suspicious outlier in a region that has almost no forest left and was located next to another outlier (polygon on the left in Fig. 2.2) that experienced tremendous levels of deforestation. By visiting this site and talking to smallholders living nearby, it became clear that the highly forested area was a Teak plantation (this was obvious just by driving by) despite the fact that on satellite imagery it looked (visually and spectrally) just like native forest throughout the study. Field verification is essential to accurate analysis of land change, as is actual investigation of the history of outliers. In addition, during the same field visit we learned about the highly deforested adjacent largeholding, which was overrun by opportunists converting forest into charcoal for use by local smelters processing iron from the enormous *Carajás* mine just down the road.

Despite the difficulty of making a dataset like the one I describe, and verifying its information with fieldwork, it is the type of work that political ecologists, human-environment geographers, and mixed-method researchers have recognized as important for quite some time and continue to call for (Le Billon and Duffy 2018; Doolittle 2015; Forsyth 2015). While many projects require a combination of some of the methods I describe above (interviews, remote sensing, surveys, analysis of existing data, archival and historical research, coding archival data into spatial and temporal location), many studies could benefit from more mixed-methods analysis, if only in order to confirm (or further refine) their primary hypotheses.

3 Primary Findings

With greater understanding of the historical, economic, and social context of Southeastern Pará, it became clear that contentious land change is not the only process at work in the region. While previous work had indicated that contention was among the most significant drivers of deforestation (Aldrich et al. 2012b), when the analysis is extended to cover a broader time-period (1984 to 2010 instead of only 1984 to 2003) covering a few years when forest cover in the region was almost totally removed and land conflict became less intense, it turns out that the passage of time, and its associated economic and policy-driven deforestation, are also important.

By employing pooled regression and panel analysis (fixed and random effects) on the spatial time-series data that we (Aldrich et al. 2020) developed, we were able to show that forest cover and conflict were associated differently during three different periods shown in Fig. 2.3. The first period, corresponding to Fig. 2.3a (1984 to 1994) maintained a low level of conflict and a high level of forest cover. However, forest declined more rapidly after 1992 (area between Fig. 2.3a and b), and a dramatic rise in conflict, peaking in 1996 with the infamous massacre of 19 landless movement members. During the final period (to the right, Fig. 2.3b) forest continued to decline, leveling out by the late 2000s, with conflict also varying dramatically from year to year. Our spatial time-series allows us to confirm that these time periods are statistically distinct from one another using fixed effects panel regression

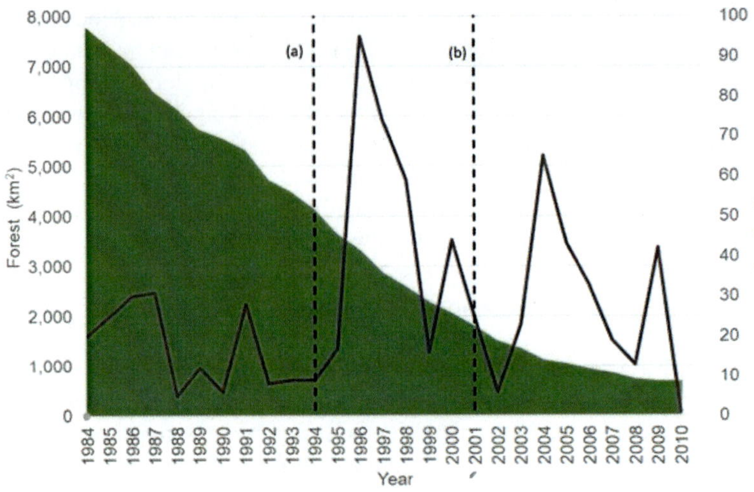

Fig. 2.3 Forest cover and conflict are episodic, concentrated in multiple time periods, and associated with deforestation

(Aldrich et al. 2020, S4 Table). This suggests that during the first period (to the left, Fig. 2.3a), deforestation was primarily driven by agronomic (economic and policy) drivers. During period two (between Fig. 2.3a and b) conflict was a strong driver of deforestation, while in the final period (to the right, Fig. 2.3b) deforestation was caused by a mix of drivers, including conflict and agronomic drivers. Indeed, categorical variables in panel regression indicate that conflict is a significant driver of deforestation during the period 1994 to 2001.

The spatial time-series also allowed us to examine the role of agrarian reform settlements on deforestation. This is an important distinction given the arguments that still abound in Brazil (and more broadly) about who is responsible for deforestation (de Paula Pereira et al. 2022; Fearnside 2017; Godar et al. 2012; Alston et al. 2000), with different researchers, and the public, blaming smallholders or largeholders, and sometimes other groups, for extensive deforestation. To some degree the history of deforestation on settlements hinges heavily on the flurry of settlement-formation activity that took place leading up to, and after, the 1996 massacre mentioned above. While later-period deforestation on settlements overwhelms deforestation on largeholdings, blaming smallholders for elimination of remaining forests on already abused former largeholdings is unfair. Additionally, though it cannot be seen in Fig. 2.4, largholding deforestation continues at approximately 1999 levels all the way through 2010, meaning that largeholders are still responsible for an average of more than 3500 km^2 of deforestation in the region each year after 1999.

Additionally, we show that deforestation is higher on properties with conflict than on those that spend the entire 27-year study period without significant conflict. This difference is significant ($p = 0.000$, using a categorical variable in a fixed effects panel regression) and substantial, with uncontentious properties having >800 hectares more forest. These trends grow over time, too, as shown in Fig. 2.5.

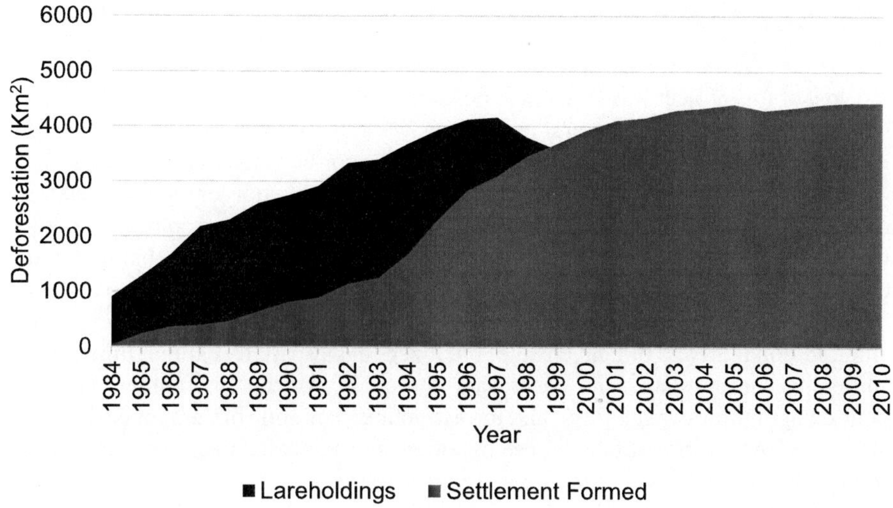

Fig. 2.4 Deforestation on agrarian reform settlements (former largeholdings) and contemporary largeholdings

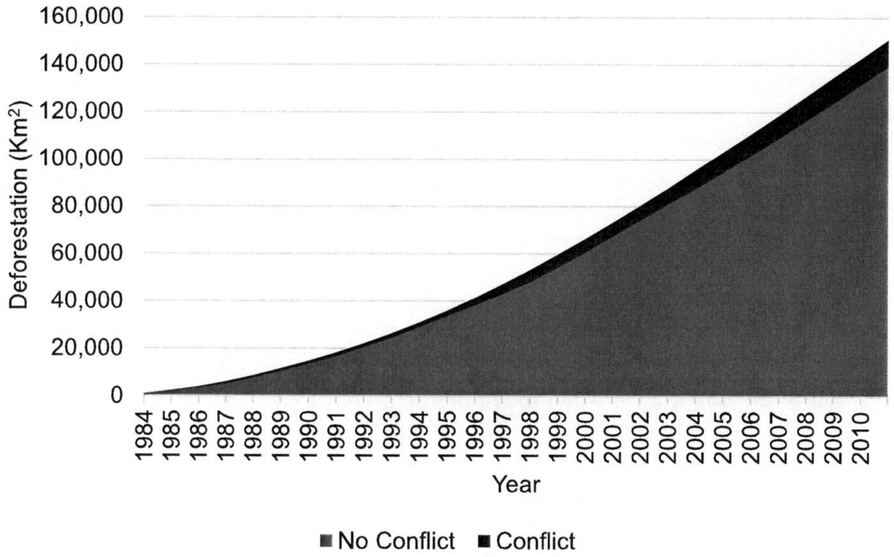

Fig. 2.5 Deforestation (in km^2) on Properties with Conflict and Without Conflict. Properties with Conflict have higher overall deforestation

4 Conclusions

The goal of this project was to understand how conflict and land change were related in Southeastern Pará. The general hypotheses were that social movements have different ideas about what makes a property attractive for occupation than other stakeholders, and that largeholders actively deforest their land to bolster their land claims in the face of conflict. Finally, I also hypothesized that deforestation does not stop when landholdings are expropriated for agrarian reform settlements, but that contentious properties have higher amounts of total deforestation. In order to address these hypotheses, and also to better understand how cultural drivers may interact with traditional drivers of land change, I realized that a spatially explicit time-series combining conflict data, land cover data, the quality of the environment (with a focus on agricultural suitability), and infrastructure/economic drivers for each property was necessary. I addressed these hypotheses statistically, using pooled regression and fixed and random effects panel regression.

The results suggest that there is broad disagreement among stakeholders regarding why properties get occupied. There is also significant complexity among drivers of land change in Southestern Pará. While conflict certainly drives deforestation, a result that has been replicated in this region (and more broadly in the Amazon) before (Aldrich et al. 2020; Simmons et al. 2002, 2007, 2019; Brown et al. 2016a; Araujo et al. 2009; Alston et al. 2000; Hall 1989; Foweraker 1981a), it would be wrong to suggest that it is a constant driver of change, or that largeholders are the only ones responsible (although they certainly *are* responsible for significant deforestation). Instead, land conflict is episodic and concentrated in certain points in time at a given place and seems to serve as a temporary exacerbator of deforestation. However, ignoring episodic exacerbators of land change is unwise as they can change the intensity and trajectory of land change processes in a very short amount of time.

Without a spatially-explicit time-series for each landholding in the study area, these analyses would not have been possible. While the cost of assembling these data can be expensive in terms of time and effort, the insights they can provide are well worth the costs, even if the mounting pressure of environmental change and human welfare combined with a desire for quick results may push researchers to employ cross-sectional techniques. In addition, taking the time to understand any region, its history, and the livelihoods and struggles of those who live within it, is always worth doing (Walker et al. 2011). This is not a new insight, but it is worth emphasizing; sometimes slower science is more fulfilling, useful, and interesting than quick analyses.

Research on cultural drivers of land change in Latin America is particularly important given the history of weak rule of law, overlapping institutions charged with land management and land tenure, and ongoing conflict hotspots surrounding land access, environmental degradation, and political strife. Conflict is a particularly difficult underlying driver of land change because it is often episodic and tough to empirically isolate when rhetoric intentionally muddies the situation. To isolate the effects of a cultural driver of land change like conflict researchers have to approach the problem both quantitatively (in a spatial and temporally aware mode) and also qualitatively in order to avoid missing important episodes, nuances, and categorizations. For example, without substantial fieldwork, we never would have learned about teak plantations, illegal charcoal operations, or the fact that social movements were active in agitating for land reform even during Brazil's dictatorship. Cross-sectional, and remote-only treatments of cultural drivers are not likely to be authoritative given these issues, and a spatial time-series approach, coupled with substantial field research efforts, are likely to provide a much more complete view of the influence of cultural drivers on land change.

Acknowledgements I would like to acknowledge funding support for this research from the National Science Foundation (NSF-GSS # 1157521). The landless of the Southeastern Pará, their largeholder rivals, and government officials must be acknowledged for their willingness to speak with me, and for their patience. The full dataset necessary to reproduce these findings is available at Sycamore Scholars, the institutional repository at Indiana State University, http://hdl.handle.net/10484/12380.

References

Acker A (2017) Volkswagen in the Amazon: the tragedy of global development in modern Brazil. Cambridge University Press, Cambridge

Aldrich SP (2009) Land conflict and deforestation in the Eastern Brazilian Amazon. Dissertation, Michigan State University, East Lansing

Aldrich SP (2015) Decision-making and the environment in the Amazon land war. J Land Use Sci 10(1):38–58

Aldrich SP, Walker RT, Arima E et al (2006) Land-cover and land-use change in the Brazilian Amazon: smallholders, ranchers, and frontier stratification. J Econ Geogr 82(3):265–288

Aldrich S, Walker R, Simmons C, Caldas M, Perz S (2012a) Contentious land change in the Amazon's arc of deforestation. Ann Am Assoc Geogr 102(1):103–128

Aldrich SP, Walker RT, Simmons CS et al (2012b) Contentious land change in the Amazon's arc of deforestation. Ann Am Assoc Geogr 102(1):103–128

Aldrich SP, Simmons CS, Arima E et al (2020) Agronomic or contentious land change? A longitudinal analysis from the eastern Brazilian Amazon. PLoS ONE 15(1):e0227378. https://doi.org/10.1371/journal.pone.0227378

Alston LJ, Libecap GD, Mueller B (2000) Land reform policies, the sources of violent conflict, and implications for deforestation in the Brazilian Amazon. J Environ Econ Manage 39(2):162–188

Araujo C, Bonjean CA, Combes J-L et al (2009) Property rights and deforestation in the Brazilian Amazon. Ecol Econ 68(8–9):2461–2468. https://doi.org/10.1016/j.ecolecon.2008.12.015

Browder JO (1988) The social costs of rain forest destruction: a critique and economic analysis of the 'Hamburger Debate'. Interciencia 13(3):115–120

Brown D, Brown JC, Brown C (2016a) Land occupations and deforestation in the Brazilian Amazon. Land Use Policy 54:331–338. https://doi.org/10.1016/j.landusepol.2016.02.003

Brown DS, Brown JC, Brown C (2016b) Land occupations and deforestation in the Brazilian Amazon. Land Use Policy 54:331–338. https://doi.org/10.1016/j.landusepol.2016.02.003

Bunker SG (1985) Underdeveloping the Amazon : extraction, unequal exchange, and the failure of the modern state. University of Chicago Press, Chicago

Caldas M, Walker R, Arima EY et al (2007) Theorizing land cover and land use change: the peasant economy of Amazonian deforestation. Ann Am Assoc Geogr 97(1):86–110

Chander G, Markham B (2003) Revised Landsat-5 TM radiometric calibration procedures and postcalibration dynamic ranges. IEEE Trans Geosci Remote Sens 41(11):2674–2677

Costa LM (1999) Discurso e Conflito: Des Anos de Disputa pela Terra em Eldorado do Carajás. Universidade Federal do Pará, Belém

CPT (2010) Conflitos no Campo Brasil – 2010 (Land conflicts in the Brazilian countryside – 2010). Commissão Pastoral de Terra (Pastoral Land Commission), Goiania, Gois

CPT (2018) Conflitos no Campo Brasil – 2018 (Land conflicts in the Brazilian countryside – 2016). Commissão Pastoral de Terra (Pastoral Land Commission), Goiania, Gois

Dale VH, O'Neill RV, Pedlowski M, Southworth F (1993) Cause and effects of land-use change in Central Rondônia, Brazil. Photogramm Eng Remote Sens 59:997–1005

DataLuta (2017) Núcleo de Estudos, Pesquisas, e Projetos de Reforma Agrária, UNESP (Nucleus for study, research, and projects on agrarian reform), São Paulo

de Almeida AVB (1990) The state and land conflict in Amazonia, 1964-88. In: Goodman D, Hall A (eds) The future of Amazonia. St. Martin's Press, New York, pp 226–244

de Paula Pereira ASA, dos Santos VJ, do Carmo Alves S, e Silva AA, da Silva CG, Calijuri ML (2022) Contribution of rural settlements to the deforestation dynamics in the legal Amazon. Land Use Policy 115:106039

Doolittle A (2015) The best of many worlds: methodological pluralism in political ecology. In: The international handbook of political ecology. Edward Elgar Publishing, Northampton

EMBRAPA, IBGE (1999) Mapa de solos do Brasil (Soil map of Brazil). Instituto Brasileiro de Geografica e Estátisticas – IBGE (Brazilian Institute of Geography and Statistics – IBGE), Brasília

Emmi M (1999) A Oligarquia do Tocantins e o Domínio dos Castanhais (The oligarchy of Tocantins and the domain of the Brazil-nut). Universidade Federal do Pará (Federal University of Pará), Belém, Pará

Fearnside PM (1979) Cattle yield prediction for the transamazon highway of Brazil. Interciencia 4(4):220–225

Fearnside PM (1982) Deforestation in the Brazilian Amazon: how fast is it occurring. Interciencia 7(2):82–88

Fearnside P (2017) Deforestation of the Brazilian Amazon. In: Oxford research encyclopedia of environmental science. Oxford University Press, Oxford

Forsyth T (2015) Integrating science and politics in political ecology. In: The international handbook of political ecology. Edward Elgar Publishing, Northampton

Foweraker J (1981a) The pioneer frontier; political violence and the peasantry. In: the struggle for land: a political economy of the pioneer frontier in Brazil from 1930 to the present day, vol 39. Cambridge Latin American studies. Cambridge University Press, New York, pp 3–27

Foweraker J (1981b) The struggle for land: a political economy of the pioneer frontier in Brazil from 1930 to the present day, vol 39. Cambridge Latin American Studies, Paperback edition edn. Cambridge University Press, Cambridge

Geist HJ, Lambin EF (2001) What drives tropical deforestation?: a meta-analysis of proximate and underlying causes of deforestation based on subnational case study evidence. LUCC International Project, Louvain-la-Neuve

Godar J, Tizado EJ, Pokorny B (2012) Who is responsible for deforestation in the Amazon? A spatially explicit analysis along the Transamazon highway in Brazil. For Ecol Manag 267:58–73. https://doi.org/10.1016/j.foreco.2011.11.046

Goodland RJ, Irwin HS (1975) Amazon jungle: green hell to red desert? An ecological discussion of the environmental impact of the highway construction program in the Amazon basin. Elsevier Scientific Publishing Co., Amsterdam

Hall AL (1989) Developing Amazonia: deforestation and social conflict in Brazil's Carajas Programme. Manchester University Press, Manchester

Hecht S (1981) Deforestation in the Amazon Basin: magnitude, dynamics and soil resource effects. Paper presented at the Semenário Regional Expansão da Fronteira Agricola e Meio Ambiente na América Latina (Seminar on Regional Expansion of the Agricultural Frontier and the Environment in Latin America), Brasilia

Hecht SB (1985) Environment, development and politics: capital accumulation and the livestock sector in eastern Amazonia. World Dev 13(6):663–684

Hecht S, Rajão R (2020) From "green hell" to "Amazonia legal": land use models and the re-imagination of the rainforest as a new development frontier. Land Use Policy 96:103871. https://doi.org/10.1016/j.landusepol.2019.02.030

Hemming J (1978) Red gold: the conquest of the Brazilian Indians. Harvard University Press, Cambridge

Jadão, PBR (ed) (1984) Marabá: a história de uma parte da Amazônia, da gente que nela vivia e da gente que a desbravou e dominou, fazendo-a emergir para a civilização. De 1892 até nossos dias. Prefeito Municipal de Marabá (Municipal Offices of Marabá), Marabá, Pará

Kröger M (2020) Deforestation, cattle capitalism and neodevelopmentalism in the Chico Mendes extractive reserve. Brazil J Peasant Stud 47(3):464–482

Kull CA (2002) Madagascar aflame: landscape burning as peasant protest, resistance, or a resource management tool? Polit Geogr 21:927–953

Lambin E (1997) Modeling and monitoring land-cover change processes in tropical regions. Prog Phys Geogr 21(3):375–393

Lauer DT, Morain SA, Salomonson VV (1997) The Landsat program: its origins, evolution, and impacts. Photogramm Eng Remote Sens 63(7):831–838

Le Billon P, Duffy RV (2018) Conflict ecologies: connecting political ecology and peace and conflict studies. J Polit Ecol 25(1):239–260

Lucas A, Warren C (2013) Land for the people: the state and agrarian conflict in Indonesia. Ohio University Press, Athens

Mausel P, Brondizio ES, You W et al (1993) Spectral identification of successional stages following deforestation in the Amazon. Geocarto Int 8(4):61–71

McCracken SD, Boucek B, Moran EF (2002) Deforestation trajectories in a frontier region of the Brazilian Amazon. In: Walsh S, Crews-Meyer K (eds) Linking people, place, and policy: a GIScience approach. Kluwer Academic Publishers, Boston, pp 215–234

Messina JP, Walsh SJ (2001) 2.5D morphogenesis: modeling landuse and landcover dynamics in the Ecuadorian Amazon. Plant Ecol 158(1):75–88

Messina JP, Walsh SJ (2005) Dynamic spatial simulation modeling of the population – environment matrix in the Ecuadorian Amazon. Environ Plann B Plann Des 32(6):835–856

Messina JP, Walsh SJ, Mena CF, Delamater PL (2006) Land tenure and deforestation patterns in the Ecuadorian Amazon: conflicts in land conservation in frontier settings. Appl Geogr 26(2):113–128

Morais T, Silva E (2012) Operação Araguaia: os Arquivos Secretos da Guerrilha (operation Araguaia: the secret archives of the little war), 5th edn. Geração Editorial, São Paulo

Moran EF, Brondízio ES, Mausel P, Wu Y (1994) Integrating Amazonian vegetation, land-use, and satellite data. BioScience 44(5):329–339

Moran EF, Siqueira A, Brondizio E (2003) Household demographic structure and its relationship to deforestation in the Amazon basin. In: Fox J, Mishra V, Rindfuss R, Walsh S (eds) People and the environment: approaches to linking household and community surveys to remote sensing and GIS. Kluwer Academic Publishers, Norwell, pp 61–89

Nascimento DM (2000) A Guerrilha do Araguaia: Paulistas e Militares na Amazônia. Universidade Federal do Pará, Belém

O'Loughlin J, Witmer FDW (2010) The localized geographies of violence in the North Caucasus of Russia, 1999–2007. Ann Am Assoc Geogr 101(1):178–201. https://doi.org/10.1080/00045608.2010.534713

Pan WKY, Walsh SJ, Bilsborrow RE et al (2004) Farm-level models of spatial patterns of land use and land cover dynamics in the Ecuadorian Amazon. Agric Ecosyst Environ 101:117–134

Percival V, Homer-Dixon T (1998) Environmental scarcity and violent conflict: the case of South Africa. J Peace Res 35(3):279–298. https://doi.org/10.1177/0022343398035003002
Pfaff ASP (1999) What drives deforestation in the Brazilian Amazon? J Environ Econ Manage 37(1):26–43
PRODES (2018) Taxas anuais de desmatamento na Amazônia Legal Brasileira (AMZ) [Annual deforestation rate in the Brazilian Legal Amazon (AMZ)]. Instituto Nacional de Pesquisas Espaciais [National Institute of Space Research]. http://terrabrasilis.dpi.inpe.br/app/map/deforestation?hl=en. Accessed 30 Aug 2022
Putzel J (1992) A captive land: the politics of agrarian reform in the Philippines. Catholic Institute for International Relations, London
Rodríguez-de-Francisco JC, del Cairo C, Ortiz-Gallego D et al (2021) Post-conflict transition and REDD+ in Colombia: challenges to reducing deforestation in the Amazon. Forest Policy Econ 127:102450
Schmink M (1982) Land conflicts in Amazonia. Am Ethnol 9(2):341–357
Schmink M, Wood CH (1992) Contested frontiers in Amazonia. Columbia University Press, New York
Serrão EA (1990) The search for sustainability in Amazonian pastures. In: Anderson AB (ed) Alternatives to deforestation: steps toward sustainable use of the Amazon rain Forest. Columbia University Press, New York, pp 196–214
Simmons CS (2004) The political economy of land conflict in the eastern Brazilian Amazon. Ann Am Assoc Geogr 94(1):183–206
Simmons CS (2005) Territorializing land conflict: space, place, and contentious politics in the Brazilian Amazon. GeoJournal 64(4):307–317
Simmons CS, Perz S, Pedlowski MM, Teixeira Silva LG (2002) The changing dynamics of land conflict in the Brazilian Amazon: the rural-urban complex and its environmental implications. Urban Ecosyst 6:99–121
Simmons CS, Walker RT, Arima EY et al (2007) Amazon land war in the south of Pará. Ann Am Assoc Geogr 97(3):567–592
Simmons CS, Walker R, Aldrich S et al (2019) Discipline and develop: destruction of the Brazil nut Forest in the lower Amazon Basin. Ann Am Assoc Geogr 109(1):242–265. https://doi.org/10.1080/24694452.2018.1489215
SUDAM (1990) Perfil da ocupação do solo e subsolo Paraenses (Profile of the occupation of the soil and subsurface claims of the State of Pará). Brasilia
UNEP (2009) Protecting the environment during armed conflict: an inventory and analysis of international law. United Nations Environmental Programme, Nairobi
Unruh JD (2009) Land rights in postwar Liberia: the volatile part of the peace process. Land Use Policy 26(2):425–433. https://doi.org/10.1016/j.landusepol.2008.05.005
Walker RT (1996) Land use dynamics in the Brazilian Amazon. Ecol Econ 18(1):1–80
Walker RT (2003) Mapping process to pattern in the landscape change of the Amazonian frontier. Ann Am Assoc Geogr 93(2):376–398
Walker R, Simmons C, Aldrich S et al (2011) The Amazonian theater of cruelty. Ann Am Assoc Geogr 101(5):1156–1170
Welch C (1999) The seed was planted: the São Paulo roots of Brazil's rural labor movement, 1924–1964. Penn State Press, State College
Williams DL, Goward S, Arvidson T (2006) Landsat. Photogramm Eng Remote Sens 72(10):1171–1178
Williams D, Goward S, Irons J et al (2018) Landsat's history: the story behind nearly a half-century of monitoring the Earth's surface with Landsat. J Earth Space Sci Open Arch. https://doi.org/10.1002/essoar.10500195.1%U
Wolford W (2004) This land is ours now: spatial imaginaries and the struggle for land in Brazil. Ann Am Assoc Geogr 94(2):409–424

Woods KM, Wang P, Sexton JO et al (2021) Integrating pixels, people, and political economy to understand the role of armed conflict and geopolitics in driving deforestation: the case of Myanmar. Remote Sens 13(22):4589

Wright A, Wolford W (2003) To inherit the earth: the landless movement and the struggle for a new Brazil. Food First Books, Oakland

Wulan YC, Yasmi Y, Purba C, Wollenberg E (2004) An analysis of forestry sector conflict in Indonesia 1997–2003. CIFOR, Bogor

Wulder MA, Loveland TR, Roy DP et al (2019) Current status of Landsat program, science, and applications. Remote Sens Environ 225:127–147. https://doi.org/10.1016/j.rse.2019.02.015

Xie P, Arkin PA (1997) Global precipitation: a 17-year monthly analysis based on gauge observations, satellite estimates, and numerical model outputs. Bull Am Meteorol Soc 78:2539–2558

Chapter 3
Crossing Boundaries: Transboundary Geographic Information in the Amazon Borderlands of Peru and Brazil

David Seward Salisbury, Claire Victoria Powell, Bertha Balbín Ordaya, Pedro Tipula Tipula, Desiree Estilita Alvarado, Miguel Alva Huayaney, Vera Reis Brown, Elaine Lopes, Antonio Willian Flores de Melo, Sidney Novoa Sheppard, Maria Luiza Pinedo Ochoa, Piero Enmanuel Rengifo Cardenas, Sonaira Souza da Silva, José Frank de Melo Silva, Stephanie A. Spera, Jorge Washington Vela Alvarado, and David Orlando González Gamarra

Abstract The Amazon borderlands' combination of remoteness, lack of state presence, high levels of biocultural diversity, abundant resources, flows of people and goods, and a lack of geographic information creates significant challenges for transboundary cooperation to reconcile conservation and development despite shared biophysical and cultural landscapes. Preliminary transboundary maps opened the

D. S. Salisbury (✉)
Department of Geography, Environment & Sustainability, University of Richmond, Richmond, VA, USA

Instituto Panamericano de Geografía e Historia, Pucallpa, Peru
e-mail: dsalisbu@richmond.edu

C. V. Powell · S. A. Spera
Department of Geography, Environment & Sustainability, University of Richmond, Richmond, VA, USA

B. B. Ordaya · P. T. Tipula
Instituto Panamericano de Geografía e Historia, Pucallpa, Peru

D. E. Alvarado
Instituto Geográfico Nacional de Perú, Lima, Peru

M. A. Huayaney
Universidad Nacional Mayor de San Marcos, Lima, Peru

V. R. Brown
Secretaria Estadual de Meio Ambiente Acre, Acre, Brazil

S. López (ed.), *Socio-Environmental Research in Latin America*, The Latin American Studies Book Series, https://doi.org/10.1007/978-3-031-22680-9_3

eyes of governmental and non-governmental organizations and civil society to the previously invisible landscapes of the neighboring countries. Increasing demand for transboundary maps and analysis led to the Transboundary Geographic Group of the Southwestern Amazon's (GTASO for its acronym in Spanish) mapping workshops in 2012, 2013, and 2019 that broadened the public's understanding of these unique and important borderland landscapes on both sides of the border. GTASO grew out of collaborations between researchers seeking to map the remote Amazon borderlands of Peru and Brazil. This chapter describes the evolution of GTASO efforts to map the Amazon borderlands shared by Acre, Brazil, and Ucayali and Madre de Dios, Peru, and to a lesser extent Pando, Bolivia. We first describe the Amazon rainforest and borderlands, PanAmazonia mapping efforts, and the national mapping approaches of Peru and Brazil, and then present a critical reflection on the GTASO transboundary mapping initiative and its outcomes.

Keywords Amazon borderlands · Mapping · GIS · Environmental change · Decision making

1 Introduction

The Southwestern Amazon region shared by Peru and Brazil spans landscapes characterized by both high levels of ecological and cultural diversity, and a broad range of rapidly accelerating threats to the peoples, flora, fauna, landforms, water, and climate of the region. The Amazon borderland location adds additional complexity due to the unique intersections, overlaps, blending, and blurring of political, cultural, and biophysical edges. Twenty years ago, regional thematic transboundary maps of the Amazon borderlands were at best, elusive and erroneous, and at worst, simply nonexistent. Since 2002, geographic information systems (GIS) have served as a common language and approach for an alliance of diverse Amazonian and Amazon-focused actors to build an improved understanding of the ecological,

E. Lopes
Universidade Federal de Lavras, Minas Gerais, Brazil

A. W. F. de Melo · S. S. da Silva
Universidade Federal do Acre, Campus Floresta, Acre, Brazil

S. N. Sheppard · P. E. R. Cardenas
Conservación Amazónica – ACCA, Madre de Dios, Peru

M. L. P. Ochoa · J. F. de Melo Silva
Comissão Pró-Índio do Acre, Acre, Brazil

J. W. V. Alvarado
Universidad Nacional de Ucayali, Ucayali, Peru

D. O. G. Gamarra
Universidad Nacional de San Antonio Abad del Cusco – Puerto Maldonado, Cusco, Peru

economic, cultural, and political dynamics of the transboundary region and to share this with Indigenous, governmental, civil society, and academic stakeholders. Soil scientists, remote sensors, anthropologists, forestry scientists, agronomists, conservation scientists, Indigenous leaders, climate scientists, geospatial analysts, and geographers have been drawn together by their concerns for the region and the power of geovisualization and geospatial analysis. More recently, GIScience and remote sensing approaches have accelerated our ability to expand, intensify, and integrate socio-environmental analysis of transboundary degradation even as new geovisualization products and processes can be adapted to a broader range of regional narratives ranging from the imperiled migratory routes of Indigenous people in isolation and initial contact to the potential climate impacts of a more expansive transboundary transportation network. As our geospatial analytical power accelerates in intensity and speed, and development pressures and changes in climate are felt in parallel, additional efforts will be required to integrate processes and share useful and relevant products with local communities and regional stakeholders living in the landscapes of inquiry. The Transboundary Geographic Group of the Southwestern Amazon (GTASO) introduced in this chapter, understands how local transboundary and borderland geographic data can help marginalized communities and landscapes to share their story and overcome their lack of consistent, accurate, and updated geographic representation at larger scales. To better understand the GTASO initiative, the chapter begins by introducing the Amazon rainforest and borderlands, PanAmazonia mapping efforts, and the national mapping approaches of Peru and Brazil.

1.1 Amazon Rainforest and Borderlands

The Amazon, the largest contiguous rainforest on the planet at 7.8 million km^2, is vital for the maintenance of regional water cycles, carbon cycles, biodiversity, and climate patterns. The 6280-kilometer-long Amazon River, by far the largest in the world by water volume, carries some 15% of the planet's running freshwater (Ghai et al. 2011). Experts estimate the Amazon's 12,000 tree species hold between 86 and 120 billion metric tons of carbon across the entire biome (Nagy et al. 2016) with one hectare (2.5 acres) potentially containing over 650 species (ter Steege et al. 2016; Bass et al. 2010). The variety of trees, abundance of water and sunlight fill the Amazon with a diversity of fauna: over 5000 types of fish, over 1500 species of birds, 400 plus mammals, 450 odd reptiles, over 1000 species of frogs, and millions of insects (Butler 2017). This biodiversity, freshwater, and carbon stocks are threatened by anthropogenic climate change and the continued expansion of extractive industries like industrial agriculture, ranching, mining, fossil fuel exploitation, and logging (Plotkin 2020). Lately, the greatest threats come from illegal activities such as illegal mining, illegal logging, coca cultivation/trafficking, and land speculation (Plotkin 2020). These activities are also associated with illegal/informal roads, tracks, and trails that fail to appear in official cartography. These roads, legal or illegal, also reduce forest cover which increases local temperatures and streamflow, decreases evapotranspiration and precipitation, and negatively affects the hydro-climatological regime that forest-based flora and fauna depend on (Walker, 2020). However, both the sustainable development of natural resources and the conservation of biodiversity across the Amazon rainforest is challenged by an Amazonian political geography characterized by nine different states: Bolivia, Brazil, Colombia, Ecuador, French Guiana (an overseas department of France), Guyana, Peru, Suriname, and Venezuela.

The Amazon rainforest is poorly understood because of its size, remoteness, and biological and cultural heterogeneity, but the Amazon borderlands are even less understood as this group of characteristics is coupled with boundaries shared between the nine countries. Together, these borderland conditions add complexity to the system due to the intersections, overlaps, blending, and blurring of political, cultural, and biophysical edges (Salisbury and Weinstein 2013). The Southwestern Amazon region shared by Peru and Brazil is no exception with landscapes characterized by both high levels of biological and cultural diversity, and a broad range of rapidly accelerating threats (logging, mining, industrial agriculture, climate change, etc.) to the peoples, flora, fauna, landforms, water, and climate of the region. Part of the challenge is associated with transboundary political ecology, in which officials and civil society from one country are often unaware of the threats coming from the other side of the boundary that have important multi-scalar impacts on shared borderland forests, rivers, watersheds, ecosystem services, cultures, and peoples on both sides of the border (Salisbury et al. 2014a, b). Recent research finds a stable forest, water, and energy balance in the Southwestern Amazon vital to not reach a critical tipping point across the entire biome by late mid-century (Walker 2020). But the Southwestern Amazon is shared across three states (Peru, Brazil, and Bolivia), and thus requires international cooperation to maintain stability, resiliency, and balanced earth systems. In addition, the lack of geographic information exacerbates the region's remoteness, tenuous state presence, and preponderance of illegal extractors and traffickers while limiting monitoring of the high levels of biocultural diversity, and abundant resources necessary for maintaining sustainable transboundary biophysical and cultural landscapes. However, for the last 20 years, state officials and non-governmental professionals from Amazonian states of Acre (Brazil), and Ucayali and Madre de Dios (Peru) have been reaching across the international boundary to work together on reconciling conservation and development in this critically important region (Perz et al. 2019; Salisbury et al. 2014a, b). The advent of GIS has provided new opportunities and challenges for transboundary cooperation. This chapter analyzes the efforts of GTASO, a collection of governmental, non-governmental, and academic GIS professionals, to coordinate across the border to improve geographic information sharing, planning, and the sustainability of the Southwestern Amazon borderlands. Before sharing the efforts of GTASO to address these challenges, how these challenges brought about GTASO, and the emerging NASA SERVIR collaboration that builds on GTASO, we introduce some of the existing GIS efforts at the Pan-Amazonian and national scales.

2 Pan-Amazonian Mapping Efforts

The following three Pan-Amazonian initiatives represent the best basin-wide collaborations to generate geographic information about Amazonia and overcome pervasive myths (“empty Amazon”, “lungs of the earth”, “homogenous biome”, “green hell”, etc.) that often led to poor management and exploitation (Nagatani 2009; CDEA 1992). The Pan-Amazonian organization most focused on developing a common understanding of the region across South American countries is the intergovernmental Amazon Cooperation Treaty Organization (ACTO). Formed in 1978 via the treaty of the same name, ACTO has facilitated cooperation between the countries of Bolivia, Brazil, Colombia, Ecuador, Guyana, Peru, Suriname, and Venezuela. ACTO has published seminal reports such as *Amazonia without Myths*, *GeoAmazonia*, and Eva and Huber’s (2005) efforts to find a common definition for the Amazon across all ACTO member states (Nagatani 2009; CDEA 1992). More recently ACTO produced repositories of Pan-Amazonian data such as the Amazonian Regional Observatory (ORA) which contains mapping tools and geographic information to contribute to policy-making and planning decisions about strategic natural resources. These types of pan-Amazonian collaborations are important because each country defines, measures, and maps the region differently.

The Large-Scale Biosphere-Atmosphere Experiment in Amazonia, better known as LBA, was a collaborative scientific investigation of the tropical rainforest of Brazil and portions of surrounding countries. LBA brought together nearly 2000 scientists from numerous scientific disciplines and countries to collaboratively marshal remote sensing techniques and field experiments to investigate the atmosphere-biosphere-hydrosphere dynamics of the Amazon rainforest. Led by the Brazilian Ministry of Science and Technology, the project had additional funding from NASA and the European Commission among other sources. While the principal focus of the project was not mapping, the endeavor produced relevant geographic information and over 1300 scientific papers. This experiment encouraged collaboration across country borders and disciplines to build a dynamic three-dimensional understanding of how Amazonia functions as a whole ecosystem (Keller et al. 2013).

The most comprehensive Amazonia wide mapping effort is the Amazon Geo-Referenced Socio-Environmental Information Network (RAISG 2021). Launched in 2009, RAISG consists of eight non-profit organizations from eight Amazonian countries and produces biome-wide analyses (such as MapBiomas) and thematic maps that are accessible from their website. RAISG basin-scale maps of Indigenous territories, protected areas, deforestation, gold mining, and carbon stocks to name a few, include graphics and tables providing quantitative results of descriptive analysis. Perhaps the biggest challenge of the RAISG initiative was aligning geographic information along and across the 12,000 km of national boundaries. Mapping the borderlands also presents challenges for the two countries with the largest portion of the Amazon: Peru and Brazil.

3 National and Regional Mapping Efforts

GTASO's efforts to map the Amazon borderlands shared by Peru and Brazil require building on the accuracy, precision, and compatibility of the Peruvian and Brazilian official cartographic systems. These systems, increasingly online and dynamic, are modernizing, but not at the same pace as imagery and data accessibility and quality. This is particularly true in peripheral regions such as the Amazon basin given that national initiatives prioritize core regions with larger populations, economies, and political power. The lack of updated and accurate official cartography in the borderlands requires initiatives like GTASO to negotiate the opportunities and challenges presented by national and regional mapping platforms so that the most effective and accurate transboundary maps, databases, and analysis can be produced in a rapidly changing and increasingly connected Amazonia. Here we introduce the Peruvian and Brazilian official cartography systems with particular attention to the Ucayali and Acre states that form the core part of GTASO's area of interest.

3.1 Peru's Spatial Data: The Proliferation of Geo Servers and the Cartas Nacionales

Peru's National Geographic Institute (IGN), formerly known as the Military Geographic Institute, has been the governing body of the official cartography of Peru for over a century. IGN's official maps or *Cartas Nacionales* consist of 500 topographic maps at a scale of 1:100,000 derived from aerial photos, optical and radar satellite images conducted by the IGN along with the Nation Geospatial-Intelligence Agency (NGA) of the United States of America, formerly Defense Mapping Agency (DMA). In 2008, the IGN switched to a digital platform, and in 2011 made the World Geodetic System 1984 (WGS84) datum the official datum of Peru in order to standardize the elaboration and updating of Peruvian cartography, thus ending the official use of the PSAD56 datum, and facilitating collaboration with neighboring countries familiar with the popular WGS84 datum (Tipula 2016).

The IGN also seeks to bring up to date and improve the *Cartas Nacionales* that form the base of Peru's GIS database. The IGN faces a diminished annual budget from the Ministry of the Economy, and significant mapping challenges due to the country's diversity of landscapes, which translates into significant costs for field checking and updating cartography. Thus, the 1:100,000 *Cartas Nacionales* continue to be outdated, particularly of the Amazon basin whose dynamic rivers create new river channels and islands relevant not only for the determination of administrative units like Indigenous territories and national parks, but also the identification of the international boundaries of Peru. IGN has a five-year goal of updating and refining the *Cartas Nacionales* to a scale of 1:25,000. Initially, the IGN prioritized the updating of any emergency areas at a 1:25,000 scale. Geographically, the IGN has prioritized the coastal region of Peru,[1] then the Andean region, with the Amazonian region coming last. Additionally, the IGN is working North to South as they move across the country. In 2021 the following department maps were available at a 1:25,000 scale: Tumbes, Piura, Ica, La Libertad, Moquegua, Tacna, Lima, and Huancavelica. Huancavelica, a poverty-stricken central Andean region was the only non-coastal department. In 2021, an IGN-NGA agreement planned to finish six 1:50,000 scale topographic maps of the department of Loreto. While Ucayali, and the other departments wait for the refined and improved *Cartas Nacionales* at 1:25,000 scale, policies will continue to be promoted on the outdated 1:100,000 national maps.

In addition, the map repository at IGN, Peru has begun distributing geographic information through a centralized online site at Peru's National Infrastructure of Fundamental Geospatial Data (IDEP) (www.idep.gob.pe). Once fully operational and consistently maintained with high quality and updated information, IDEP may eliminate some of the need for transboundary data sharing initiatives like GTASO. IDEP, part of Peru's Council of Ministers, is a conglomeration of policies, standards, organizations, human resources, and technical capacity that seeks to facilitate the production, use, and access of Peru's geographic information to support socio-economic development and timely decision making (IDEP 2021). IDEP efforts to modernize and improve the quality and accessibility of Peru's geographic information is immediately visible through the 23 links to geoportals on their website. Interestingly, of the six departmental geo servers linked to the IDEP's main page, five were of Amazonian departments: Amazonas, San Martìn, Huànuco, Loreto, and Ucayali. This Amazon focus results in part from a United States of America's Forestry Service cooperation program which has been working with the Interregional Amazonian Council (CIAM) since 2012 with more recent support from Germany's Society for International Cooperation (GIZ), the Japanese International Cooperation Agency (JICA), the Peruvian government through the Presidency of the Council of Ministers (PCM), and the National Forest and Wildlife Service (SERFOR). Although the Peruvian geo servers provide access to spatial data, the quality of data and access varies widely, with access sometimes requiring entering through the ministry or departmental website rather than IDEP or IGN. Moreover, in recent years, the inconsistency between the data managed by local governments and those of the national government has reached critical points, especially in strategic sectors such as agriculture, environment and transportation. In addition, Peru's National Commission of Aerospace Research and Development (CONIDA) also provides technical expertise in remote sensing and GIS while also operating PeruSAT-1, Latin America's most powerful Earth observation satellite capable of capturing images at a 0.7-meter resolution (Villa and Finer 2018).

Peru's border region of Ucayali's most recent efforts to manage and share geographic data began with the creation of the Spatial Data Infrastructure (http://

[1] The prioritization of the coastal region coincides with the "El Niño Costero" 2017 climate event which impacted over 500,000 people, killed 75, and collapsed 10,000 houses.

ider.regionucayali.gob.pe/) and Catalog of Geographic Objects for the Ucayali Department Base Map in 2017. In 2018, Ucayali created two more catalogues, the Catalogue of Geographic Symbols and Objects of the Native Communities and the Catalogue of Geographic Objects for Forestry Management. In 2020 Ucayali began hosting a functional webmap with data on Economic Ecological Zoning (ZEE), Native Communities, Forestry Management, Population Centers, Transport, Hydrography, and Administrative Limits and Areas. We note, however, at least part of the data is incomplete (Native Communities, Protected Areas) and other datasets (Population Centers, Native Communities, among others) have the same weaknesses as the original data sources. Of course, similar challenges exist in the neighboring country of Brazil.

3.2 *Brazil's RADAMBRASIL Project, Acre's EEZ, and Ethno-Mapping*

The GTASO initiative to construct a transboundary database builds on and complements Brazil's RADAMBRASIL Project, Acre state's Ecological and Economic Zoning (EEZ), and Indigenous territory Ethno-mapping. The RADAMBRASIL project is the foundation of Brazil's Amazonian cartography. This project started in 1970 and relied on radar images, aerial photographs, and soil samples collected over 15 years to generate thematic maps and ecological zoning of the Brazilian Amazon. After 1985, the Brazilian Institute of Geography and Statistics (IBGE in Portuguese) absorbed RADAMBRASIL. IBGE's geoportal has close to 33,000 maps available, including coverage of the entire country at 1:250,000 scale and maps of the states of the Legal Amazon at 1:100,000 scale. Other Brazilian federal institutions such as the Ministry of the Environment (MMA), National Institute of Spatial Research (INPE), and National Water Agency (ANA), also provide important map products. Some of their maps and data are available from the National Infrastructure of Spatial Data (INDE), which integrates and homogenizes geospatial data from Brazilian institutions at federal, state, and municipal scales if the corresponding Brazilian institutions follow the spatial data quality protocols established by INDE's metadata catalogue.

Acre, Brazil's westernmost state, has a longer border with Peru's Ucayali and Madre de Dios regions than it does with the rest of Brazil, and does not have their geospatial data integrated into the INDE catalog making information access and standardization more difficult. Despite this limitation, Acre used the RADAMBRASIL framework both to map land use and land cover, and to define priorities for land and resource management as part of their Ecological and Economic Zoning (ZEE for its acronym in Portuguese) project. Acre's ZEE has produced a suite of thematic maps and cartographic databases at 1:1000,000 scale (Phase I in 2001) and then at 1:250,000 scale (Phase II in 2007). Currently, Acre has a new 2021–2025 project called Phase III both to revise and bring the ZEE up to date. The Acre state entity in charge of geographic and ZEE information is the State Environmental Secretariat (SEMA), through the Integrated Center for Geoprocessing and Environmental Monitoring (CIGMA) which includes the Central Geoprocessing Unit of state of Acre (UCEGEO). A central challenge for borderland states like Acre and Ucayali and their regional mapping offices like CIGMA and IDE Ucayali is that they often cannot access the geographic information of the neighboring country despite the crucial importance of transboundary information to the sustainable management of their territory and resources due to shared rivers, forests, and Indigenous cultures, not to mention transboundary impacts of development projects.

Brazil's ethno-mapping initiatives for the management of Indigenous territories should also be highlighted, particularly because of the signing of Decree 7.747 of the National Policy on Environmental and Territorial Management (PNGATI) in 2012 that underscored the importance of ethno-cartography along borderlands for monitoring and management purposes. The Amazon borderlands contain high levels of cultural diversity and a high density of Indigenous territories that often border neighboring states even as they share cultures, resources, and migration patterns (Salisbury and Weinstein 2013). Acre's first ethno-mapping began in 2004, with the borderland Indigenous Territory of Kampa do Rio Amonea. This 2004 mapping effort provided accurate information about invasions by loggers into Indigenous territories and the Serra do Divisor National Park, facilitated dialogue with neighboring communities, and gathered information about transboundary dynamics that eventually resulted in subsidies for borderland policies. The ethno-maps prepared by Indigenous communities not only provided essential information for territorial and environmental management plans, but also served as strategic tools for territorial protection, environmental conservation, and dialogue between Indigenous people, their neighbors, and government officials.

4 Pre-GTASO Efforts

In March of 2004, two graduate students and a new geography professor at the Universidade Federal do Acre (UFAC) made a preliminary map of the Southwestern Amazon borderlands that included Acre, Brazil and Ucayali, Peru. The Preliminary Map: Borders between Ucayali, Peru and Acre, Brazil (Fig. 3.1) had only a few features: cities, localities, roads, main rivers, the international boundary, state boundaries, protected areas, proposed protected areas, titled and proposed Indigenous territories. The three cartographers produced the map because just a few days later, on March 11, 2004, the Governor of Acre, Brazil, the Governor of Ucayali and the Minister of Foreign Relations of Peru were to meet in Pucallpa for the First Binational Meeting of the Authorities and Businessmen of Acre state, Brazil and the Ucayali region, Peru. At the time, the politicians, businesspeople, and decision makers of Acre and Ucayali knew very little about each other. The cartographers hoped the map might introduce a greater geographic perspective to discussions about transportation, products, and peoples. The map contained a disclaimer beneath the

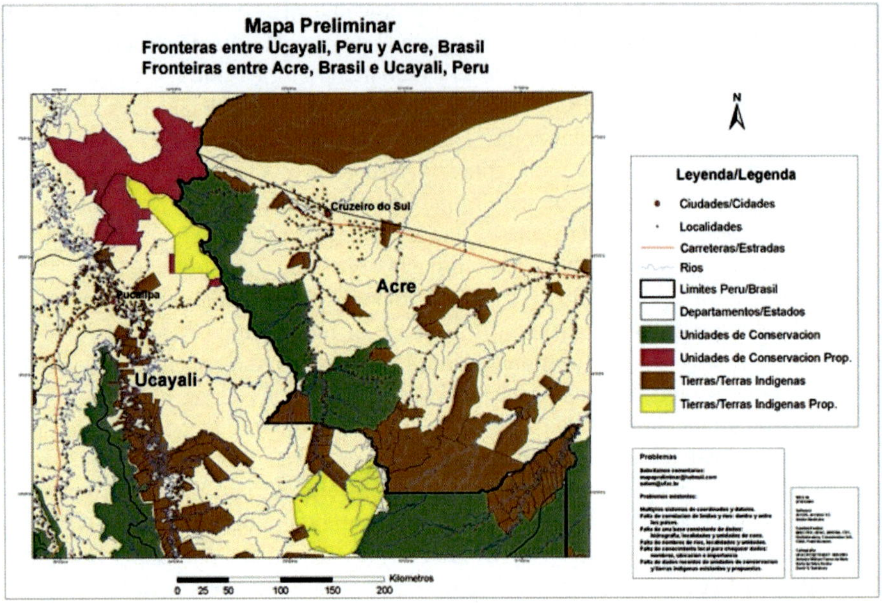

Fig. 3.1 The *Preliminary Map: Borders between Ucayali, Peru and Acre, Brazil* was created in 2004, and represents one of the first cartographic products to detail settlements, natural protected areas, and Indigenous territories on both sides of the Ucayali/Acre border. The map was made to be printed at poster scale, and thus is here shown only for illustrative purposes rather than detailed analysis

legend acknowledging five of the most important known problems with the map and thus the Brazilian and Peruvian cartographic databases (Fig. 3.1): (1) multiple coordinate systems and datums; (2) lack of congruence between borders and rivers within and between countries; (3) lack of toponyms for rivers, localities, and areas; (4) lack of local knowledge to check data (names, location, and importance); (5) lack of recent data on existing and proposed protected areas and Indigenous territories.

That same year, two authors on this publication formed the Amazon Borderlands Research Center (CIFA in Spanish) at the Universidad Nacional de Ucayali along with five professors and five research students. CIFA sought to combine scientific research, local knowledge, geographic technology, and participatory methods to provide information and recommendations to reconcile integrated sustainable development with the conservation of biological and cultural diversity in the borderlands. CIFA identified the imposition of logging, mining, and oil and gas concessions, as well as road projects onto the poorly mapped forests, rivers, and peoples of the borderlands. The imposition of development interests on the silent spaces (Harley 1988) of the "empty" Amazon (Hecht 2013) generated a great deal of conflict and environmental injustice. CIFA sought to provide insight and simultaneously develop human capacity, contribute to improving borderlands livelihoods, and improve the networking of neighboring countries and borderland populations. Following CIFA, a 2012–2015 Higher Education for Development-United States Agency for International Development (HED-USAID) project titled Building Conservation Capacity for a Changing Amazonia supported and trained 21 Universidad Nacional de Ucayali students to conduct their thesis research in the Ucayali borderland watersheds of Yurua and Purús (Salisbury et al. 2015). While the CIFA and HED-USAID projects strengthened relationships and knowledge building between future professionals and borderland populations, they still did not provide a mechanism and funding for regular transboundary interchanges of geographic information, knowledge, and expertise.

Years later, at the time of this writing, many of the Preliminary Map's known problems continue to challenge specialists mapping the borderlands shared by Acre and Ucayali. These cartographic challenges have important ramifications for borderland populations given they struggle to receive state support to address threats to their people and place when national databases and maps fail to register their village or river correctly, much less the dangers on the other side of the border. Nevertheless, geographic information now moves between the neighboring borderland states, and the maps and mapping technology are much improved. One important improvement has been the successful series of GTASO workshops that helped accelerate an improved geographic understanding in the Southwestern Amazon borderlands.

5 The GTASO Initiative

In 2012, a group of GIS professionals, geographers, and environmental scientists came together at the Universidad Nacional de Ucayali's Amazon Borderlands Research Center for a week-long workshop with a handful of objectives:(1) create an integrated plan for the updating and improvement of a transboundary Acre-Ucayali map with documented metadata; (2) sign an agreement for future collaborations between Acre, Brazil and Ucayali, Peru to integrate and update a geographic database; (3) create a joint declaration between participants for the establishment of working relationships between geographers and natural resource specialists of Acre and Ucayali; (4) create a list of technical capacity building needs for Acre and Ucayali; (5) document priorities for additional data layers; (6) list data needs for the region (Salisbury et al. 2014a). Participants represented a multidisciplinary team that included eight geographers, four foresters, one agronomist, one zootechnician, one systems analyst, and one environmental technician. The success of the workshop led to two more workshops, one in 2013, and one in 2019. Despite the changing names for the group and workshop, these later workshops shared many of the same objectives and were also increasingly diverse and multidisciplinary.

A quick comparison of the three workshops reveals several notable trends (Table 3.1). First, the presence of female participants increased with each workshop. Second, the workshops had a large percentage of Peruvian participants. This is in part due to two of the three workshops being held in the Peruvian Amazon. The

Table 3.1 Demographics of the three Transboundary Geographic Group of the Southwestern Amazon (GTASO) workshops

Title	2012 workshop for data integration and development of technical capacity to mitigate environmental challenges in the Peruvian and Brazilian Amazon	2013 socio-environmental corridors in the Southwestern Amazon workshop	2019 workshop to mitigate environmental challenges in the Peruvian and Brazilian Amazon- transboundary geographic group of the Southwestern Amazon
Location	Universidad Nacional de Ucayali (UNU), Pucallpa, Ucayali, Peru	Centro de Formação dos Povos da Floresta, Comissão Pró-Índio do acre, Rio Branco, acre, Brazil	Universidad Nacional de San Antonio Abad del Cusco (UNSAAC), campus Puerto Maldonado, Madre de Dios, Peru
# of participants	17	27	25
% women	17.6%	29.6%	36.0%
% Brazilian	23.5%	40.7%	24.0%
% Peruvian	64.7%	44.4%	52.0%
% Bolivian	0.0%	3.7%	0.0%
% USA	11.8%	11.1%	24.0%
%NGOs	41.2%	40.7%	28.0%
%GOs	29.4%	22.2%	16.0%
%education	29.4%	37.0%	56.0%

1-1.5-hour flight from Lima, the capital city of Peru, to the Southwestern Amazon allows for greater participation of professionals from governmental and non-governmental offices in the capital, whereas the 3.5-hour flight from Brasilia to the southwestern Amazon, and the sheer number of Amazonian sites available in Brazil, means the workshops had no representation from Brazil's capital city. Acre is often considered the most remote state in the entire country of Brazil. That said, the workshops had excellent participation from Acre. Third, the Non-Governmental Organization (NGO) sector provided the greatest percentage of participation across the three employment sectors when students were removed from the education sector (Table 3.1). However, we also note the fluidity with which GIS professionals move between governmental, non-governmental, and academic sectors within the Southwestern Amazon. GIS professionals in the Amazon are thus not only interdisciplinary in their approach, but also inter- and multi-institutional throughout their professional career.

6 Challenges

The three workshops identified a series of basic GIS data challenges such as official cartographic databases that require updating, variable scales (national, regional, local) and methodologies (field, remote sensing, etc.) used in geographic data generation, and the existence of incompatible toponyms between shared data classes. Other challenges were related to the specific suite of maps each workshop focused on. For example, the 2013 corridor workshop struggled with the multidisciplinary challenge of finding the best way to define a socio-environmental corridor in terms of defining variables and extent. In addition to the basic and detailed challenges above, two general groups of challenges were shared by all three workshops, namely those related to transboundary issues and those associated with the remoteness of the Amazon borderlands.

6.1 *Transboundary Challenges*

A major challenge continues to be the lack of agreement between each country's national boundary dataset. Also, the lack of congruence between international borders and the borders of overlapping administrative units created a great deal of noise that had to be overcome to produce clean cartography and perform GIS analyses. The first workshop used a neutral border dataset from the Humanitarian Information Unit (HIU), rather than showing the lack of agreement between the Peruvian and Brazilian border datasets. Unfortunately, the HIU border also failed to agree with the borders of Indigenous territories, natural protected areas, and other overlapping administrative units between and within countries. The lack of alignment of borders resulted from the layers being based on different datums, coordinate systems, and scales. The lack of a common datum is becoming less of a problem given Brazil's legal transition in 2015 to using SIRGAS2000 as the sole geodetic official reference system after a decades-long process of moving from Córrego Alegre, Astro Datum Chuá, SAD69, with concurrent use of SIRGAS2000 with SAD69 starting in 2005. Unfortunately, many older layers must be carefully transformed from SAD69 to SIRGAS2000 to minimize error (Borges et al. 2017), and this includes layers from Acre. Perhaps, the most visible example of non-alignment of layers/coordinate systems is that the Brazilian and Peruvian databases contain large transboundary rivers like the Juruá and the Purús which fail to line up due to the Brazilian layers' basis in SAD69 and the Peruvian in WGS84. This underscores the larger challenge that neither physical geography nor ecological, biological, and cultural diversity respect political geography and international boundaries. Moreover, each country defines, measures, and maps their features differently.

6.2 *Amazon Borderlands Challenges*

The foremost challenge in the remote Amazon borderlands recognized by the spatial analysts present at the three workshops is the limited quantity and poor quality of borderland geographic information. This challenge both follows and precedes the lack of a robust state presence in the borderlands. The state lacks the significant physical presence in the borderlands necessary to conduct rigorous fieldwork and produce the needed quantity and quality of geographic information. A common approach is to fill these poorly understood borderland landscapes and blank spaces on the map with state extractive administrative units (state forests, forestry concessions, mining concessions, hydrocarbon blocks). To a lesser degree, conservation scientists, usually affiliated with both NGOs and GOs, have also promoted the establishment of a variety of national and regional scale protected areas although often with a greater degree of fieldwork and geographic information. The placement of extractive and conservation administrative units on the borderland forests, fields, rivers, and population centers results in overlapping and contested claims that exacerbate territorial conflict and complicate effective planning and land use policy making. Yet, these, and other borderland resource and territorial conflicts can be silenced by the lack of state institutions able to intervene, the invisibility of local people on the map, and most importantly the lack of political will to engage with local conflicts, furthering marginality in the borderlands. In Ucayali and Madre de Dios, the situation is further aggravated by the increasing number of migrants from the Selva Central and Andean slopes. Many arrive with an interest in pursuing informal and illegal activities such as coca cultivation, drug trafficking, illegal gold mining, illegal logging, land trafficking or other pursuits that often provoke additional socio-environmental conflicts.

Neighboring countries in the Amazon borderlands also have different ways of defining, organizing, and mapping the characteristics of human geography, such as roads, that complicate the creation of transboundary databases without overlaps or gaps. Brazil and Peru also have different understandings of land tenure. Land tenure is contested within each respective country, and particularly in the remote borderlands. Examples of diverse land tenure regimes include the INCRA settlement projects in Brazil, the UMAR settlement projects in Peru, the Extractive Reserves of Brazil, the illegal coca farmers in Peru, the extensive Brazilian Indigenous Territories with their communal territory, the smaller fragmented Peruvian Native Communities, the expansive Brazilian cattle ranches, and the abundance of household farm lots of Peru. We can add to these the variety of concessions, lots, and units for the extraction of resources such as timber, gold, hydrocarbons, and non-timber forest products. The diversity of land tenure regimes, lack of tenure documentation, inadequate geographic information, and management are complicated by frequent land speculation, corruption, and generation of fake titles.

After the 2013 workshop, GTASO participants from Ucayali witnessed an increase in international funding to facilitate the titling of Indigenous territories. However, Indigenous titling proposals have to contend with other overlapping demands for territory and resources such as forestry concessions. The GTASO transboundary database is well placed to inform land use policy decisions like titling Indigenous territories given that transboundary geographic information provides additional context and rationales. Nevertheless, to maximize transboundary analysis, the Ministries (culture, environment, agriculture) of the respective countries also need to become more comfortable using GIS data. When working near the borderlands, they must learn to use data from outside their country and official borders, but these data do come with a learning curve as articulated previously and below. Additional challenges regarding Indigenous territoriality also include the lack of alignment between Indigenous territories and international borders. Regarding Indigenous peoples who live in a state of isolation or initial contact, workshop participants decided not to share information about transboundary and borderland migratory routes. The geographic information is sensitive given the intense curiosity of the media and public about these vulnerable people.

7 Data Needs

At each workshop the transboundary database was updated according to the data provided by the participants of the different institutions attending. The first workshop presented the greatest challenge given the initial construction of a transboundary database. Subsequent databases could build on that first effort. The thematic focus of the workshop also dictated in part the type of data needed to create the suite of thematic maps. For example, the second workshop focused on producing maps of transboundary socio-environmental corridors and watersheds. This focus led to a heightened interest in geographic data that facilitated the definition and presentation of corridors and watersheds (Salisbury et al. 2014b). Table 3.2 includes the various layers lacking or needing significant improvement during the workshop. In some cases, the data from one country would be available but not that of the other. For display purposes, we list the variable of concern, even if only one country's data were missing or inadequate.

First, participants from states bordering Bolivia (Acre, Brazil and Madre de Dios, Peru) asked for more data and participation from GIS professionals of the neighboring Bolivian state of Pando. Other common interests between the workshops included a desire to improve on access to and quality of shared deforestation and degradation data. By 2019, the workshop was able to take advantage of the Pan-Amazonian MapBiomas deforestation database (https://amazonia.mapbiomas.org/), but still searched for rigorous forest degradation data.

Mapping degradation is inherently complicated due to the myriad definitions and spatial and temporal scales at which 'forest degradation' can be defined. Rivers also proved a significant challenge across all three workshops given the lack of alignment between Brazilian and Peruvian river databases due to different coordinate systems, datums, and scales, not to mention the Peruvian official database's polygonal representation of the larger rivers versus the linear representation used in Brazil. SRTM derived hydrography data like HydroSheds showed promise but lacked the river names and detailed metadata most useful for analysis and representation at this regional transboundary scale. While still not perfect, the quality of and access to official roads data improved across the three workshops and continued to be of high interest due to the range of socio-environmental impacts associated with transportation infrastructure. That said, an interest in rigorously documented databases of unofficial roads also increased for the same reasons, and particularly as recent

Table 3.2 Data inadequacies identified during the 2012, 2013, and 2019 GTASO workshops

2012	2013	2019
No Bolivian data	Pando, Bolivia: Limited data	No Bolivian data
Deforestation: Poor quality	Deforestation: No shared database	
Degradation data unavailable		Degradation data unavailable
Hydrography: Lacks toponyms	Watersheds: Need editing	Water: No use and access data
Roads: Inconsistent coverage and quality	Roads: Need editing	
	Unofficial roads: No metadata	Unofficial Roads: No 2019 data
Urban zones insufficient	Urban zones insufficient	
	Migratory routes lacking	Migratory routes lacking
Hydrocarbon lots: Need updates		
No fauna: Terrestrial, aquatic, and birds	No fauna biodiversity data	
Vegetation and flora incomplete	No flora biodiversity data	
Rural settlements: Incomplete		Demography data lacking
Indigenous territories: Incomplete		
Climate data lacking		
SRTM elevation data only at 30 m		
		No land use types
		No seasonal crop data
		No carbon storage
		Lack local conflicts

official documents, maps, and road proposals use illegal and informal road initiatives as rationales for their creation (DRTC 2017; Briceño Ampuero 2010). Other important shared needs across at least two of the workshops included a desire for data on species diversity (flora and fauna), demography (settlements and Indigenous territories), land use, carbon stocks, and other environmental data such as climate change scenarios.

8 Human and Technical Capacity Building

The principal capacity building challenge identified by workshop participants was language. Despite the utility of GIS as a common dialect, higher-level communication between stakeholders suffered due to participants' uneven command of Spanish, Portuguese and/or English. The uneven language skills also limited workshop participants' goal of improving inter-institutional relationships to learn from their peers from neighboring countries to obtain more and better results from the workshop and transboundary GIS analysis and cartography. Interestingly, despite the twin goals of improving language and transboundary relationships, during meals, breaks, and transit, workshop participants tended to group with their regional and national peers rather than with participants from other countries and regions. The rigors of the workshop and foreign language fatigue likely contributed to the tendency of GTASO members to spend downtime with those with whom they had the most in common. Despite these challenges, GTASO members in all three workshops demonstrated a shared vision, exemplary work ethic, and can-do attitude that not only allowed members to reach our goals, but also led to strengthened relationships across borders, disciplines, and scales.

In terms of technical capacity, participants voiced the need for higher level training in GIS data analysis, remote sensing (passive and active image processing, classification, and analysis), cartography, GIS programming, and modeling. In addition to these desired technical capacities, participants also expressed an interest in obtaining improved hardware with high resolution and greater processing power and storage capacity that would allow them to use the latest software versions. Indeed, participants recognized that the workshop would work better if participants shared the same level of technical ability, commanded the same software types and versions, and had hardware of similar capacities. Cloud computing appears to be the direction of choice given access to software platforms such as Google Earth Engine (GEE) and high-resolution imagery products (e.g., NICFI) can transcend some hardware limitations.

9 Workshop Results

The three workshops each ended with a strengthened organization, an updated transboundary database, a new suite of transboundary maps, and a greater understanding of the Amazon borderlands through conference presentations and working groups. Our analysis not only finds commonalities in planning but also an evolution in how participants thought future workshops and collaborations should proceed. Common elements between the three workshops included the creation and strengthening of the transboundary network of professionals. The network originally called the Technical Geographic Transboundary Acre-Ucayali Group (GTGTA-U) became the Transboundary Geographic Group of the Southwestern Amazon (GTASO). A second result was the creation and updating of a transboundary GIS database between all participants. This database served as the basis for the third result, the creation of 14 total maps through the sum of the suite of maps produced in each workshop: 2012 (4), 2013 (5), 2019 (5) (Table 3.3). The sum total of 14 do not include all of the maps elaborated afterwards by an individual or a subset of participants using the GTASO transboundary database and network: there are likely over 100. GTASO distributed digital maps through online repositories and listservs, but also distributed printed maps to borderland schools, universities, and both governmental and non-governmental offices. These maps provided the geographic

Table 3.3 Transboundary maps of the Amazon borderlands resulting from the three GTASO workshops

2012	2013	2019
Natural protected areas and indigenous territories	Indigenous territories and natural protected areas	Indigenous territories and natural protected areas
Threats to natural protected areas	Threats to watersheds	Threats to indigenous territories and natural protected areas
Threats to indigenous territories	Socio-environmental corridors	Indigenous territories and cultural diversity
Ethno-geographic	Transport infrastructure and areas of influence	Vulnerable zones
	Cartographic challenges	Climate changes until 2040

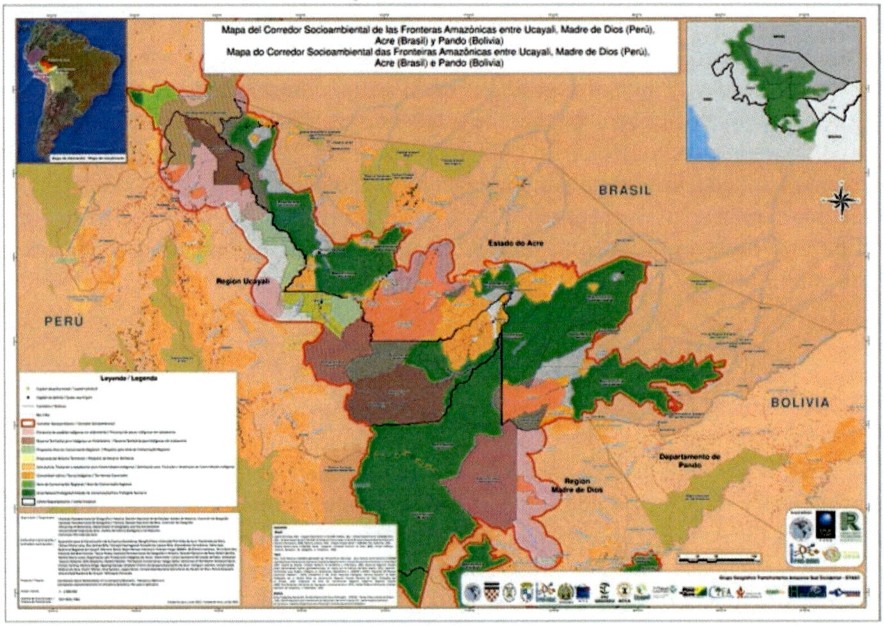

Fig. 3.2 One of the 14 GTASO workshop maps, this 2013 map shows a transboundary corridor of protected areas and Indigenous territories shared across the borderlands of Acre, Ucayali, and Madre de Dios. This map, originally produced at poster size, is here shared for illustrative purposes only and is not meant to be legible at this scale. The GTASO maps can be seen at their appropriate resolution here: https://scholarship.richmond.edu/geography-maps/5/

information and analysis to analyze critically important Amazon borderland characteristics like cultural geographies that transcend political boundaries, shared transboundary watersheds, socio-environmental corridors that straddle multiple countries (Fig. 3.2) as well as multiple threats like roads, deforestation, extraction zones, and climate change (Table 3.3).

Conferences, the fourth outcome, also served as tools for generating awareness of the importance of transboundary collaboration and workshops using GIS. The three GTASO conferences had 290 participants across over 20 institutions: governmental, non-governmental, academic, and other. The 2012 conference was the most diverse with 128 attendees from 28 institutions. The next year, the Acre conference was held in a smaller venue that had only 53 attendees, but across 27 institutions. In 2019, the conference had fewer institutions, but 109 total attendees. Local media covered the conferences and interviewed workshop participants and conference presenters in order to share the workshop goals and approach in print, radio, television, and internet mediums. For example, in 2013 GTASO members from Brazil facilitated multiple press releases and the publication of six newspaper articles. A fifth result included workshop declarations and protocols for work relations and data sharing across the GTASO network.

These protocols culminated in 2019 with the creation of an ESRI ArcGIS hub, https://gtaso-data-urichmond.hub.arcgis.com/, where users could view, and with permission, download GIS data concerning the transboundary region. This hub also provides access to a video and a StoryMap of the workshop experience. Publications, reports, posters, and presentations also resulted from the workshops. In addition to the technical report describing each workshop, GTASO published in outlets such as *El Geógrafo*, *ArcNews*, *Revista Cartográfica*, and *Revista Geográfica* (Salisbury et al. 2012a, b, 2014a, b). The 2019 workshop resulted in six academic posters presented by students in venues like the Conference of Latin American Geography (3), Geography2050: Borders and a borderless world (2), and the PanAmerican Institute of Geography and History's technical meetings. Finally, at least nine conference presentations or invited talks built on the material generated by the workshops.

Perhaps, even more important than shared products, were shared processes such as the multi-disciplinary and inter-institutional spatial analysis of a threatened transboundary region rich in ecological and cultural diversity. This analysis not only lists in the tables above the workshop products, the missing and desired GIS layers, and the technical skills needing development, but also served as the basis for a powerful interchange of knowledge, ideas, experiences, and friendships across borders. Today, GTASO maintains a vibrant WhatsApp group for the interchange of information about what is happening in Amazonia, as well as questions, concerns, and solutions, but this group also has strengthened morale during challenging times in the COVID-19 pandemic and the loss of one of our leaders, Bertha Balbín Ordaya, who until her death was the principal member of the Geography Commission for Peru's National Section of the PanAmerican Institute of Geography and History.

10 Future Strategies

Each of the three GTASO workshops expressed goals and objectives for the future. Shared goals across all three workshops included (1) improving human and technical capacity building; (2) maintaining and strengthening a transboundary and multidisciplinary GTASO network for data interchange and transboundary problem solving; and (3) holding regular workshops to update, grow, and standardize geographic information in a transboundary context. Initially, participants thought an annual workshop ideal, but now the focus has been on once every 2–3 years.

Similarly, the primary target audience for GTASO products has also expanded. Initially, GTASO targeted regional policy makers as the primary audience for maps and analysis, but in the second workshop, GTASO also distributed maps to borderland municipal authorities, NGOs, Indigenous communities, and schools. The 2020–2021 COVID-19 pandemic created challenges for 2019 map distribution efforts to all local venues, but GTASO was able to establish the aforementioned ArcGIS Hub for data sharing amongst members. Another 2019 workshop objective was a desire to see the data and cartographic products stimulate more research opportunities and academic publications. These goals are realized in this book chapter as well as through a NASA SERVIR grant "Quantifying the Effects of Forest Cover Changes on Provisioning and Regulating Ecosystem Services in the Southwestern Amazon", which includes a selection of GTASO participants from Brazil, Peru, and the USA. After three workshops, GTASO participants became increasingly aware that mitigation of socio-environmental problems, impacts, and challenges will require multi-disciplinary, integrated, and coordinated actions between the Southwestern Amazon's multiple regions. Participants have also noted increasing recognition within Brazil and Peru of each country's interconnectivity with neighboring countries and states.

Along these lines, GTASO envisions future workshops that can expand the GTASO area of interest to incorporate Pando, Bolivia, and perhaps even Loreto, Peru and Amazonas, Brazil. While growth in geographic area and institutions would be of interest to many participants, they also recognize the challenges such growth presents to the sharing, cleaning, and distribution of transboundary data and metadata (another shared goal).

One partial solution to the transboundary data quality control problem would be for GTASO to incorporate more of the increasingly robust global and regional datasets focused on forest cover, forest change, watersheds, river systems, and other basin wide variables (Linke et al. 2019; Kalacska et al. 2020; RAISG 2020). These larger scale initiatives prove easier to incorporate into the transboundary database rather than reconciling the numerous idiosyncrasies of the burgeoning databases of neighboring Amazonian states, municipalities, and non-governmental institutions. Burgeoning databases are a good problem to have given the dearth of geographic information when GTASO began.

GTASO participants are also aware that technical advances in GIS, remote sensing, and quantitative analysis are greatly expanding the possibilities, contributions, and reach of a group of professionals like GTASO. Yet, the exponential expansion and acceleration of geospatial data, analytical power and geospatial technology, may also pose challenges for meaningful integration with local actors and landscapes. Here, GTASO, a knowledge network founded to engage with the complex local, borderland, and transboundary dynamics of the Southwestern Amazon, can make a significant contribution even as our technical ability expands. Perhaps no other institution better understands the power of local transboundary and borderland geographic data in helping marginalized communities and landscapes that lack consistent, accurate, and updated geographic representation at larger scales to share their story. GTASO analysis and mapping of transboundary watersheds, cultural landscapes across borders, biocultural conservation corridors, and threats have allowed borderland residents, governmental officials, NGO representatives, professors, and the general public better understand the opportunities and challenges that exist in these remote, dynamic and bio culturally diverse borderlands.

GTASO participants believe future workshops should revolve around themes that combine our local transboundary approach with the expanding capacity of geospatial technology, and new challenges, such as those presented by the global pandemic. This could be similar to the 2013 workshop's thematic focus on transboundary socio-environmental corridors. Outside experts would ensure a more robust capacity building element and train GTASO members in geo-analytical and geovisualization techniques specific to the chosen theme. Thus, future GTASO workshops would add new skills to our consistent goals of producing a suite of rigorous and effective transboundary maps and geovisualizations, an updated and expanded transboundary GTASO database, a strengthened and expanded GTASO network, and an integrated GTASO plan forward with identified needs in data, metadata, and training. To do this, GTASO will continue to use GIS as a bridge between diverse languages, cultures, disciplines, institutions, and countries. To underscore this point, at the time of this writing, GTASO members are working across boundaries to respond to requests from government ministries, Indigenous communities, NGOs, and universities to produce timely transboundary maps and geospatial analysis of two road projects threatening the biological and cultural integrity and diversity of the Amazon borderlands shared by Ucayali and Acre (AACAPPY et al. 2021, APIB et al. 2021). Future GTASO workshops might also lead to a goal articulated best by Bertha Balbin Ordaya, who argued for the institutionalization of GTASO as a formal and sustainable transboundary institution combining multiple disciplines and advanced technical capacity to address Amazonian challenges.

In conclusion, we note three important points relevant to GTASO's role in the future of the Amazon borderlands. First, GTASO's ability to share data across boundaries and provide informed borderland context is crucial to collaborative transboundary scientific inquiries about the Southwestern Amazon's most important challenges: climate change, hydrological systems, ecosystem services, pandemics, extractive industries, road proposals, and forest degradation to name a few. Similarly, Latin America's increase in GIS technical capacity and open data geoportals, while helpful for data sharing, does not provide the transdisciplinary, transboundary, and scientific discussions necessary to conduct research and create analytical maps in the dynamic and poorly understood borderlands. Second, that while techno-scientific approaches can influence political will, GTASO's methodology of incorporating borderland knowledge into GIS to present human-environment dynamics on both sides of the border accelerates the scaling up/down of geographic information to inform decision makers and land managers at multiple scales to collaborate across borders. Finally, continued investment in GTASO workshops is required to further build human capacity and to ensure the continued production of useful geospatial analysis, geovisualizations, and updated transboundary geodatabases. This is particularly meaningful to human groups living in a dynamic, poorly understood borderland region threatened by illegal extractors, and land-use and climate change. Participatory mapping grounded in local knowledge will be exceptionally useful to protect these landscapes still replete with the biological and cultural diversity so important for a just and sustainable future. In June of 2022, a fourth GTASO workshop benefited from just such an exchange of GIScience and traditional ecological knowledges along the transboundary course of the Juruá River before convening at the Universidad Nacional de Ucayali, Pucallpa, Peru with support from SERVIR Amazonia, UNU, and the University of Richmond's Amazon Borderlands Spatial Analysis Team (ABSAT-www.absatrichmond.com).

References

AACAPPY, ACONADYISH, ACC-Yurúa, Apiwtxa, CPI-Acre, Sawawo, Saweto, ORAU, OPIRJ, UAC (2021) A Estrada illegal "Nuevo Italia-Puerto Breu": Uma grande ameaça para os povos Indígenas do Yurua, Alto Tamaya e Alto Juruá. Dossiê. https://apiwtxa.org.br/wp-content/uploads/2021/08/Dossie%CC%82-Estrada-Ilegal-Nueva-Italia-%E2%80%93-Puerto-Breu_ok.pdf. Accessed 9 Sep 2021

APIB, Apiwtxa, AACAPPY, AAPBI, ACONADIYSH, Associação Floresta Viva, AIN, AKARIB, AMAAIAC, Ascema Nacional, Associação do Povo Indígena Jaminawa Arara, Associação do Povo Arara do Rio Amonia, Associación ProPurús, SOS Amazonia, ASIBAMA/AC, CTI, CPI-Acre, Comite Chico Mendes, COIAB, DAR, IEPÉ, Instituto Yorenka Tasorentsi, OPI, ORAU, OPIRE, OPIRJ, RCA, UAC/Conservación Alto Amazonas, UMIAB, WWF-Brasil (2021) Carta Abierta: A defesa dos direitos dos povos indígenas e das comunidades tradicionais e a conservação das florestas como perspectiva de desenvolvimento na Fronteira Acre-Ucayali. CPI-Acre: https://bit.ly/2V5BsvB. Accessed 12 Sep 2021

Bass MS, Finer M, Jenkins CN et al (2010) Global conservation significance of Ecuador's Yasuní National Park. PLoS One 5(1):e8767. https://doi.org/10.1371/journal.pone.0008767

Borges AF, Timbó MA, Nero MA, et al. (2017) Sistemas geodésicos de referência adotados no Brasil e a conversão dos dados geográficos para o sistema oficial SIRGAS2000: transformações e avaliação de erros. Revista Geografias 45–63

Briceño Ampuero L (2010). Gestión de desarrollo por el lado Peruano de la zona de integración Fronteriza Peru-Brasil. Presentation given at the Binational Workshop, Hacia un desarrollo sostenible mediante la integración fronteriza Acre-Ucayali. Pucallpa

Butler R (2017) Amazon wildlife. Mongabay.com, Accessed 26 Jan 2018

CDEA – Commission on Development and Environment for Amazonia (1992). Amazonia without myths. IDB/UNDP/OTCA

Dirección Regional de Transportes y Comunicaciones Ucayali (DRTC) (2017) Plan Vial Departamental Participativo – Ucayali 2017–2026. Pucallpa

Eva HD, Huber O (2005) A proposal for defining the geographical boundaries of Amazonia: Synthesis from the results from an expert consultation workshop organized by the European Commision in collaboration with the Amazon Cooperation Treaty Organization. Office for Official Publications of the European Communities

Ghai R, Rodríguez-Valera F, McMahon KD, Toyama D, Rinke R, de Oliveira TCS, Garcia JW, de Miranda FP, Henrique-Silva F (2011) Metagenomics of the water column in the pristine upper course of the Amazon river. PLoS ONE 6(8):23785

Harley JB (1988) Silences and secrecy: the hidden agenda of cartography in early modern Europe. Imago Mundi 40(1):57–76

Hecht SB (2013) The scramble for the Amazon and the "lost paradise" of Euclides da Cunha. University of Chicago Press

IDEP (2021) Que es la IDEP. https://www.geoidep.gob.pe/que-es-la-idep. Accessed 7 Sep 2021

Kalacska M, Arroyo-Mora JP, Lucanus O et al (2020) Deciphering the many maps of the Xingu River basin–an assessment of land cover classifications at multiple scales. Proc Acad Nat Sci 166(1):1–55

Linke S, Lehner B, Dallaire CO et al (2019) Global hydro-environmental sub-basin and river reach characteristics at high spatial resolution. Science 6(1):1–15

Nagatani K (2009) GeoAmazonia: Environment outlook in Amazonia. United Nations Environmental Program, Amazon Cooperation Treaty Organization, University of the Pacific, Stockton

Nagy L, Forsberg BR, Artaxo P (eds) (2016) Interactions between biosphere, atmosphere, and human land use in the Amazon Basin (227). Springer, Switzerland

Perz SG, Selaya G, Muñoz-Carpena R et al (2019) Scientists and stakeholders, data and diagnostics: crossing boundaries for modeling the impacts of highway paving in a tri-national frontier in the Amazon. In: Perz SG (ed) Collaboration across boundaries for social-ecological systems science. Palgrave Macmillan, Cham, pp 327–359

Plotkin MJ (2020) The Amazon: what everyone needs to know. Oxford University Press, New York

Red Amazónica de Información Socioambiental Georreferenciada (RAISG) (2020). Amazonia Under Pressure.www.amazoniasocioambiental.org. Accessed 7 Sep 2021

Red Amazónica de Información Socioambiental Georreferenciada (RAISG) (2021). http://raisg.socioambiental.org. Accessed 7 Sep 2021

Salisbury DS, Flores de Melo AW, Vela Alvarado J, Balbín Ordaya B (2012a) Amazonian states map threatened borderlands. ArcNews 34(3):33. http://www.esri.com/news/arcnews/fall12articles/amazonian-states-map-threatened-borderlands.html

Salisbury DS, Flores de Melo AW, Balbín Ordaya B (2012b) Taller transfronterizo para la Amazonía Peruana y Brasileña. El Geógrafo 8:25–26

Salisbury DS, Weinstein BG (2014) Cultural Diversity in the Amazon Borderlands: Implications for Conservation and Developments. Journal of Borderlands Studies. 29(2): 217–241, https://doi.org/10.1080/08865655.2014.916462

Keller M, Bustamante M, Gash J, Dias, PS (eds) (2013) Amazonia and global change. John Wiley & Sons. ISBN 978-0-87590-476-4

Salisbury DS, Flores de Melo AW, Tipula Tipula P (2014a) Transboundary political ecology in the Peru-Brazil borderlands: mapping workshops, geographic information, and socio-environmental impacts. Rev Geogr 152:105–115

Salisbury DS, Leal DB, Chávez Michaelsen AB et al (2014b) Cartografía, corredores, y cooperación: La búsqueda de soluciones transfronterizas en las fronteras Amazónicas. Revista Cartográfica 89:131–143

Salisbury DS, Anderson E, Bilsborrow R et al (2015) Transformando la Educación Superior para una Amazonía Cambiante. Investigación Universitaria, Edición Extraordinaria, Revista de la Universidad Nacional de Ucayali, pp 7–9

ter Steege H, Pitman NCA, Sabatier D et al (2016) The discovery of the Amazonian tree flora with an updated checklist of all known tree taxa. Sci Rep 6:29549. https://doi.org/10.1038/srep29549

Villa L, Finer M (2018) Introducing PeruSAT-1, Peru's new high-resolution satellite. MAAP 91

Walker RT (2020) Collision course: development pushes Amazonia toward its tipping point. Environ Sci Policy Sustain Dev 63(1):15–25

Chapter 4
Territorial Implications of Economic Diversification in the Waorani Ancestral Lands

Rodrigo Sierra, Sylvia Villacís, Javier Vargas, Oscar Calva, Abraham Boyotai, Gilberto Nenkimo, Aurelia Ahua, and Ana Puyol

Abstract At the beginning of the 2020s, roughly 80% of the original forest cover of the Western Amazon remained in a relatively compact form. Immersed in these forests lived a growing Indigenous population, increasingly concentrated in an expanding number of multi-family permanent settlements, each using a stable, rarely overlapping territory, mostly composed by natural forests, lakes, and rivers, to produce all the material necessities they traditionally required. Most were also going through important demographic and economic transformations, with direct implications on their particular territories and on the aggregate use and condition of the region's forests. In this chapter, we examine if and how the adoption of a commercial crop alters the traditional use of the territory of 10 communities in the Waorani Ancestral Territory of the Ecuadorian Amazon. We propose and implement a spatial model of such a territory and compare it with the cultivation patterns of a common and widespread commercial crop: cacao. Results strongly suggest that the cultivation of cacao absorbs a scarce resource needed to use the territory. Furthermore, most changes are associated with the management of older productive trees, pointing at activities such as harvesting, processing, and the transportation of products, as important draws of this resource. We hypothesize that this resource is mostly 'uptime' (the proportion of time a community member is working or avail-

This study was carried out in collaboration with EcoCiencia and NAWE. It intended to provide NAWE with a broad understanding of the territorial budget of the WAT.

R. Sierra (✉) · O. Calva
GeoIS Consulting, Austin, TX, USA
e-mail: rsierra@geo-is.com

S. Villacís · J. Vargas · A. Puyol
EcoCiencia, Quito, Ecuador

A. Boyotai · G. Nenkimo · A. Ahua
Waorani Nationality of Ecuador (NAWE), Puyo, Ecuador

S. López (ed.), *Socio-Environmental Research in Latin America*, The Latin American Studies Book Series, https://doi.org/10.1007/978-3-031-22680-9_4

able for work), resulting in a reduction of the area used for traditional production. These results illustrate the situation of the productive territories of most permanent communities in the Waorani Ancestral Territory, and possibly other Indigenous communities of the Western Amazon where similar demographic and economic transformations are taking place.

Keywords Land use · Indigenous territories · Waorani · Western Amazon · Ecuador

1 Introduction

Does the adoption of commercial activities by Indigenous communities in the Western Amazon affect their traditional demand for a productive territory, or the territory used to produce all the material necessities a community customarily requires? In the early 2010s, in the Ecuadorian Amazon alone, there were approximately 260 isolated permanent Indigenous communities immersed in approximately 6.6 million hectares of a relatively continuous forest matrix, only accessible by foot, canoe or airplane. Each used a relatively well defined, rarely overlapping territory to produce most of those necessities.[1] By the early 2020s, there were possibly more than 300.[2] The majority probably appeared between the early 1980s and the late 1990s, likely driven by declining productivity in accessible areas of the existing communities' territories due to rapidly increasing populations, itself a demographic feedback of the establishment of permanent communities, and a large forest reserve available for territories of new communities.[3] In the Waorani Ancestral Territory of the Upper Curaray River (WAT) in the central Ecuadorian Amazon, the focus of this chapter, the number of permanent multi-family communities went from one in the 1950s to 31 in 2021, only three of which were accessible by road. To the south, in the territory of the Achuar Nationality of Ecuador, the first permanent community was also established in the 1950s. By 2021 there were 88 Achuar communities. On the other hand, in both nationalities the formation of new communities declined rapidly after the 1980s and 1990s (Fig. 4.1), at a time when many started to engage in some type of commercial exchange with the so-called "national" economy, through the sale of crops, cattle, wildlife, credit, and paid work. Today probably most do so (Vasco et al. 2017).

A rapidly declining number of new communities point to changes in the conditions that initially contributed to their rapid formation: high population rates of

[1] From a wall-to-wall survey of official aerial photography (SIGTIERRAS) for the region by the first author for this study. Approximately 85% of the region's original forest cover remained by the end of the 2010s (Sierra et al. 2021).

[2] Extrapolated from a non-random subsample of 23 communities with year of establishment information.

[3] In a few cases, communities have been established for geopolitical reasons (i.e., to establish property rights), but in our experience these are rare.

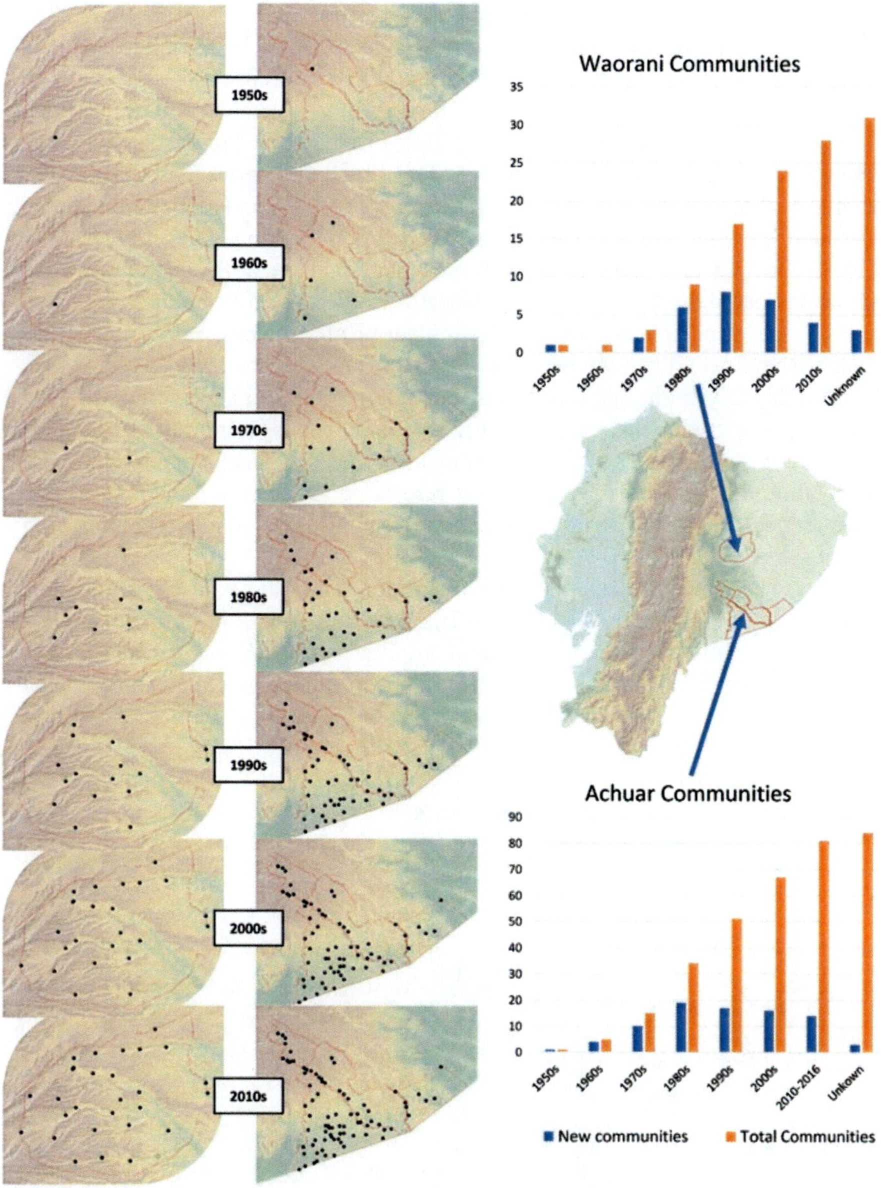

Fig. 4.1 Creation of permanent communities in the Achuar (right of left panel) and Waorani (left of left panel) Ancestral Territories, 1950–2019. Does not include communities that were abandoned

growth and territory availability, the former being probably more important than the latter, as suggested by the fact that in some areas the territories of new communities overlap those of older communities. External factors, including education, health services, and work opportunities outside of the household, especially for women, contributed to lower fertility rates and populations rates of growth (Carr et al. 2006). The desire to have access to infrastructure, services (e.g., schools, airstrips, health centers), and substitute goods (e.g., food, building materials) also contributed to community permanency. On the other hand, most of these effects will probably not be fully felt for some time. To begin with, while dropping, fertility rates remain high among Indigenous women. Davis et al. (2015) found that between 2001 and 2012 the number of families increased an average of 40%, and populations in 32 Indigenous communities in the Northern Ecuadorian Amazon, an average of 50%. Also, the population is very young yet to have children. In 2010, one of every two Indigenous people in the Ecuadorian Amazon was under 15 years of age, 6 of 10 were younger than 19 years (INEC 2022). In terms of availability of territory in the Ecuadorian part of the biogeographic Amazon (i.e., below 1300 m.a.s.l), there were roughly 20,000 ha per community including protected areas, around 16,000 excluding protected areas, and 12,000 if only Indigenous territories were considered. As this study suggests, this means that, on the aggregate, existing communities continue to potentially have a forest reserve to support the functions of their active productive territories, and, if needed, to establish new non-overlapping communities in certain areas.

In this chapter, we examine the general question stated above directly through the spatial lens of the productive territories of 10 communities in the WAT and their widespread adoption of a commercial perennial crop, cacao (*Theobroma cacao*, Fig. 4.2). We propose that the analysis is relevant for the rest of the communities in the WAT,

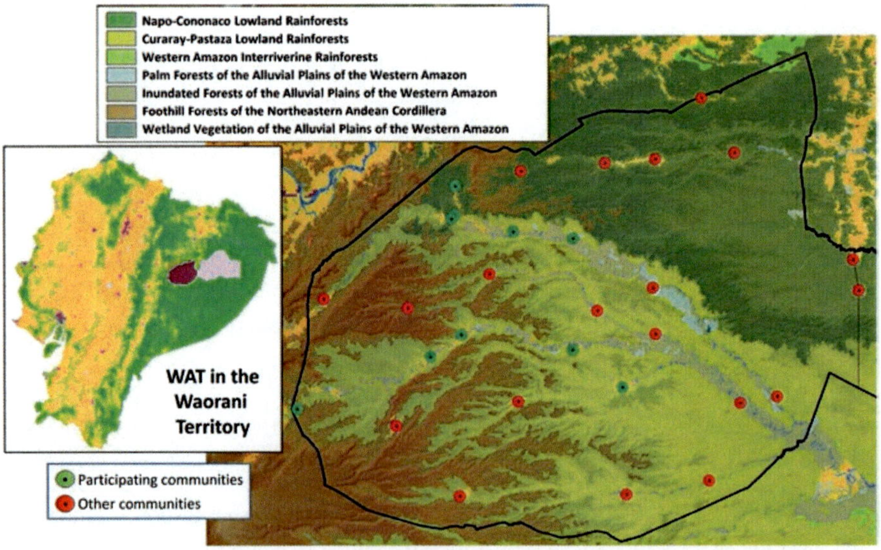

Fig. 4.2 Waorani Ancestral Territory of the Upper Curaray River (WAT) and the communities in this study

and could be an indication of broader land use changes in many areas of the Western Amazon where similar demographic and market-engagement arrangements are becoming increasingly common. Specifically, we ask the following questions:

1. What are the spatial characteristics of the productive territories of the Waorani communities in the WAT?
2. What are the patterns of cacao cultivation in these communities? And,
3. Does cultivation of cacao affect the size of their productive territories?

In the analysis that follows, the territory is expected to correspond to a permanent community with significant dependence on horticulture and agriculture, a territorial structure that is increasingly common in the Western Amazon. We define a productive territory as the territory used for: (a) living space in general (homes, schools, airstrips, etc.) and the nearby cultivation of subsistence crops (e.g., manioc, fruits; active and fallow), (b) the outlying cultivation of subsistence and commercial crops (e.g., cacao, vanilla), animal husbandry (e.g., cattle) and the harvesting of resources in a matrix of semidomesticated forests in different stages of succession from long-term horticultural cycles, (c) periodical harvesting of wild and semi-domesticated plants (e.g., fibers, roots, fruits, nuts) and hunting, fishing and harvesting of foods of animal origin (e.g., honey, beetle larvae) in a matrix of old growth forests. Communities may also have a reserve territory (d), used occasionally and to which only community members have use rights (Fig. 4.3). The concentration of soils that have been improved by human activity (anthrosols) and the higher abundance and richness of domesticated and semi-domesticated plants and trees around these sites suggests that similar arrangements, without the commercial components, were

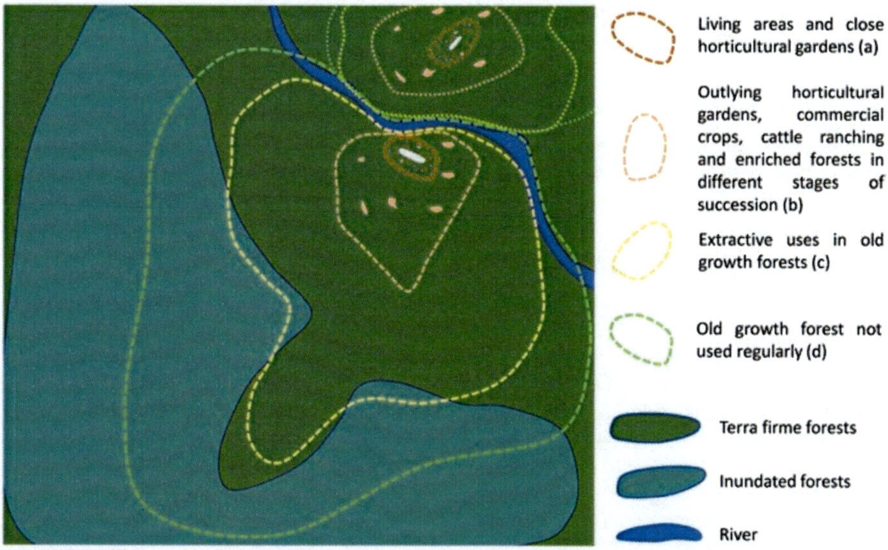

Fig. 4.3 Productive territory model for a permanent Indigenous community in the Western Amazon

common in many areas of the Amazon dating as far as 2500 ybp until shortly after the arrival of Europeans (Sombroek et al. 2002; Levis et al. 2017).[4] To be clear, not all traditional productive territories in the Amazon are structured in the same way. Beckerman (1987) found great variation in the dependence on horticulture in the diet of 34 traditional Amazonian Indigenous groups, ranging from over 80% among the Bari and Kuikuru to equal or less than 50% among the Amahuaca and the Secoya. Some nationalities, like the Akuriyo of Suriname and the Waorani of the northern Ecuadorian Amazon, persisted until recently without significant agriculture, mainly from hunting, fishing, and gathering (Beckerman et al. 2009; Kloos 1977), which probably meant that their territories were very large compared to communities more dependent on horticulture. In riverine environments, fishing may be the most important animal component of the diet, with families consuming wild fish at least once a day (Dorea et al. 2003), while in intra-riverine communities, forests for hunting may be more important.

The adoption of new commercial land uses should be expected to have different effects on the extent of a productive territory depending in the nature of the activity. For instance, Sierra et al. (1999) found that the closer to a market Indigenous families were in the WAT during their early stages of economic integration, the greater the time dedicated to hunting at the cost of less time dedicated to social activities. Similar tradeoffs have also been documented among the Kichwa in the northern Ecuadorian Amazon (Macdonald 1981) and the Shipibo of the Peruvian and Brazilian Amazon (Behrens 1992). It is reasonable to assume that increased time dedicated to hunting meant increased pressure and a larger productive territory, which in turn would be an incentive for some Indigenous households to move to a new area where resources are available at desired levels. In contrast, the adoption of commercial crops or income from labor is generally expected to reduce the pressure and area of productive territories by requiring more time to be spent in agricultural or paid-work (e.g., Engel et al. 2008; Vasco and Siren 2019).[5] Income also contributes to the substitution with purchased goods and services of some of the resources and services available in the territory. For example, the progressive replacement of traditional palm-leaf thatched with tin roofs reduces the extractive pressure on palm populations near a community and increases the availability of work-time for commercial activities. De la Montaña et al. (2015) estimated that a 10% increase in a family's income from paid-labor in two Cofán and Secoya communities in the northern Ecuadorian Amazon resulted in a reduction of 16–20% in the animal

[4] The first European account by Gaspar de Carvajal, c. 1540–1542, describe large areas along major rivers with what appeared to be high-density settlements, with horticulture and keeping of wild animals (e.g., turtles for meat and eggs), separated by depopulated areas between them.

[5] Perhaps, the greatest interest in this relation originated in policymakers and managers needing to understand the effectiveness of conservation strategies using economic and social development interventions. A common criticism has been the lack of evidence of these effects (Znajda 2014; Blom et al. 2010; Mistry et al. 2010).

(hunting and fishing) biomass extracted, and that an increase of 50% resulted in a reduction of >50%, which would suggest a contraction of their productive territories.

A thorough review of the literature yielded no empirical assessments of the extent and condition of the territory Indigenous communities use in the Amazon or anywhere else. Land use and territorial analysis have focused on its horticultural and agricultural component probably because these are observable through mapping or socioeconomic surveys (e.g., Denevan 2001; Gray and Bilsborrow 2020). From a methodological point of view, this chapter describes a spatial modeling effort using data collection and modelling techniques that allow to quantitatively and objectively study land uses that are hard to measure remotely or through socioeconomic surveys. It is possible to observe the forest, but not the people using it. A second innovation is the measure of distance as time to reflect more closely short-distance decision making by Indigenous families about how to use their territory.

2 Conceptual Framework

2.1 Characterization of Cacao Cultivation Among Waorani Communities in the WAT

Data on cacao cultivation was collected through personal interviews and field mapping of all cacao sites of all producers in nine of the 10 communities included in this study (N = 121 plots). The analysis differentiates broadly between the high-value, low-output aroma variety, and the low-value, high-output CCN51 variety and between productive (older, >4–6 years for CCN51 and > 5–7 for aroma) and not-productive trees (recent, <4–6 years and < 5–7 years, respectively). Cultivation of the CCN51 variety is widely supported by State agencies for economic development purposes. Cultivation of aroma cacao is promoted by national and international non-governmental and cooperation organizations for biodiversity and climate regulation purposes. Waorani families also cultivate wild cacao (*Theobroma bicolor*), which is not promoted actively by State agencies or NGOs, but has a niche market to which Indigenous producers have access to. To estimate the distances of the plots to locations of interest, such as the nearest cocoa market, we generated a movement-cost model within a geographic information systems (GIS) framework that considers the average speed (m/seg) in open spaces (horticultural and agricultural, settlements), roads and rivers using standard speeds applied in previous studies in the region (e.g., Sierra and Calva 2019). In forested areas, a slope-weighted binned time-cost model was estimated from a random sample of 100,000 speed observations (Fig. 4.4). Cultivation intensity is measured as trees per plot or per family, not as trees per hectare, because cacao is cultivated alone or in combination with other crops.

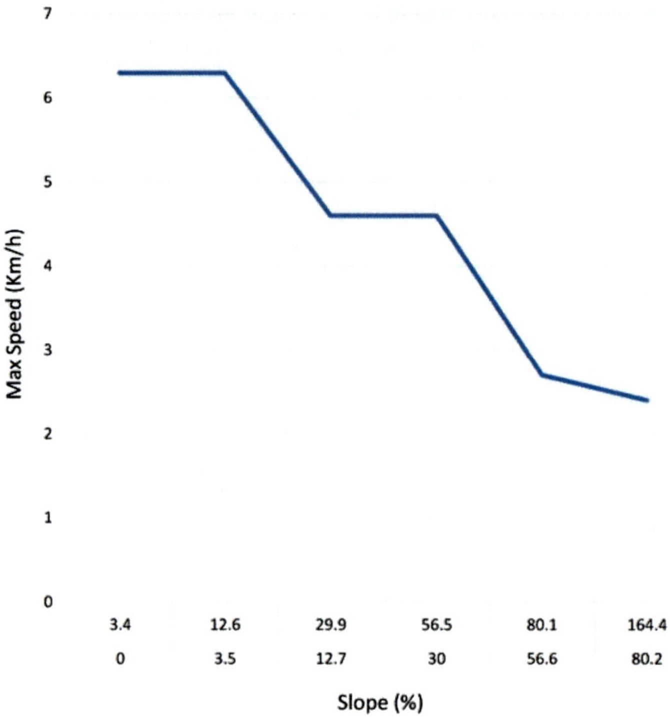

Fig. 4.4 Binned model of the maximum speed of movement through forested areas based on terrain (slope) conditions

2.2 A Model of the Productive Territory of Traditional Waorani Communities in the WAT

Traditionally, the Waorani are an egalitarian social group, living in close groups of one to three communal homes occupied by extended multigenerational families without a clear power structure. The traditional Waorani productive territory had two layers: one short-cycle active territory, used for 1 to 2 years, with small horticultural gardens dispersed inside an old-growth forest matrix, up to 30–60 min, or 3–5 km, from their homes, and a larger long-cycle reserve territory, used over several years, inside of which several short-cycle territories were rotated (Beckerman et al. 2009; Cardoso et al. 2012). The periodic replacement of short-cycle territories mainly composed of extractive areas, within the larger long-cycle reserve territory, was a strategy to resolve an active territories' low carrying capacities, ensuring that there was a minimum acceptable return per unit of effort, rapidly moving extractive areas once their productivity fell below a local optimum. Two results of this mobility were less dependence on agricultural production and small populations.

Although dependence on local natural resources remains high (Macía 2004; Lu 2007), most contemporary Waorani territories are different from the traditional ones. Perhaps the most significant difference is the transformation of a semi-sedentary into a fundamentally sedentary life system with a relatively constant productive territory; similar to most Indigenous communities in the Western Amazon that still depend on the surrounding forests, rivers, etc., for their livelihood. This transformation is illustrated, for example, by the adoption of permanent housing, increasingly made of sawn wood and metal roofs, single-family, monogamous homes with fewer children, and the construction of infrastructure that requires large investments of time and funds (e.g., airstrips, schools – Fig. 4.5). A sedentary life also helped reverse the Waorani tendency to lose population and increased their life expectancy, creating pressure for additional territories for a growing number of communities. Over the past five decades, population grew on average >3% per

Fig. 4.5 Selected Waorani settlements in the WAT, c.2010: (**a**) Ñaibecaro (1–2 families, access by river), (**b**) Kenaweno (12 families, access by river and small plane), (**c**) Gareno (18 families, access by road), (**d**) Tiweno (20–30 families, access by plane and river). (Aerial photography: SIGTIERRAS)

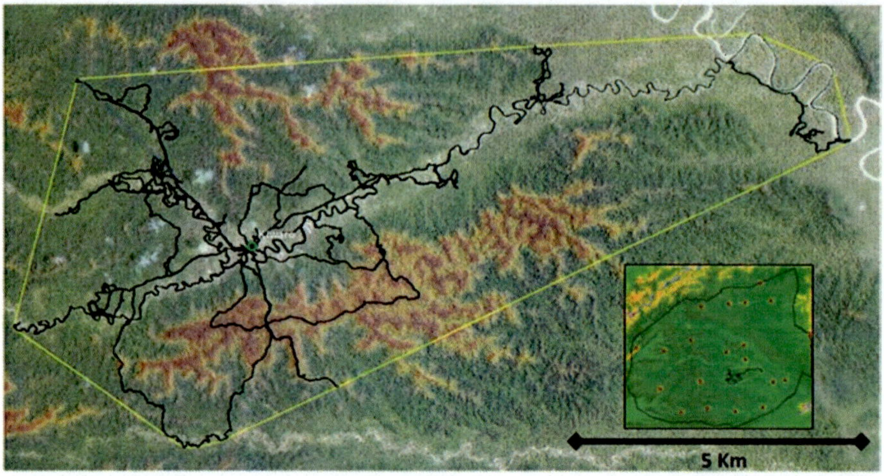

Fig. 4.6 Spatial model of the productive territory of the Waorani Community of Kiwaro developed in a geographic information system (GIS) using a polygon envelope (in yellow) connecting maximum travel distances from the focus community

year.[6] Although there are indications the rate is slowly falling (Davis et al. 2015), compared to other Indigenous women in Amazonian Ecuador, Waorani women maintain high fertility rates, suggesting the population will continue to grow rapidly for a few more decades.

We propose a model of the contemporary Waorani productive territory that extends to the maximum travel time that allows community members to go and return home the same day, with a strong preference to minimize travel time per unit of production; a pattern widely observed in other sedentary Indigenous communities in the Western Amazon. Sirén et al. (2004), for example, observed among Kichwa communities a maximum hunting distance of 15 km, with almost 70% of the hunting effort taking place less than 6 km and 90% less than 12 km from the community center (see also Peres and Lake 2003). We implement this territorial model by mapping the daily displacement of members of 10 communities in the WAT. Operationally, each community's productive territory is equal to the convex polygon that includes all displacements from that community (Fig. 4.6).

We collected forest use information in an opportunistic way in the period September 2018 to May 2019 (Table 4.1). We assume these territories are structured as zones a, b and c in the proposed territorial model following a theoretical use-intensity gradient with a maximum at the center of the community and a minimum at the edge of the territorial model defined by the aggregate daily displacement area of all community members. A simplified pressure model was created by reprojecting each community's territory using a central point of the living areas. Pressure is equivalent to the frequency a location is used or visited by a community given its linear distance and access time to the community's center (Fig. 4.7).

[6] Estimated from an initial population of 500 in 1960 (Cardoso et al. 2012) and a late-2010s population of 3500 (Per. Comm., G. Nenkimo 2020).

Table 4.1 Sample of transects per community used to estimate a community's territorial demand

Community	Transects (N)
1	11
2	20
3	27
4	31
5	20
6	33
7	9
8	7
9	25
10[a]	59
Total	183

[a]No cacao cultivation data

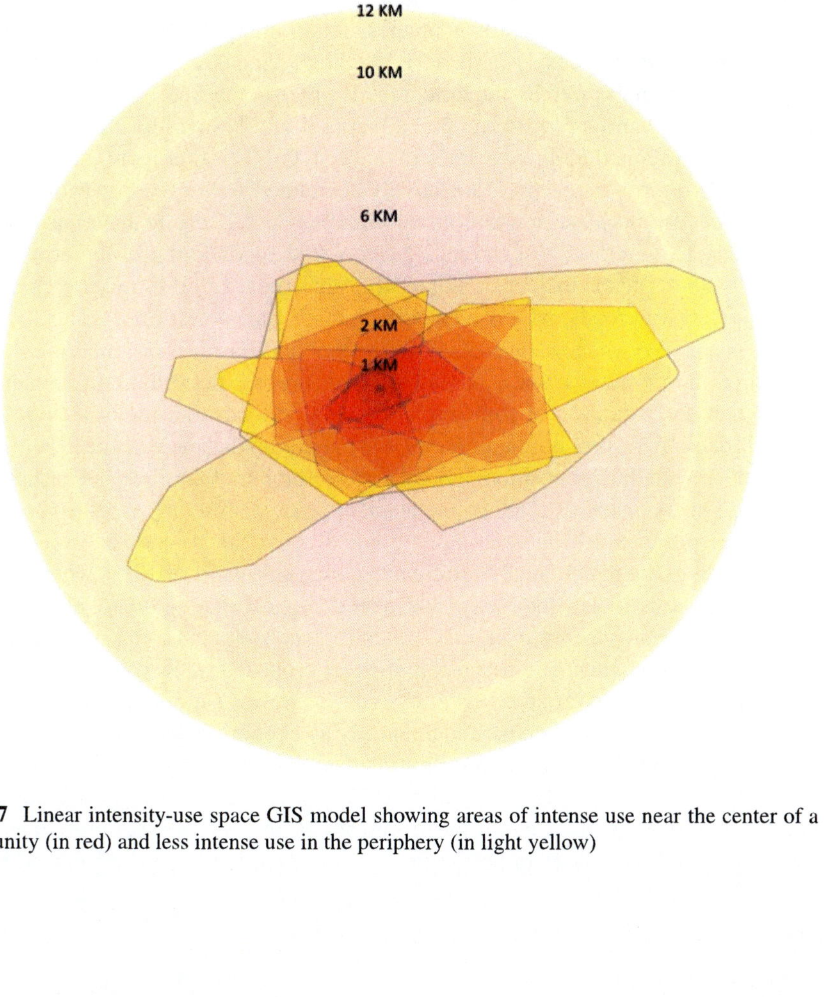

Fig. 4.7 Linear intensity-use space GIS model showing areas of intense use near the center of a community (in red) and less intense use in the periphery (in light yellow)

Due to the small sample size, the analysis of the relationships between the territory and cocoa production at the community level is primarily based on a visual interpretation of their correspondence. We assume that there is a relationship when the distribution of the values shows an obvious trend. We use parametric statistics (correlations, linear regression) only to offer a complementary empirical approximation of this observable relationship.

3 Patterns of Cacao Cultivation Among WAORANI Communities in the WAT

Table 4.2 summarizes the plot-level patterns of cacao cultivation in nine of the 10 communities included in this study. Each family managed an average 145 cacao trees on one 0.3 ha plot, with cacao planted alone or in combination with other crops. Density varied between 1.7 and 21.5 trees per 100 m^2, with an average of 7.8 trees per 100 m^2. The level of investment in commercial production, measured as the number of trees each family manages, is significantly affected by local and regional transportation costs (Table 4.3). The more expensive it gets to sell products or buy inputs, as measured by the time it takes to reach markets or agricultural input suppliers, the lower the probability that a Waorani family invest in cacao cultivation (i.e., less area planted and fewer trees managed). On the other hand, having an airstrip, which improves access, is negatively correlated with cacao investment. This probably occurs because newer communities are less likely to have airstrips but more likely to be closer to markets by rivers and roads and are, therefore, more likely to adopt cacao. In fact, it is possible that most trade is mainly channeled through rivers and roads, which is cheaper than air transportation.

The probability of new investment in commercial production, measured as the number of non-productive trees, increases with distance to the community center, suggesting that the expansion of commercial production tends to avoid affecting subsistence production. At the community level, the maximum distance for cacao cultivation is consistent with this proposition: the greater the number of families, the farther away cultivation takes place, as more of the available closer areas are used or needed for subsistence production (Fig. 4.8). Potential and active subsistence cultivation areas closer to homes are maintained, pushing new cacao cultivation to areas where competition for space is lower. Indirectly, this would suggest that commercial productive capacity is not drawn from subsistence agricultural capacity but by transferring capacity from other productive and social activities, as proposed by Sierra et al. (1999).

The patterns observed at the level of cacao varieties support the argument that adoption of cacao cultivation is fundamentally an economic decision and that transportation costs probably are a determining factor in a family's decision invest in commercial production, its magnitude and form. Since families in a community tend to experience similar conditions, transportation costs for example, they tend also to respond to market opportunities in similar manners. Producers with low

Table 4.2 Cacao adoption patterns by the families of 9 communities of the WAT

Community	Year established	Families 2018	Cacao plots				Cacao trees							
			N	Average has	Plots/ family	Has/ family	Total	Trees/ plot	% CCN51 productive	% CCN51 no productive	% Aroma productive	% Aroma no productive	% Wild cacao productive	% Wild cacao no productive
1	2003	12	9	0.2	0.8	0.18	822	91.33	0.7	0.1	51.7	16.2	27.7	3.5
2	2001	7	4	0.9	0.6	0.52	1442	360.50	30.9	21.2	32.9	11.2	3.2	0.8
3	1998	33	41	0.4	1.2	0.48	9118	222.39	7.7	14.6	31.4	11.9	5.2	18.3
4	1970	14	14	0.1	1.0	0.08	882	63.00	2.3	0.6	74.3	8.4	11.7	2.8
5	2005	10	10	0.3	1.0	0.27	1461	146.10	12.0	2.9	64.8	10.3	2.9	7.0
6	1999	18	21	0.3	1.2	0.38	4756	226.48	63.5	6.0	24.2	2.6	0.7	3.0
7	1995	8	8	0.2	1.0	0.20	965	120.63	0.6	0.3	59.3	25.9	11.4	2.5
8	1998	7	7	0.2	1.0	0.21	1481	211.57	4.3	0.9	93.3	1.5	0.0	0.1
9	1980	51	43	0.1	0.8	0.12	3462	80.51	1.5	0.5	64.8	21.5	7.1	4.7
Total 2020		**160**	**157**	**0.3**	**1.0**	**0.3**	**24,389**	**169.2**	**13.7**	**5.2**	**55.2**	**12.2**	**7.8**	**4.7**

Table 4.3 Key factors determining cacao cultivation at the family level in 9 Waorani communities in the WAT

Factor	Area planted (m^2)	Total trees per plot without bicolor	Total trees per plot with bicolor	Number of trees per plot								
				CCN51 productive	CCN51 not productive	CCN51 total	Aroma productive	Aroma not productive	Aroma total	Bicolor productive	Bicolor not productive	Bicolor total
Time to community center				–**	+*			+*			+**	+**
Time to closest cacao collection center	–**	–**	–**	–**	–*	–**					–*	–*
Time to closest local market	–**	–**	–**	–**	–*	–**					–*	–*
Time to closest regional market	–**	–**	–**	–**	–*	–**					–*	–*
Time to closest paved road	–**	–**	–**	–**	–*	–**					–*	–*
Time to closest airstrip	+**	+**	+**				+**		+**			
Time to closest navigable river								+**				+*
Time to closest agricultural input supplier	–**	–**	–**	–**	–*	–**					–*	–*

Time to closest health center	+*	+**	+**				+**		+**			
Time to closest education center							+**		+**			
Terrainw	−**	−**	−**			−*			−*			
Elevation							−**	+*	−*		+*	+*

*Significant 0.00 level, **significant 0.000 level

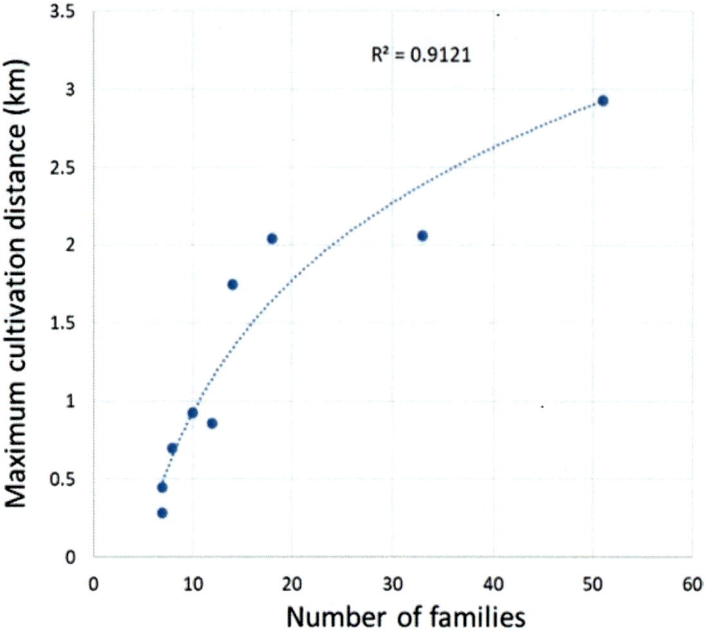

Fig. 4.8 Distance of furthest cacao cultivation plot vs number of families in the community

transportation costs prefer to produce more volume of CCN51, low priced but high yield, to optimize their returns, while producers with high transportation costs prefer the lower productivity but high value of the aroma variety. Consistently, higher transportation costs reduce the interest in the CCN51 because of its lower price-to-transportation costs ratio. In contrast, cultivation of the aroma variety increases with distance because, for a given transportation capacity and cost, it is more efficient to transport high value commodities.

4 The Productive Territories of 10 Communities in the WAT

Figure 4.9 and Table 4.4 summarize the characteristics of the productive territories of the 10 communities included in this study. These are a probable description of the other communities in the WAT as well. At the beginning of the 2020s, an average community had 17 families, had existed for approximately 3 decades, used 3300 hectares or around 200 hectares per family, and was covered mostly by natural forests in various stages of succession. It is evident that territories of nearby communities can overlap, but the tendency is to avoid competition with other communities by extending the territory to areas located away from other communities' productive territories. This results in territories that do not extend radially and uniformly away from a community center. A territory can be elongated and the community can be located at one

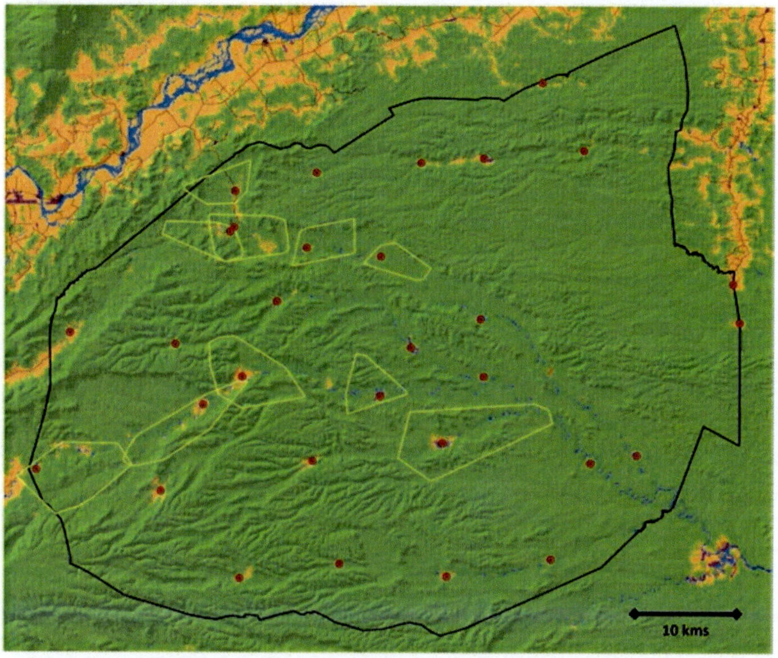

Fig. 4.9 Productive territories of the communities in this study (Yellow polygons)

Table 4.4 Summary findings

Community	Established	Families 2019	Active territory (has)	Active territory per family (has)
1	2003	12	5055	421
2	2001	7	2051	293
3	1998	33	1742	53
4	1970	14	3734	267
5	1970	8	7120	890
6	2005	10	3160	316
7	1999	18	2105	117
8	1995	8	2061	258
9	1998	7	1359	194
10[a]	1980	51	4397	86
Average	1992	17	3278	195
Total		185	36,063	195

Productive territories of the communities in this study
[a]No cacao cultivation data

extreme of the territory. Its extent is variable, being affected by various endogenous factors, such as community age and terrain (Fig. 4.10). For instance, older communities tend to have larger complementary territories, probably because of decreasing productivity in closer areas caused by long-term use – which is consistent with the

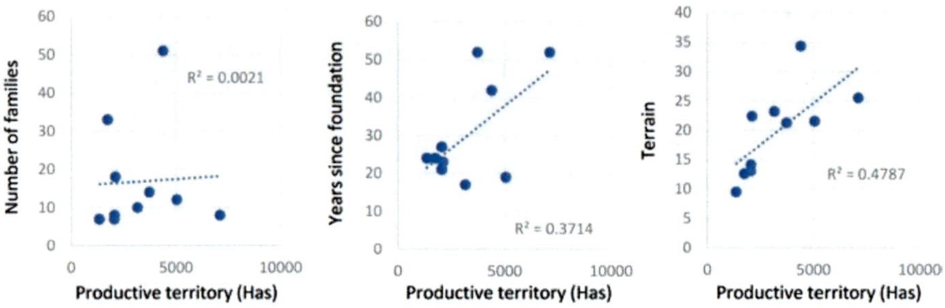

Fig. 4.10 Relationships between selected endogenous control factors (number of families, years of existence, terrain) and territorial demand (area) among nine communities in the WAT

use-intensity gradient model proposed. Territories tend to be also larger in more rugged areas, probably because a larger total area is needed to provide the required usable productive territory (i.e., steep areas are not easily used).

Empirically, it is possible to identify two management zones within a territory. A primary productive zone, where productive activities are frequent and within quick reach from homes. In all communities studied, areas within ~3 km or 3 h of a community center are actively used (Fig. 4.11), regardless of community size, age, location and level of commercial engagement. This is a minimum constant territory in all cases, suggesting it is an access-defined critical area for family livelihood. This area probably corresponds to uses (a) and (b) in the proposed model. A complementary productive zone occurs beyond the primary zone, corresponding to zone (c) in the proposed model. Interestingly, the extent of the complementary active area is not affected by the number of families in a community.

5 The Territorial Outcome of Cacao Cultivation in the WAT

Figure 4.12 illustrates the relationships between investment in cacao cultivation, measured as the average number of trees per family in a community, and the spatial demand for a productive territory by that community: the more trees a family manages in a community, the smaller the productive territory demanded by that community. Most of this effect corresponds to the relationship with established cultivation of aroma cocoa, which is by far the preferred variety – almost 70% of all cacao trees are aroma, 80% of which are productive (i.e., >5–7 years old). This suggests that: (a) there is a structural bottleneck in some capacity needed to use the territory, and (b) the territory needed to produce the minimum required resources is smaller through some level of resource substitution facilitated by commercial production. An empirical model linking these two factors, transportation costs and investment in cacao production, produces an almost perfect estimate of the area of a community's productive territory (Fig. 4.13). In a stepwise model, investment in cacao show a stronger association than transportation costs with the area of the

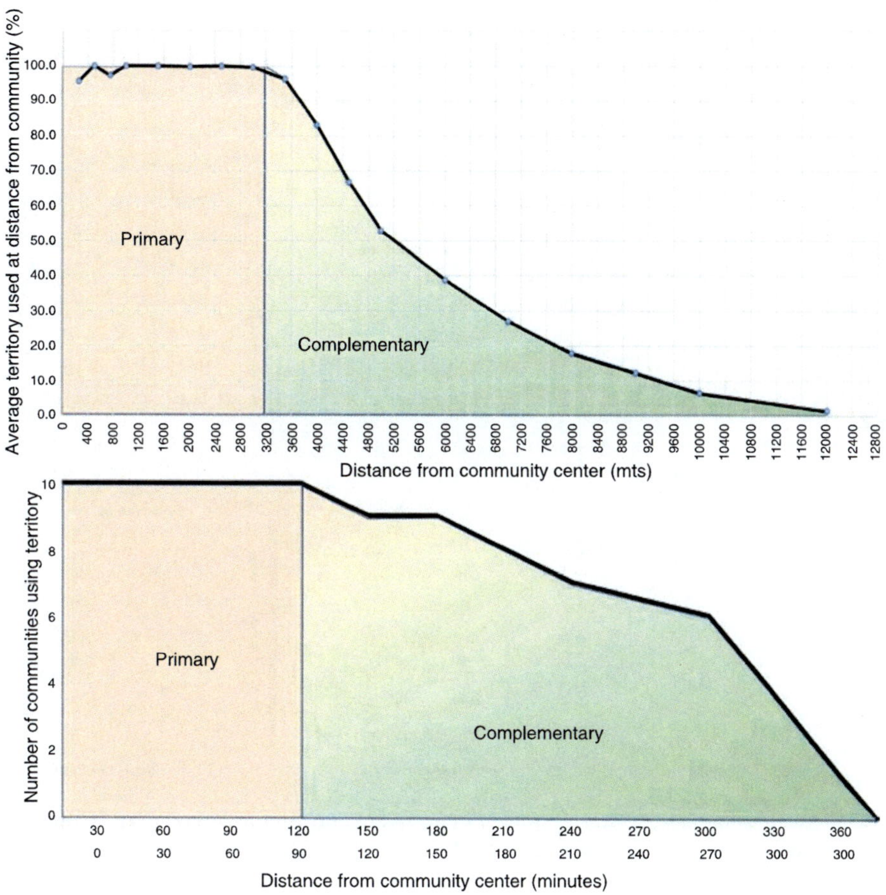

Fig. 4.11 Space and time model of the management zones of the productive territory of the communities in the WAT

territory, suggesting that the mechanism (a) plays a key role in this change and that (b) may be complementary. If the number of trees per family term is removed from the model, transportation costs lose significance, which suggests that these alone would probably not have a significant effect. Of the two, ability to allocate a key resource or factor to commercial production is the most important. Only once this factor is potentially available for commercial farming, its economic return is considered. To expand commercial production, more land (i.e., fallow gardens, forest in different stages of succession) are brought into cultivation. This requires labor to establish and maintain it, which is likely drawn from social and productive activities that take place far from homes (i.e., in the secondary zone). As cultivation grows outward from the communities, related nonproductive time demands increase too, which also likely reduce the ability of households to use the older, larger territory. For example, longer time is needed for reaching newer plots and bringing harvest to

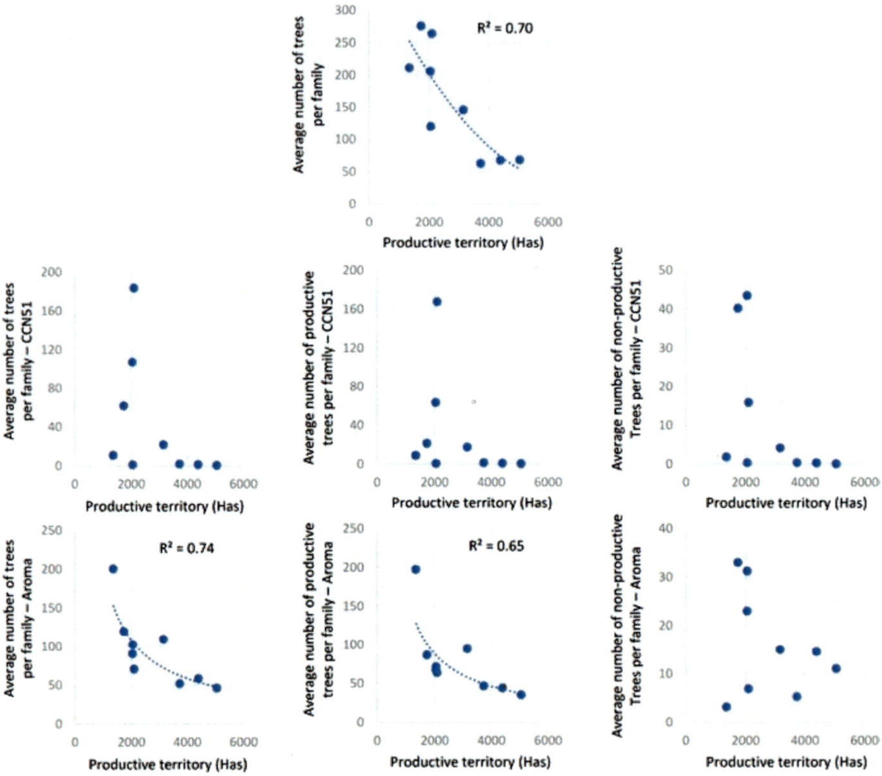

Fig. 4.12 Relationships between different dimensions of cacao farming and territorial demand (area) among nine communities in the WAT

markets. In fact, each additional tree seems to absorb a substantial amount of the capacity needed to use the old territory, probably the equivalent of several hectares; 17.1 hectares according to the empirical model. It is reasonable to assume that similar relationships can be expected in other communities in the WAT.

Once capacity to produce is secured, transportation costs become a determining factor relative to the value of the cacao produced. Families from communities with lower transportation costs and smaller territories invest more on commercial production than distant and labor-poor communities. Results vary slightly depending in what market definition is used (i.e., cacao collection center, local markets, regional markets, nearest road, etc.), but all are important (and highly correlated of course) to determine the effect of cacao cultivation in the area of a territory. Furthermore, it is reasonable to assume that, from the point of view of the intensity of use of the territory, the reduction in the area is reflected by a shortening of the distance and a contraction of the intensity gradient in the complementary zone of the territory shown in Fig. 4.3. This translates to patterns at the family level: the number of CCN51 and aroma productive trees per family, and transportation costs determine the size of their productive territories (Fig. 4.14). More cacao plots, more productive

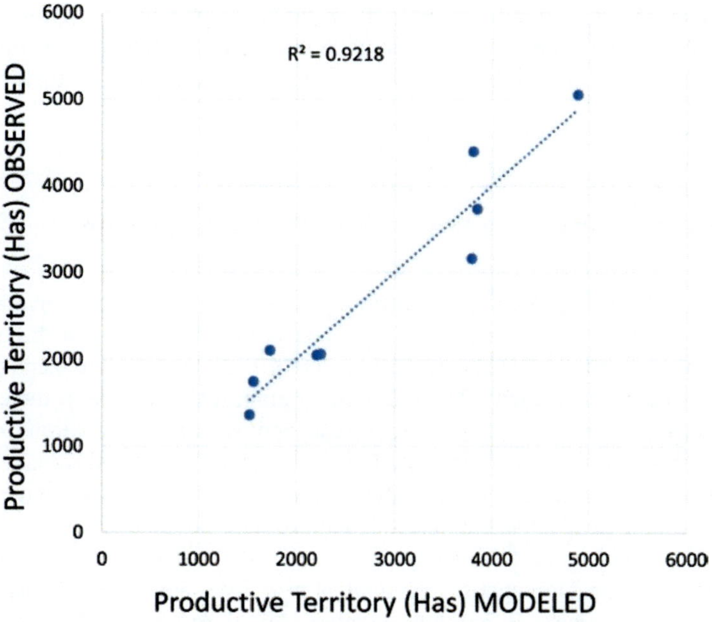

Fig. 4.13 Observed vs modelled productive territories' areas of nine communities in the WAT

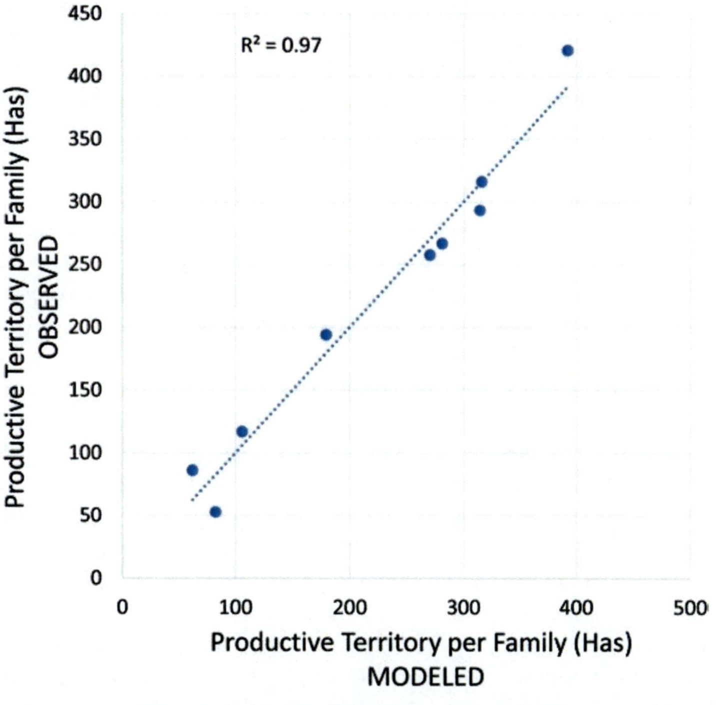

Fig. 4.14 Observed vs modelled productive territories' areas per family in nine communities in the WAT

trees, and lower transportation costs result in smaller productive territories per family. Interestingly, the empirical coefficients in the model for the number of CCN51 and aroma trees are almost the same (~ − 1 ha/tree), suggesting that their effect is similar and additive.

6 Expanding from the WAT to the Western Amazon

Definition and measurement of integrated productive territories of Amazonian Indigenous communities have been mostly limited to descriptions of the distance traveled for harvesting, primarily hunting, and to the location, structure and composition of horticultural gardens. There is limited empirical evidence of the changes such a territory experiences as demographic and economic conditions change. Indirect evidence, such as that presented by Sierra et al. (1999) and De la Montaña et al. (2015), and the results presented here suggest that these changes have important effects in the productive territories of Amazonian Indigenous communities. In practice, it is probably a combination factors that reshape the structure of a productive territory among Indigenous communities in the Western Amazon, but uptime (the proportion of time a community member is working or available for work)is likely a key bottleneck. To spend more time on commercial production it is necessary to spend less time in social and subsistence production in the territory – and probably harvesting in the complementary zone of the territory. Other factors that should be considered are the positive feedback from the reduction of the frequency of use of the complementary territory (i.e., the extractive pressure) and changes in knowledge that affect the usefulness of traditional resources. Lower extractive pressure from less labor availability itself contributes to reducing the maximum distance at which required resources can be found, reducing the time-distance needed to obtain them. Traditional knowledge is also likely lost by reduced use and its substitution with non-Indigenous knowledge due to demographic and cultural changes. With this loss, the use of those territories may be also reduced as it may be connected with that knowledge or resource. There is evidence that older Waorani know more about the resources available in the forest than younger Waorani (Weckmüller et al. 2019), so it is possible that younger populations take less advantage of their territory, and therefore use it less (See also Gray et al. 2015).

The fact that the reduction in the demand for territory is mainly associated to aroma cacao trees of productive age, planted 5–7 years earlier (2012–2014 in this case), also suggests that a key mechanism is the absorption of labor in harvesting, product preparation, and delivery to markets; activities that are not required for non-productive trees, although both require planting and care. On the other hand, the expected increase in new cocoa production, represented by the increasing importance of non-productive trees in the most distant plots, suggests that the shift from non-productive to productive trees has the potential to further reduce the demand of territory in each community. Therefore, actions that promote sustainable commercial production could contribute to reducing extractive pressures in the rest of the communities in the WAT and possibly the rest of the Western Amazon, where similar traditional land use patterns are common.

It is reasonable to expect that the ongoing demographic and economic transformations taking place among existing Indigenous communities throughout the Amazon will have also an effect on the selection of places to establish new communities, and therefore on the distribution of productive territories inside larger Indigenous territories in general, and in the WAT specifically. Among the Waorani communities in the WAT, newer communities tend to be closer to markets and roads, will likely use a smaller productive territory, and are closer to each other in comparison to older communities. A simple calculation put the aggregate productive territory of the 31 communities at the end of 2010s at approximately 100,000 hectares, or 30% of the WAT.

References

Beckerman S (1987) Swidden in Amazonia and the Amazon rim. In: Turner BL II, Brush SB (eds) Comparative farming systems. Guilford Press, New York, pp 55–94

Beckerman S, Ericksonb P, Yost J, Regaladod J et al (2009) Life histories, blood revenge, and reproductive success among the Waorani of Ecuador. Proc Natl Acad Sci 106:8134–8139

Behrens C (1992) Labor specialization and the formation of markets for food in a Shipibo subsistence economy. Hum Ecol 20:435–462

Blom B, Sunderland T, Murdiyarso D (2010) Getting REDD to work locally: lessons learned from integrated conservation and development projects. Environ Sci Pol 13:164–172

Cardoso S, Alfonso M, Valverde L et al (2012) Genetic uniqueness of the Waorani tribe from the Ecuadorian Amazon. Heredity 108:609–615

Carr D, Pan W, Bilsborrow R (2006) Declining fertility on the frontier: the Ecuadorian Amazon. Popul Environ 28:17–39

Davis J, Bilsborrow R, Gray C (2015) Delayed fertility transition among Indigenous women in the Ecuadorian Amazon. Int Perspect Sex Reprod Health 41:1–10

De la Montaña E, Moreno R, Maldonado J, Griffith D (2015) Predicting hunter behavior of Indigenous communities in the Ecuadorian Amazon: insights from a household production model. Ecol Soc 20:30–41

Denevan W (2001) Cultivated landscapes of native Amazonia and the Andes. Oxford University Press, New York, 432pp

Dorea J, Barbosa A, Ferrari I, De Souza J (2003) Mercury in hair and in fish consumed by riparian women of the Rio Negro, Amazon, Brazil. Int J Environ Health Res 13:239–248

Engel S, Pagiola S, Wunder S (2008) Designing payments for environmental services in theory and practice: an overview of the issues. Ecol Econ 65:663–674

Gray C, Bilsborrow R (2020) Stability and change within Indigenous land use in the Ecuadorian Amazon. Glob Environ Change 63:102116. https://doi.org/10.1016/j.gloenvcha.2020.102116

Gray C, Bozigarb M, Bilsborrow R (2015) Declining use of wild resources by Indigenous peoples of the Ecuadorian Amazon. Biol Conserv 182:270–277

INEC (2022) Censo de Población y Vivienda del Ecuador, 2010. Available at http://redatam.inec.gob.ec/cgibin/RpWebEngine.exe/PortalAction?&MODE=MAIN&BASE=CPV2010&MAIN=WebServerMain.inl Accessed 4 Apr 2022

Kloos P (1977) The Akuriyu of Surinam: a case emergence from isolation, 36p. International Working Group on Indigenous Affairs. Document 27. Denmark, Copenhagen

Levis C, Costa C, Bongers F et al (2017) Persistent effects of pre-Columbian plant domestication on Amazonian Forest composition. Science 355:925–931
Lu F (2007) Integration into the market among Indigenous peoples: a cross-cultural perspective from the Ecuadorian Amazon. Curr Anthropol 48:593–602
MacDonald T (1981) Indigenous responses to an expanding frontier: jungle Quichua economic conversion to cattle ranching. In: Whitten NE Jr (ed) Cultural transformations and ethnicity in modern Ecuador. University of Illinois Press, Urbana, pp 356–384
Macía M (2004) Multiplicity in palm uses by the Huaorani of Amazonian Ecuador. Bot J Linn 144:149–159
Mistry J, Berardi A, Simpson M et al (2010) Using a systems viability approach to evaluate integrated conservation and development projects: assessing the impact of the North Rupununi Adaptive Management Process, Guyana. Geogr J 176:241–252
Peres C, Lake R (2003) Extent of nontimber resource extraction in tropical forests: accessibility to game vertebrates by hunters in the Amazon Basin. Conserv Biol 17:521–535
Sierra R, Calva O (2019) Análisis de la Dinámica y los Factores de Deforestación en la Amazonia del Ecuador, 1990–2017. Plan Estratégico de Intervención Territorial para la Reducción de la Deforestación en la Región Amazónica del Ecuador. Reporte 2. Ministerio de Ambiente del Ecuador y FAO, Quito
Sierra R, Rodríguez F, Losos E (1999) Forest resource use change during early market integration in tropical rain forests: the Huaorani of upper Amazonia. Ecol Econ 30:107–119
Sierra R, Calva O, Guevara A (2021) La Deforestación en el Ecuador, 1990–2018. Factores promotores y tendencias recientes. Ministerio de Ambiente y Agua, Ministerio de Agricultura y Programa Integral Amazónico de Conservación de Bosques y Producción Sostenible, Quito, Ecuador, 216pp.
Sirén A, Hamback P, Machoa J (2004) Including spatial heterogeneity and animal dispersal when evaluating hunting: a model analysis and an empirical assessment in an Amazonian community. Conserv Biol 18:1315–1329
Sombroek W, Kern D, Rodrigues T et al (2002) Terra Preta and Terra Mulata: pre-Columbian Amazon kitchen middens and agricultural fields, their sustainability and their replication. Presented at 17th world congress of soil science, 14–21 August, Bangkok, Thailand
Vasco C, Siren A (2019) Determinants of wild fish consumption in Indigenous communities in the Ecuadorian Amazon. Soc Nat Resour 32:21–33
Vasco C, Tamayo G, Griess V (2017) The drivers of market integration among Indigenous peoples: evidence from the Ecuadorian Amazon. Soc Nat Resour 30:1–17
Weckmüller H, Barriocanal C, Maneja R, Martí M (2019) Factors affecting traditional medicinal plant knowledge of the Waorani, Ecuador. Sustainability 11:4460–4472
Znajda S (2014) What is 'successful development' in conservation and development projects? Insights from two Nicaraguan case studies. Conserv Soc 12:318–328

Chapter 5
New Insights on Water Quality and Land Use Dynamics in the Napo Region of Western Amazonia

Santiago López and Adolfo Maldonado

Abstract Environmental degradation including water quality decline, soil loss, biodiversity loss, and forest reduction are at the center of current socio-environmental concerns in the Amazon region. This chapter re-examines the factors that contribute to the likelihood of forest loss and provides a baseline characterization of water quality in the Napo Region in the Ecuadorian Amazon – a megadiverse area threatened by increased human pressure. We focus primarily on issues of land use/land cover transformations and water quality since changes in the lithosphere and hydrosphere currently affect ecosystem services critical to sustain biodiversity and human populations. We rely on a land change science approach, combining remotely sensed data and multinomial regression analysis, and borrow elements of watershed science to evaluate land cover and water quality changes in the past decade. Our results indicate that although deforestation rates have remained relatively unchanged the reduction of secondary and successional forests due to the expansion of African palm cultivation, pastures, and other forms of agriculture is problematic. The likelihood of deforestation is associated with the socio-spatial conditions of production areas. Our water quality analysis results show that dissolved oxygen, pH, and conductivity measurements are outside the optimal ranges suggested by national and international guidelines for drinking water, with direct consequences on public health. By focusing on LULC change and water quality, this study sheds new light into the complexity of human-environmental dynamics and resource use systems along the Andean foothills of Western Amazonia.

Keywords Remote sensing · Forest degradation · Land use and land cover · Water quality · Western Amazonia

S. López (✉)
School of Interdisciplinary Arts and Sciences, University of Washington Bothell, Bothell, WA, USA
e-mail: cslopez@uw.edu

A. Maldonado
Clínica Ambiental, Acción Ecológica, Quito, Ecuador

S. López (ed.), *Socio-Environmental Research in Latin America*, The Latin American Studies Book Series, https://doi.org/10.1007/978-3-031-22680-9_5

1 Introduction

The Amazon region, an area of about 7 million Km^2, is characterized by dynamic resource-use systems that could modify the structure and composition of native ecosystems, leading to high risks of irreversible change (Nobre et al. 2016). Although the Amazon region has experienced significant environmental changes and human modifications for millennia (Morcote-Ríos et al. 2020), contemporary transformations have significantly increased in recent decades due to processes of colonization (i.e., in-migration and settlement of landless peasants) (Barbieri et al. 2009; Laurance et al. 2002), expansion of the agricultural frontier and the replacement of forest with monocultures (Vijay et al. 2018), and extraction of natural resources (e.g., logging, crude oil production, mining) (Sonter et al. 2017; Mena et al. 2006). Environmental degradation including water quality decline, soil loss, biodiversity loss, and forest degradation are at the center of current socio-environmental concerns. Studies on the factors that contribute to these changes together with baseline characterizations of environmental change are needed for informing national and international resource management policies, conservation initiatives, and future environmental impact assessments. This study aims at assessing the spatio-environmental factors that contribute to the likelihood of deforestation and providing a baseline characterization of water quality in the Napo River basin in the Ecuadorian Amazon – a megadiverse area threatened by increased human pressure (Myers 2003; Myers et al. 2000). In this study, we defined deforestation as the probability that trees will be completely or partially removed from land and used for purposes other than forests.

Most socio-environmental change studies in the Amazon basin have focused on the quantification of land use and land cover (LULC) transformations. In the Napo region, such studies have concentrated on the quantification of forest loss (cf. Santos et al. 2019; Mena et al. 2006; Pichón 1997; Bilsborrow et al. 2004; Sierra 2000), and the associated factors including population dynamics, labor and market indicators, tenure conditions, and spatio-environmental constraints to agricultural production. Most studies usually rely on some form of remotely sensed data to measure the extent and intensity of deforestation. Sierra (2000), for example, used Landsat MSS and TM data resampled at a spatial resolution of 100 m to quantify forest loss. This study showed that forests are likely cleared at higher rates in the core of the Napo deforestation front than in the periphery. The author reported an average deforestation rate of <1% ha $year^{-1}$ between 1986 and 1996 for an area of about 5 times the size of the area analyzed in this study. Mena et al. (2006) used Landsat TM data at a resolution of 30 m and obtained a deforestation rate of 2.5% $year^{-1}$ between 1986 and 1996 and a reduced rate of approximately 1.8% between 1996 and 2002 for an area of about two times the extent analyzed here. These results demonstrate that the quantification of forest loss and its rates vary depending on the size of the study area and spatial resolution of the sensors. Other studies have shown that land cover change patterns vary depending on particular local demographic, socio-economic, and environmental contexts (Gray and Bilsborrow 2013; Bilsborrow et al. 2004).

Mena et al. (2006) and Santos et al. (2019) showed, for example, that forest areas close to urban centers or administrative areas with high population densities, and secure land tenure, usually experience very intense deforestation. However, less populated areas with relatively insecure land tenure could also experience intense deforestation if enough financial incentives exist to clear forests.

It is well known that deforestation processes are non-linear (Santos et al. 2019; Walsh et al. 2008) and changes in socio-economic conditions, sometimes triggered by national-level processes associated with national development policies, affect households' behavior (Barber et al. 2014; Pichón 1997). The expansion of road infrastructure and land distribution policies, for example, facilitate resource extraction (Laurence et al. 2009) and it is well known that most new forest clearings in the Amazon region occur in the proximity of roads (Asner et al. 2006; Alves 2002; Chomitz and Thomas 2001). The creation of settlements along navigable river routes opens new possibilities to reaching forested regions, which could accelerate deforestation in previously inaccessible areas (Verissimo et al. 1998). In the Napo region, the substantial increase and improvement of transportation routes and consequently the provision of better access to local and national markets and forested lands are conspicuous characteristics of place affecting forest extent and health (Santos et al. 2019; Bilsborrow et al. 2004).

Topographic and soil characteristics are well-known land conditions affecting LULC change. Areas with steep slopes, where on-site forest conversion costs are higher due to restricted accessibility, and poor soils are less used and more often remain forested (López and Sierra 2010; Pan et al. 2004). In general, the conversion of forests to other land uses including pastures, new settlements, roads and other forms of public infrastructure, and agriculture, depends on a series of proximate and underlying conditions affecting the extent and rate of LULC change (Geist and Lambin 2002). In this chapter, we evaluate current deforestation processes in the Napo region in Ecuador through the quantification of the probability of forest loss, particularly caused by agriculture and pastures expansion. The approach includes a multinomial logistic regression (MNLR) modeling exercise within a geographic information systems (GIS) framework to identify the factors associated with forest clearing and the determination of the probability of forest change.

Besides its bio and cultural diversity, the Amazon region is also well known for its abundant water resources, currently threatened by decreased water supply due to tropical glacier recession and degradation of land cover in water catchment areas (Buytaert and Bièvre 2012). Major threats to Amazonian freshwater resources include agricultural expansion, point and nonpoint source pollution, hydroelectric dams, oil extraction, mining and road building, and climate change (Mena et al. 2017; Guimberteau et al. 2017; Castello and Macedo 2016; Swenson et al. 2011; Potes 2010). Such threats have severe repercussions on water's physico-chemical parameters, with implications for ecosystem function and public health (Azevedo-Santos et al. 2016; Finer and Jenkins 2012). In the Ecuadorian Amazon, for example, Jocknick et al. (1994) found high levels of polycyclic aromatic hydrocarbons (PAH) in a sample of water quality observations collected in major oil fields in Sucumbíos and Orellana likely associated with high cancer incidence in the region.

Although the idea of studying coupled hydro-terrestrial systems is not new and some agreement exist on how land use changes and water resource use have altered hydrological regimes throughout history (Savenije et al. 2014), scarce empirical studies exist that document the physico-chemical properties of water in areas of the Amazon basin heavily affected by crude oil operations and agricultural expansion with a few exceptions (cf. Velázquez-Gómez and Lacorte 2020; Maurice et al. 2019; Barraza et al. 2018). This type of work is particularly relevant for areas in Latin America like Ecuador or Peru where about two-thirds of their Amazon regions are covered by oil and gas leases, many of which overlap protected areas, Indigenous territories, and agricultural lands (López and López-Sandoval 2016; Finer et al. 2008). As energy demands grow, the impacts on water resources of extractive projects such as the controversial drilling operation at Yasuní/Ishipingo-Tambococha-Tiputini (Yasuní ITT) in Ecuador or the Camisea gas pipeline in Peru are likely to become more common, particularly in the Andean Amazon, where some relatively large oil and mineral reserves still exist (López and López-Sandoval 2016). Although it is beyond the scope of this study to formally (statistically) evaluate the linkages between water quality and land degradation, this study aims to provide a baseline characterization of water quality changes in two sub-catchments of the Napo River basin in the last decade and offer some intervention recommendations.

Due to increased human activity, largely connected to expansion of extractive industries and the rise of commercial agriculture, the Napo region in Northeast Ecuador has experienced accelerated LULC transformations and environmental deterioration in recent years. In this context, we address three main research questions that encapsulate our scholarly interests on human environment interactions: What significant LULC transitions have occurred in this region over the last 10 years? What socio-environmental drivers account for the probability of deforestation? How has water quality varied in areas impacted by land degradation and contamination in the past decade? Since changes in the lithosphere and hydrosphere are currently affecting ecosystem services necessary to maintain biodiversity and human populations, we primarily focus on land use/land cover (LULC) and water quality (Schielen and Börner 2018; Brauman 2015; Elmhagen et al. 2015). We follow a land change science approach (i.e., the systematic study of the changing landscapes as mediated through human land use) and borrow elements of watershed science (i.e., the study of water quality and the factors that affect it) to evaluate environmental change in an area of the Amazon region that has witnessed accelerated environmental degradation.

2 Study Area

The study area comprises 3400 km^2 and is divided into 15 parishes (*parroquias*) in the Ecuadorian provinces of Sucumbíos and Francisco de Orellana in Ecuador (Fig. 5.1). Our assessment of water quality concentrates on three sub-catchments of the Napo and Aguarico rivers (Huamayacu, Sacha, and Pacayacu rivers), which

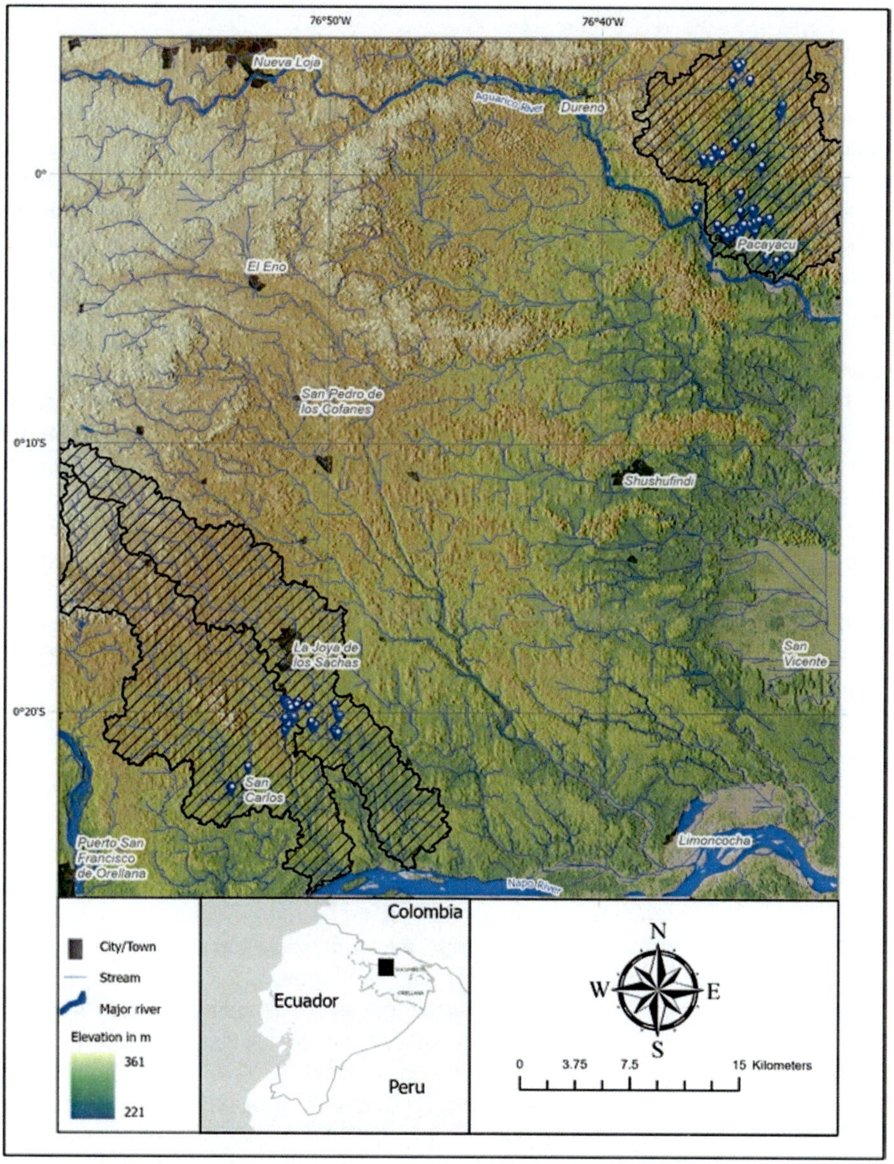

Fig. 5.1 Study area: Subcatchments of the Napo River basin in the Ecuadorian Amazon. Blue markers on the map represent the locations of water quality observations

together cover an area of 540 km^2. These catchments are home to two communities representative of the *mestizo* (mixed race or non-Indigenous) population of Ecuador: Pimampiro and Pacayacu, each of which has about 50 and 450 families respectively. The region has a tropical rainforest climate with heavy to very heavy rainfall year-round reaching approximately 3000 mm year^{-1} of precipitation, with an average yearly temperature of 25.3 °C (INAMHI 2014).

Due to concerns about accelerated environmental degradation caused by economic development and migration processes such as colonization, agricultural expansion, the replacement of native forests with monocultures, and the growth of the oil and gas industry, the Napo River basin in the Ecuadorian Amazon has drawn significant attention from the international community since the early 1980s. Before then, traditional land uses with minimal human impact dominated the land (Beristain et al. 2009; Mena et al. 2006). The discovery of petroleum reserves, road construction, and the 1964 agricultural reform contributed to a massive exodus of primarily landless peasants from other regions of Ecuador and the settlement of these new migrants in the Sucumbíos and Orellana provinces. Although some of this in-migration was unplanned, the government nevertheless encouraged it to promote the occupation and development of what government officials often saw as "empty lands" (Vasco et al. 2018). At the moment, the landscape is a patchwork of mainly Amazon lowland evergreen forests, non-native land uses mixed with traditional ones, and densely populated urban centers, arranged in spatial patterns consistent with road networks, crude oil extraction projects, agricultural activities, and cattle raising operations. Since the early 1990s, the total population of the NorthernEcuadorian Amazon has reportedly increased at rates of around 5% year^{-1}, one of the highest rates in the country (Bilsborrow 2003). For instance, the 2010 census shows that the population in Lago Agrio, the main town and market found in the study area, increased from roughly 30,000 people in 2000 to nearly 58,000 people in 2010 (almost a two-fold increase in one decade) (INEC 2010).

Although deforestation rates have slowly decreased since the early 1990s, forest degradation remains an issue of concern (Santos et al. 2019; Sierra 2000). Most forest changes in the region are due to colonist farmers' interventions who established land claims along roads originally developed to aid the crude oil industry (Pichón 1997). In addition to a continued conversion of forests to agriculture and pastures, the conversion from extensive land uses to more intensive ones has been clearly documented among Indigenous and non-Indigenous groups (Lu et al. 2010; Rudel et al. 2002). Although vast evidence exists that not all Amazonian soils are poor and some areas, especially in the lower parts of the Amazon River basin, are in fact extremely rich in nutrients (e.g., Amazonian dark earths or *Terra Preta de Índio* in the *várzea* ecosystem) (Hilbert and Soentgen 2020), most of the study area lies in non-*Terra Preta*, primary forest soils (oxisols) that are generally extremely nutrient poor and not adequate for intensive industrial agricultural use without external inputs (e.g., fertilizers, draining systems, machinery). Despite these environmental constraints and the lack of robust agrobiodiversity knowledge of the region, large agroindustry complexes promoting the use of non-native species, such as African palm (*Elaeis guineensis*), expanded quickly in the region since their introduction in the 1970s. A large palm plantation, for example, can be identified to the east of the study area (Palmeras del Ecuador S.A.) covering an area of about 41 km^2. Given the rising demand for palm oil worldwide, African palm monoculture is likely the most lucrative land use in the moist forests of Western Amazonia (Castellanos-Navarrete et al. 2021); however, this type of land use has also been linked to increased tropical deforestation (Vijay et al. 2018) and widespread soil and water pollution because of the intensive use of chemical fertilizers and pesticides (Dislich et al. 2017).

Water quality is also affected by the leaching of naturally occurring heavy metals like mercury, exacerbated by the clearing of forests (Webb 2005). In addition, large concentrations of tannins from decaying vegetation and sediment from erosion processes contribute to the high acidity levels in surface and subsurface waters (Castello and Macedo 2016). Gas flare emissions and spills of petroleum and toxic waste from oil operations into freshwater systems are common in the region and affect large areas beyond wells and oil camps (Lessman et al. 2016). The area is also impacted by chemical compounds from pesticides and fertilizers used in agriculture (Velázquez-Gómez and Lacorte 2020). These issues exacerbate the need for water quality monitoring since most families in peri-urban areas rely on rain and well water as their cleanest alternative to local rivers and streams since there is limited centralized water supply. River and stream sources of water are sometimes used for consumption, sustaining livestock, bathing, and recreation (Rivera-Parra et al. 2020).

3 Methods and Research Approach

3.1 Statistical Modeling of Land Use and Land Cover Change

We propose that the socio-environmental context of a region correlates with the likelihood of forest reduction, influencing individual landholder decisions regarding the allocation of land use (Lopez 2022). We employed a MNLR framework to assess the likelihood of potential LULC transitions given a collection of independent socio-environmental parameters. MNLR has been used to predict the likelihood of different possible outcomes when the dependent variable is categorical, and the independent co-variates are categorical or interval (Millington et al. 2007; dell'Olio et al. 2011). One advantage of MNLR in relation to other regression frameworks is that there is no need for the independent variables to be statistically independent from each other; however, collinearity is assumed to be low (Belsley 1991). Lopez (2022) formally define the MLNR model as:

$$P_{i,u,v} = \begin{cases} \dfrac{1}{1+\sum_{k\in\{1,\dots,U\},k\neq u} e^{\left(\beta_{0,u,k}+\beta_{1,u,k}X_{1,k}+\dots+\beta_{p,u,k}X_{p,i}\right)}}, for\ v = u \\ \dfrac{e^{\left(\beta_{0,u,v}+\beta_{1,u,v}X_{1,i}+\dots+\beta_{p,u,v}X_{p,i}\right)}}{1+\sum_{k\in\{1,\dots,U\},k\neq u} e^{\left(\beta_{0,u,k}+\beta_{1,u,k}X_{1,i}+\dots+\beta_{p,u,k}X_{p,i}\right)}}, for\ v \neq u \end{cases} \tag{5.1}$$

Where $P_{i,u,v}$ is the probability that pixel i will change from LULC type u to type v. Observations are split into U groups based on their initial LULC change states as identified by cross-comparison of satellite data. The independent covariates $X_1 \ldots X_p$ represent a series of socio-spatial factors. The regression parameters β are estimated via maximum likelihood (Hosmer and Lemeshow 2000). Based on a thorough revision of existing literature, Lopez (2022) identified the following potential factors affecting the probability of deforestation in the area: cost-distance to cities and towns (CstDstTwn), distance to rivers and waterways (DstStrm), distance to principal roads (DstMnRds), population density proxied by light conditions (PopDen), road density (RdDen), and terrain characteristics: slope (Slope), elevation (Elev), and ruggedness (TRI). LULC transition data were derived from intercomparison analysis of remotely sensed data. Distance factors were derived from official cartographic sources generated by the Ecuadorian Geographic Military Institute at a scale of 1:50,000. Topographic factors were obtained from 30-m Shuttle Radar Topography Mission (STRM) V3.0 elevation data. Population density estimates were derived from a 2013 cloud-free U.S. Airforce Defense Meteorological Satellite Program Operational Line Scan (DMSP/OLS) lighting dataset assuming that population concentrates in well-lit areas as measured by satellite sensors. Please see Lopez (2022) for a detailed description of the GIS methods employed in the operationalization of socio-spatial co-variates. We used a stratified random technique to generate robust statistical pixel samples comprising 1% of the total number of pixels ($N \approx 5200$) for periods T1(2013–2015) and T2 (2015–2019) based on the results of LULC change detection analysis. To reduce the issue of spatial autocorrelation and given the size of some land cover transition areas, pixels that fell within 100 m from another potential location were not considered in the evaluation.

3.2 *Time Series of Land Use and Land Cover*

We followed the pre-processing and processing remote sensing protocols detailed in López (2022) to analyze three subsets of Landsat 8 OLI scenes (USGS 2020) from years 2013, 2015 and 2019. This protocol included second order polynomial geometric and relative radiometric rectifications based on the pseudo-invariant features method to correct for sensor specific distortions. After corrections, we performed object-oriented random tree (RT) classifications in ArcGIS Pro (ESRI 2020) since evidence suggests that they perform better than conventional and unconventional machine learning classification techniques (Breiman 2001; Liu et al. 2016; Thanh Noi and Kappas 2018; Oo et al. 2022). From data collected from the interpretation of images, the Global Positioning System (GPS), and expert knowledge of the region amassed by the research team over the previous 10 years, we produced 50 (about 1250 pixels) training areas per class scene^{-1}. We quantified LULC change using a post-classification, inter-comparison technique (Singh 1989), for the periods T1 and T2. We used Sierra et al. (2002) and our knowledge of the area to differentiate seven LULC classes: (1) Amazon lowland and inundated evergreen forests

(LINEV); (2) barren areas or exposed soils (BARR); (3) mixed development areas that include roads, infrastructure, and buildings (MIXDEV); (4) monocultures with predominance of African palm plantations (PALM); (5) pastures and agricultural areas (PASAGR); (6) water bodies such as rivers and lakes (WBDY); and (7) secondary and successional forest (SECFOR). We used Maxar imagery as reference (years 2014 and 2018) (cell size ≈ 0.5 m) for validation of our classified data with confusion matrixes weighted by the proportion of class area (Olofsson et al. 2014) and following the approach described by Congalton and Green (1999).

Since we were interested in investigating forest reduction processes, we concentrated on the following LULC changes: (a) LINEV to BARR, (b) LINEV to PALM, (c) LINEV to PASAGR, (d) LINEV to SECFOR, (e) SECFOR to BARREN, (f) SECFOR to PALM, and (g) SECFOR to PASAGR. As "pivot" category (i.e., a possible LULC transition to which the other changes are compared against), we used the transition from LINEV to SECFOR since the emphasis was on changes from dominant land cover types to major land uses. Based on these transitions, deforestation rates were calculated for the entire study area and individual administrative units.

3.3 *Water Quality Multi-Temporal Analysis*

We hypothesize that water quality as characterized by water temperature (Temp), dissolved oxygen (DO), potential of hydrogen (pH), and conductivity (Cond) significantly declined between 2011 and 2017 given communities' reports about the slow progress made by local authorities over the past decade to improve water quality, collection, and distribution infrastructure. To determine whether significant changes existed from year to year in the observed water quality parameters, we relied on a series of independent sample T-tests. We used Cohen's d statistic to determine the magnitude and direction of the differences.

Water quality information was obtained from shallow wells (less than 8 m deep) (93% sample), piped water (4%), and stream water (3%) observations in the *parroquias* of Joya de los Sachas, San Carlos, and Pacayacu collected between 2011 and 2017 (Fig. 5.1). Our colleagues at *Accion Ecológica's Clinica Ambiental* (i.e., a local environmental organization that has worked in the area since the early 1990s) helped with the selection of communities where the water quality sampling took place and coordinated the work with community leaders. The data were obtained at the end of the month of August (less wet season) in 2011 (N = 56), 2013 (N = 39), and 2017 (N = 17), between 9:00 in the morning and 5:00 in the afternoon. We obtained the samples systematically following the order of farms and houses along the main road in both communities (Pimampiro near the Sacha Oil Field and Pacayacu near the Pichincha Oil Field). In 2017, we selected a random subset of the original samples due to severe weather conditions and time limitations in our field campaign.

The sample observations were georeferenced using a Trimble GPS receiver unit (Juno 3B) with a nominal accuracy of ±5 m in a geographic coordinate system (WGS84). We collected three-minute GPS recordings and averaged them to reduce positional errors. The areas we visited were photographed with a Nikon camera to keep a record of the observations (Fig. 5.2).

To quantify water quality, the research team used a YSI Prometer multiparameter meter, calibrated daily with neutral standard solutions and according to the equipment specifications. Our team measured Temp (°C), DO (mg/l), pH (−log [H +]), and Cond (μS/cm). Although there are no minimum standard requirements for drinking water in the Ecuadorian legislation and most international guidelines, temperature provides information on the characteristics of surface aquifers and streams. Water temperature also influences the rates of chemical and biological processes and affects other measured parameters like DO (EPA 1986). DO provides information about the amount of oxygen present in the water and helps evaluate the health of organisms such as fish, invertebrates, bacteria, and aquatic plants. DO is also linked to the presence or absence of aerobic bacteria in the water and generally shares an inverse relationship with temperature. The amount of DO in water polluted with contaminants is often low as microbes consume oxygen to process certain organic chemicals such as hydrocarbons (Joye et al. 2011). The indication of basicity or acidity of water can be measured with a pH scale. Because pH is measured on a logarithmic scale, a pH of 7 (basic), for instance, is much more hospitable to biological organisms compared to a pH of 4 (acidic). Although pH does not have a direct impact on human health, this parameter is one of the most important parameters to measure in water quality assessments since it usually correlates with chemical contamination levels (WHO 2018). Waters heavily contaminated with chemicals and/or heavy metals usually exhibit low pH values (WHO 2008). Conductivity is the ability of water to pass electrical current, which is directly related to the concentration of ions in the water (Rusydi 2018). These ions usually come from dissolved salts and inorganic materials and are known as electrolytes. Distilled (demineralized) water, for instance, will have very low (if not negligible)

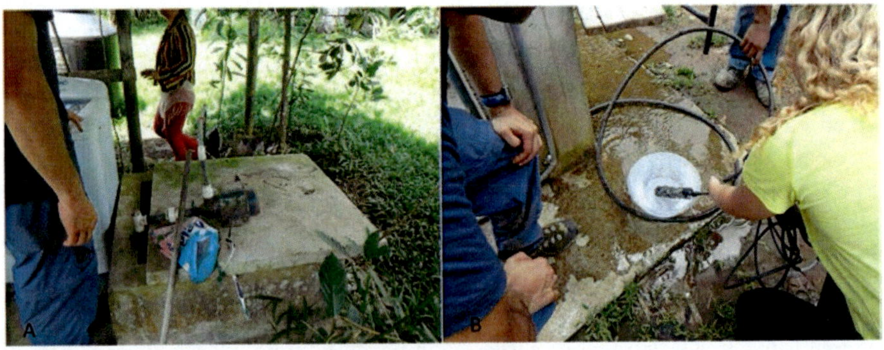

Fig. 5.2 Sampling of well water in the community of Pacayacu by the research team. (**a**) Concrete water well with pumping system. (**b**) Water quality testing with multiparameter probe

Table 5.1 Water quality guidelines generated by international and Ecuadorian agencies for temperature (Temp), dissolved oxygen (DO), potential of hydrogen (pH), and conductivity (Cond)

Parameter	US EPA[a] – WHO[b] – The Council of the European Union[c] – Canadian Guidelines[d]	TULSMA (Ecuador)
Temp (°C)	≤ +15[d]	+5 (aquatic life preservation)
DO (mg l^{-1})	> 6.5 (cold water) or 5.5 (warmwater)[a]	> 6 (human consumption)
pH	(6.5–8.5)[a,b] or (6.5–9.5)[c]	6.0–9 (human consumption) or 6.5–9 (aquatic life preservation)
Cond (μS cm^{-1})	<1500[b] or (200 < Cond <2500)[c]	–

conductivity. National and international environmental and health agencies have established permissible limits for some of these physico-chemical parameters for drinking water (Table 5.1).

4 Results

4.1 Land Use/Land Cover Characterization and Deforestation Assessment

The confusion matrices weighted by class area showed that our pre-processing and processing protocols helped us attain an overall accuracy of at least 78% for our classifications. All accuracies are presented with a 95 percent confidence interval (Table 5.2). We did not derive the Kappa index, as recent studies have discouraged its use and it is considered a redundant accuracy measurement (Olofsson et al. 2014). Despite the relatively overall high accuracies some classes were particularly hard to delineate. Specifically, MIXDEV, PASAGR and BARR had large commission or omission errors.

The assessment of deforestation revealed that the current average forest loss rate is 3.33% year $^{-1}$, which is comparable to a previous report by Mena et al. (2006) for an area twice the size of the area analyzed here. In comparison to earlier studies, the majority of parishes in the study area's central region, where deforestation has historically been most severe, had a minor increase of roughly 1% year^{-1} on average. Interestingly, the parish of Joya de Los Sachas gained forest at a rate of 0.15% year^{-1}, a significant reversal from the previous loss rate of 7.63% year $^{-1}$ in the early 2000s (Table 5.3). The rate of forest loss also decreased about 0.2% in the San Carlos parish.

Results of the MNLR analysis show that in T1 approximately 3% (43 km^2) of lowland and inundated forest cells transitioned to barren, 1% (14 km^2) to African palm cultivation, 2% (28 km^2) to pastures and agriculture, and 18% (224 km^2) to secondary and successional forest (Fig. 5.3). When quantifying changes from secondary forest to other LULC types in the same time period, our results showed that

Table 5.2 Overall and partial accuracies of random forest classifications of Landsat 8 OLI data based on a seven-class scheme. Cells entries are expressed as the estimated proportion of area. Classified categories are the rows while reference classes are the columns. Accuracy measures are presented with a 95 percent confidence interval

	Class	*LINEV*	*BARR*	*MIXDEV*	*PALM*	*PASAGR*	*WBDY*	*SECFOR*	*Total*	*Areas (ha)*	*Wi*	*U acc*	*P acc*	*Overall*
A	***LINEV***	0.28	0.01	0.01	0.00	0.01	0.01	0.06	0.05	124,500	0.37	0.76 ± 0.06	0.95 ± 0.03	0.78 ± 0.04
	BARR	0.00	0.08	0.01	0.01	0.03	0.00	0.01	0.08	44,038	0.13	0.60 ± 0.12	0.83 ± 0.11	
	MIXDEV	0.00	0.00	0.03	0.00	0.00	0.00	0.00	0.22	10,324	0.03	0.93 ± 0.13	0.56 ± 0.20	
	PALM	0.00	0.00	0.00	0.05	0.01	0.00	0.01	0.16	22,504	0.07	0.72 ± 0.15	0.86 ± 0.13	
	PASAGR	0.00	0.00	0.00	0.00	0.11	0.00	0.01	0.09	45,471	0.14	0.83 ± 0.09	0.66 ± 0.10	
	WBDY	0.00	0.00	0.00	0.00	0.00	0.03	0.00	0.36	8873	0.03	1.0 ± 0.00	0.72 ± 0.21	
	SECFOR	0.01	0.01	0.00	0.00	0.01	0.00	0.20	0.02	79,132	0.24	0.85 ± 0.07	0.69 ± 0.08	
	Total	0.30	0.10	0.05	0.06	0.17	0.04	0.29	1.00	334,842	1.00			
B	***LINEV***	0.29	0.00	0.00	0.00	0.01	0.00	0.02	0.33	109,536	0.33	0.89 ± 0.06	0.89 ± 0.03	0.85 ± 0.04
	BARR	0.00	0.08	0.00	0.00	0.02	0.00	0.00	0.12	39,624	0.12	0.71 ± 0.12	0.93 ± 0.11	
	MIXDEV	0.00	0.00	0.03	0.00	0.00	0.00	0.00	0.04	12,148	0.04	0.94 ± 0.13	0.76 ± 0.20	
	PALM	0.00	0.00	0.00	0.08	0.00	0.00	0.00	0.09	28,672	0.09	0.95 ± 0.15	0.97 ± 0.13	
	PASAGR	0.00	0.00	0.00	0.00	0.15	0.00	0.01	0.17	58,038	0.17	0.86 ± 0.09	0.78 ± 0.10	
	WBDY	0.00	0.00	0.00	0.00	0.00	0.02	0.00	0.02	8023	0.02	1 ± 0.00	0.79 ± 0.21	
	SECFOR	0.04	0.00	0.00	0.00	0.01	0.00	0.19	0.24	78,822	0.24	0.81 ± 0.07	0.82 ± 0.08	
	Total	0.33	0.09	0.04	0.08	0.19	0.03	0.23	1.00	334,864	1.00			
C	***LINEV***	0.24	0.00	0.00	0.00	0.00	0.00	0.01	0.25	109,536	0.25	0.94 ± 0.05	0.93 ± 0.06	0.82 ± 0.04
	BARR	0.00	0.06	0.00	0.00	0.03	0.00	0.01	0.11	39,624	0.11	0.52 ± 0.15	0.83 ± 0.15	
	MIXDEV	0.00	0.00	0.03	0.00	0.00	0.00	0.00	0.04	12,148	0.04	0.69 ± 0.23	0.72 ± 0.23	
	PALM	0.00	0.00	0.00	0.07	0.01	0.00	0.01	0.10	28,672	0.10	0.74 ± 0.14	0.93 ± 0.09	
	PASAGR	0.01	0.01	0.00	0.00	0.19	0.00	0.03	0.23	58,038	0.23	0.81 ± 0.08	0.66 ± 0.09	
	WBDY	0.00	0.00	0.00	0.00	0.00	0.02	0.00	0.02	8023	0.02	0.9 ± 0.20	1 ± 0.00	
	SECFOR	0.02	0.00	0.00	0.00	0.02	0.00	0.20	0.24	78,822	0.24	0.84 ± 0.07	0.77 ± 0.08	
	Total	0.26	0.07	0.04	0.08	0.26	0.02	0.27	1.00	334,864	1.00			

LINEV Lowland and inundated evergreen vegetation, *BARR* barren areas, *MIXDEV* mixed development, *PALM* African palm cultivation, *PASAGR* pastures and agriculture, *WBDY* water bodies, *SECOR* secondary and successional forest

Table 5.3 Deforestation rates in nine parishes in the Napo region and the whole study area between 2013 and 2019. Rates were calculated using the approach suggested by (Ochoa-Gaona and Gonzales-Espinosa 2000) and compared with the estimates obtained by Mena et al. (2006) for period 1996–2002

Area of analysis	Year 2013 (km^2)	Year 2019 (km^2)	r^a (% $year^{-1}$)	r^b (% $year^{-1}$)
La Joya de los Sachas	59	60	−0.15	7.63
San Pedro de los Cofanes	28	17	8.17	7.52
San Carlos	65	48	4.96	5.13
Siete de Julio	79	52	6.81	5.44
Enokanqui	101	66	6.93	5.90
Pompeya	118	98	2.98	0.37
Shushufindi	265	201	4.50	2.97
Dureno	207	188	1.63	1.35
El Eno	253	215	2.73	2.26
Whole study area	2036	1661	3.33	–
$r = 1 - \left[1 - \frac{A1 - A2}{A1}\right]^{\frac{1}{T}}$		*A1* = area of forest at the beginning of the period, *A2* is the area at the end, *T* is the number of years in the period.		

[a]This study
[b] Mena et al. (2006)

11% (88 km^2) of the cells in this category changed to barren, 4% (28 km^2) to palm cultivation, and 15% (115 km^2) to pastures and agriculture. In period T2, about 3% (30 km^2) of forest cells changed to barren, 1% (8 km^2) to palm cultivation, 6% (62 km^2) to pastures and agriculture, and 16% (173 km^2) to secondary and successional forest (Table 5.4). During this period, 9% (73 km^2) of secondary forest cells changed to barren, 3% (27 km^2) to palm cultivation, and 21% (164 km^2) to pastures and agriculture. Our data show that, in general, deforestation (i.e., complete removal of forests including secondary forests) rates decreased from 3.88% $year^{-1}$ in T1 to 2.07% $year^{-1}$ in T2 (Lopez 2022).

4.2 *Multinomial Statistical Analysis of Land Use and Land Cover Change*

We first statistically described the data (Table 5.5) and looked for multicollinearity among the covariates using an iterative regression technique. We calculated variance inflation factors (VIF) by taking a predictor and regressing it against every other predictor. Because all VIFs fell between 1 and 2, the results indicated that multicollinearity was not a problem. Most of the parameters in the model are significant ($p < 0.01$) for periods T1 and T2, except for slope in period T1 and terrain

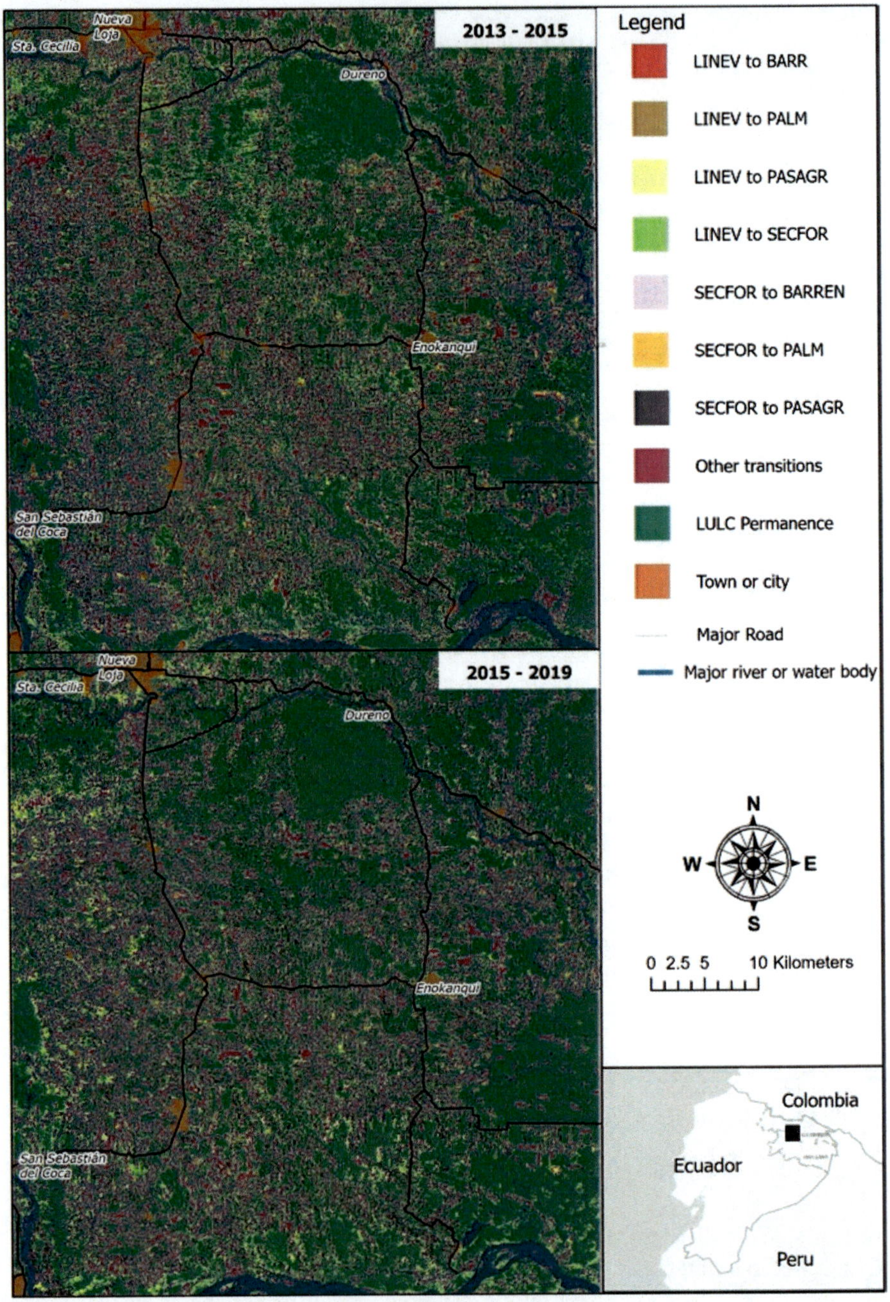

Fig. 5.3 Land use and land cover change and persistence between 2013–2015 and 2015–2019. (Modified from Lopez 2022)

Table 5.4 LULC transition matrixes showing areas (ha) and percent changes (in parenthesis) at the class level for three periods of analysis: (A) 2013–2015, ((B) 2015–2019, and (C) 2013–2019

A	*Year 2015*								
Year 2013	***LULC class***	**LINEV**	**BARR**	**MIXDEV**	**PALM**	**PASAGR**	**WBDY**	**SECFOR**	**Total**
	LINEV	932 (75)	43 (3)	2 (0)	14 (1)	28 (2)	3 (0)	224 (18)	1243 (100)
	BARR	28 (6)	152 (35)	9 (2)	26 (6)	129 (29)	1 (0)	96 (22)	440 (100)
	MIXDEV	0 (0)	0 (0)	102 (99)	0 (0)	0 (0)	0 (0)	0 (0)	103 (100)
	PALM	2 (1)	18 (8)	1 (1)	200 (89)	1 (0)	0 (0)	3 (1)	225 (100)
	PASAGR	32 (7)	92 (20)	4 (1)	18 (4)	307 (67)	1 (0)	1 (0)	455 (100)
	WBDY	7 (8)	3 (4)	1 (1)	1 (1)	1 (1)	72 (0)	3 (0)	89 (100)
	SECFOR	95 (12)	88 (11)	2 (0)	28 (9)	115 (15)	2 (0)	461 (58)	791 (100)
B	***Year 2019***								
Year 2015	***LINEV***	818 (75)	30 (3)	0 (0)	8 (1)	62 (6)	0 (0)	174 (16)	1095 (100)
	BARR	10 (2)	121 (31)	6 (2)	17 (4)	130 (33)	1 (0)	111 (28)	396 (100)
	MIXDEV	0 (0)	0 (0)	121 (99)	0 (0)	0 (0)	0 (0)	0 (0)	121 (100)
	PALM	2 (1)	27 (9)	1 (0)	247 (86)	2 (1)	1 (0)	7 (3)	287 (100)
	PASAGR	7 (1)	113 (19)	2 (0)	27 (5)	425 (73)	0 (0)	6 (1)	580 (100)
	WBDY	2 (2)	2 (2)	2 (3)	0 (0)	1 (1)	69 (86)	4 (5)	80 (100)
	SECFOR	11 (1)	73 (9)	1 (0)	27 (3)	163 (21)	2 (0)	511 (65)	788 (100)
	Year 2019								
Year 2013	***LINEV***	771 (62)	47 (4)	3 (0)	22 (2)	99 (8)	4 (0)	299 (24)	1245 (100)
	BARR	12(3)	118 (27)	13 (3)	40 (9)	160 (36)	10 (0)	96 (22)	440 (100)
	MIXDEV	0 (0)	0 (0)	102 (99)	0 (0)	0 (0)	0 (0)	0 (0)	103 (100)
	PALM	2(1)	22 (10)	2 (1)	187 (83)	6 (2)	0 (0)	7 (3)	225 (100)
	PASAGR	10 (2)	80 (18)	6 (1)	31 (7)	287 (63)	1 (0)	40 (9)	455(100)
	WBDY	4 (5)	2 (2)	3 (3)	1 (1)	2 (3)	68 (77)	8 (9)	89 (100)
	SECFOR	50 (6)	97 (12)	4 (1)	46 (6)	229 (29)	2 (0)	362 (46)	791 (100)

LINEV Lowland and inundated evergreen vegetation, *BARR* barren areas, *MIXDEV* mixed development, *PALM* African palm cultivation, *PASAGR* pastures and agriculture, *WBDY* water bodies, *SECOR* secondary and successional forest

Table 5.5 Descriptive summary of the independent covariates for periods *T1* = 2013–2015 and *T2* = 2015–2019

Covariate	*Period*	*N*	*Mean*	*Median*	*Min*	*Max*	*Std Dev*
CstDstTwn (h)	*T1*	5336	7.60	6.65	0.00	26.90	4.68
DstMnRds (m)	*T1*		4868.70	4145.00	5.60	17209.80	3616.32
DstStrm (m)	*T1*		427.76	364.69	0.00	8403.67	406.21
PopDen (%)	*T1*		55.67	55.76	0.55	100.67	29.94
TRI (m)	*T1*		6.00	4.81	0.00	480.05	14.96
Slope (°)	*T1*		3.17	2.51	0.00	18.68	2.74
Elev (m)	*T1*		284.72	281.31	226.00	355.20	22.72
RdDen (km × km^{-2})	*T1*		0.71	0.64	0.00	5.88	0.55
CstDstTwn (h)	*T2*	5171	7.42	6.50	0.00	26.50	4.57
DstMnRds (m)	*T2*		5062.28	4398.04	5.60	16841.30	3721.60
DstStrm (m)	*T2*		446.23	374.81	0.00	8613.92	471.79
PopDen (%)	*T2*		55.83	55.57	0.26	100.88	30.52
TRI (m)	*T2*		5.83	4.82	0.00	490.64	12.87
Slope (°)	*T2*		3.12	2.56	0.00	18.56	2.60
Elev (m)	*T2*		285.46	281.12	228.00	353.00	23.87
RdDen (km × km^{-2})	*T2*		0.70	0.64	0.00	5.48	0.48

roughness in T2 ($p > 0.05$) (Table 5.6). The model also exhibits good fit to the data (Table 5.7). In period T1, when comparing the transition from evergreen forest (LINEV) to barren soils (BARR) against the reference category (LINEV to SECFOR), the model revealed that cost-distance to towns and distance to main roads are significant predictors of this type of change ($p < 0.01$). The signs of the coefficients suggest that forest locations with a lower cost-weighted distance but farther away from main roads are more likely to be cleared. In fact, the odds ratios indicate that for every one unit increase in CstDstTwn and DstMnRds the likelihood of forest clearing decreases and increases by factors of 0.99 and 1.01 respectively (Table 5.8). In period T2, the model shows that areas with higher population densities and elevations are more likely to experience this type of transition. For every one unit increase in population density and elevation, the likelihood of change increases by factors of 1.03 and 1.01 respectively.

When assessing the transition from evergreen forest (LINEV) to palm cultivation (PALM), the model revealed that elevation and road density are significant predictors of the likelihood of change ($p < 0.001$). For every one unit increase in elevation and road density, the likelihood of this type of LULC change decreases and increases by factors of 0.98 and 2.26 respectively. In period T2, the model shows that for every one unit increase in DstMnRd, the likelihood slightly decreases by a factor of 0.99. The model shows that in period T1, cost-distance to towns and distance to main roads are significant predictors of the likelihood of change from LINEV to PASAGR ($p < 0.001$). For every one unit increase in CstDstTwn and DstMnRds the likelihood of forest clearing decreases and increases by factors of 0.99 and 1.01

Table 5.6 Likelihood ratio tests of the overall contribution of each independent variable to the overall model for periods *T1* = 2013–2015 and *T2* = 2015–2019

Variable	*Period*	*-2 Log Likelihood of Reduced Model*	*Likelihood Ratio Tests (Chi-Square)*	*df*	*Sig.*	
Intercept	***T1***	16321.02	99.25	6	0.000	*
CstDstTwn		16480.05	258.28	6	0.000	*
DstMnRds		16273.40	51.63	6	0.000	*
DstStrm		16241.58	19.81	6	0.003	*
PopDen		16376.33	154.56	6	0.000	*
TRI		16246.91	25.14	6	0.000	*
Slope		16226.39	4.62	6	0.593	
Elev		16484.89	263.12	6	0.000	*
RdDen		16330.16	108.39	6	0.000	*
Intercept	***T2***	16490.77	106.61	6	0.000	*
CstDstTwn		16397.17	13.00	6	0.043	*
DstMnRd		16398.41	14.24	6	0.027	*
DstStrm		16402.11	17.94	6	0.006	*
PopDen		16407.49	23.32	6	0.001	*
TRI		16387.67	3.51	6	0.743	
Slope		16411.58	27.42	6	0.000	*
Elev		16472.87	88.71	6	0.000	*
RdDen		16405.64	21.48	6	0.002	*

*Significant at a 0.05 level

Table 5.7 Log likelihood test showing the goodness of fit of the models for periods *T1* = 2013–2015 and *T2* = 2015–2019. Non-significant results for Pearson and Deviance indicate that model fits the data well

	Period	*-2 Log Likelihood*	*Chi-Square*	*df*	*Sig.*	*Nagelkerke (Pseudo R-square)*
Full model	***T1***	16221.769	702.353	48	0.000	0.130
Pearson		–	32442.893	31,962	0.059	
Deviance		–	16221.769	31,962	1.000	
Full model	***T2***	16384.164	339.905	48	0.000	0.120
Pearson			30753.601	30,972	0.810	
Deviance			16384.164	30,972	1.000	

respectively. In period T2, the model shows that for every one unit increase in elevation the odds of this type of transition slightly increases by a factor of 1.01.

The model shows that all factors except terrain slope played a significant role ($p < 0.001$) in explaining the likelihood of change from SECFOR to BARR in period T1. The odd ratios show that for every one unit increase in these variables, the likelihood of change decreases by a factor of 0.9 for CstDstTwn, DstStrm, PopDen, TRI, and Elev. The signs of the coefficient and the odd ratios of DstMnRds and RdDen,

Table 5.8 Results of the multinomial regression analysis for periods T1 = 2013–2015 and T2 = 2015–2019 when comparing agains the reference category LINEV to SECFOR

LULC TRANSITION[a]		B^{T1}	*Std. Error*^{T1}	*Wald*^{T1}	*df*$^{T1 \text{ and } T2}$	*Sig.*^{T1}	*Exp(B)*^{T1}		B^{T2}	*Std. Error*^{T2}	*Wald*^{T2}	*Sig.*^{T2}	*Exp(B)*^{T2}	
LINEV TO BARR	Intercept	−2.8086	0.8660	10.5176	1.0000	0.0012			−5.8106	1.0620	29.9339	0.0000		
	CstDstTwn	0.0000	0.0000	21.2957	1.0000	0.0000	0.9900	*	0.0000	0.0000	0.0911	0.7627	1.0000	
	DstMnRds	0.0001	0.0000	8.5767	1.0000	0.0034	1.0100	*	0.0000	0.0000	1.3451	0.2461	1.0000	
	DstStrm	−0.0002	0.0002	1.3230	1.0000	0.2501	0.9900		0.0001	0.0001	1.0387	0.3081	1.0001	
	PopDen	−0.0016	0.0021	0.5406	1.0000	0.4622	0.9900		0.0077	0.0026	8.9044	0.0028	1.0077	*
	TRI	0.0022	0.0020	1.2174	1.0000	0.2699	1.0100		−0.0035	0.0076	0.2132	0.6443	0.9900	
	Slope	−0.0061	0.0190	0.1022	1.0000	0.7492	0.9900		0.0382	0.0256	2.2251	0.1358	1.0390	
	Elev	0.0053	0.0027	3.7320	1.0000	0.0534	1.0100		0.0121	0.0033	13.6482	0.0002	1.0121	*
	RdDen	0.2130	0.1537	1.9198	1.0000	0.1659	1.2400		0.0477	0.1861	0.0658	0.7976	1.0489	
LINEV TO PALM	Intercept	2.1079	1.4605	2.0831	1.0000	0.1489			−1.1121	1.6518	0.4533	0.5008		
	CstDstTwn	0.0000	0.0000	1.7396	1.0000	0.1872	0.9900		0.0000	0.0000	0.4273	0.5133	1.0000	
	DstMnRds	0.0000	0.0000	0.8910	1.0000	0.3452	0.9900		−0.0001	0.0000	6.7861	0.0092	0.9900	*
	DstStrm	−0.0004	0.0003	1.6551	1.0000	0.1983	0.9900		0.0000	0.0002	0.0000	0.9979	1.0000	
	PopDen	0.0021	0.0034	0.3872	1.0000	0.5338	1.0100		−0.0038	0.0043	0.7963	0.3722	0.9962	
	TRI	−0.0167	0.0654	0.0648	1.0000	0.7991	0.9800		−0.0074	0.0358	0.0427	0.8363	0.9926	
	Slope	−0.0032	0.1029	0.0010	1.0000	0.9750	0.9900		−0.0246	0.0694	0.1253	0.7233	0.9757	
	Elev	−0.0174	0.0049	12.5327	1.0000	0.0004	0.9800	*	−0.0046	0.0053	0.7576	0.3841	0.9954	
	RdDen	0.8159	0.1511	29.1692	1.0000	0.0000	2.2600	*	0.2429	0.2895	0.7039	0.4015	1.2750	
LINEV TO PASAGR	Intercept	−2.2567	1.0218	4.8771	1.0000	0.0272			−3.9507	0.7625	26.8448	0.0000		
	CstDstTwn	0.0000	0.0000	19.0231	1.0000	0.0000	0.9900	*	0.0000	0.0000	0.1752	0.6755	1.0000	
	DstMnRds	0.0001	0.0000	13.2120	1.0000	0.0003	1.0000	*	0.0000	0.0000	0.2964	0.5862	1.0000	
	DstStrm	0.0000	0.0002	0.1016	1.0000	0.7500	1.0000		0.0000	0.0001	0.0395	0.8424	1.0000	

(continued)

	PopDen	−0.0008	0.0025	0.0942	1.0000	0.7589	0.9900		0.0019	0.0019	0.9588	0.3275	1.0019	
	TRI	−0.0854	0.0488	3.0640	1.0000	0.0800	0.9200		−0.0001	0.0029	0.0016	0.9683	0.9999	
	Slope	0.0612	0.0749	0.6661	1.0000	0.4144	1.0600		0.0093	0.0185	0.2522	0.6155	1.0093	
	Elev	0.0025	0.0032	0.6095	1.0000	0.4350	1.0100		0.0095	0.0024	16.1003	0.0001	1.0095	*
	RdDen	0.0664	0.1925	0.1190	1.0000	0.7301	1.0700		0.0376	0.1458	0.0666	0.7963	1.0383	
SECFOR TO BARR	Intercept	2.2692	0.6346	12.7842	1.0000	0.0003			0.6115	0.6936	0.7773	0.3780		
	CstDstTwn	0.0000	0.0000	52.0976	1.0000	0.0000	0.9900	*	0.0000	0.0000	0.0002	0.9887	1.0000	
	DstMnRds	0.0001	0.0000	18.7655	1.0000	0.0000	1.0100	*	0.0000	0.0000	1.1176	0.2904	1.0000	
	DstStrm	−0.0004	0.0001	11.3119	1.0000	0.0008	0.9900	*	−0.0002	0.0001	4.5181	0.0335	0.9900	*
	PopDen	−0.0031	0.0015	4.0263	1.0000	0.0448	0.9900	*	0.0048	0.0017	7.6489	0.0057	1.0100	*
	TRI	−0.0868	0.0305	8.0897	1.0000	0.0045	0.9200	*	−0.0077	0.0145	0.2830	0.5947	0.9923	
	Slope	0.0436	0.0475	0.8393	1.0000	0.3596	1.0400		−0.0473	0.0283	2.7873	0.0950	0.9538	
	Elev	−0.0085	0.0021	17.0730	1.0000	0.0000	0.9900	*	−0.0057	0.0022	6.5042	0.0108	0.9944	*
	RdDen	0.5079	0.1028	24.4164	1.0000	0.0000	1.6600	*	0.4161	0.1128	13.6031	0.0002	1.5160	*
SECFOR TO PALM	Intercept	6.3404	1.0236	38.3678	1.0000	0.0000			3.2425	1.1079	8.5662	0.0034		
	CstDstTwn	0.0000	0.0000	13.7105	1.0000	0.0002	0.9900	*	0.0001	0.0000	5.0584	0.0245	1.0001	*
	DstMnRds	0.0001	0.0000	14.1349	1.0000	0.0002	1.0000	*	−0.0001	0.0000	5.0570	0.0245	0.9900	*
	DstStrm	0.0002	0.0001	2.4544	1.0000	0.1172	1.0000		−0.0005	0.0002	5.3239	0.0210	0.9901	*
	PopDen	−0.0032	0.0023	1.9123	1.0000	0.1667	0.9900		0.0091	0.0028	10.7381	0.0010	1.0091	*
	TRI	−0.1437	0.0551	6.7924	1.0000	0.0092	0.8700	*	−0.0732	0.0512	2.0412	0.1531	0.9294	
	Slope	0.1549	0.0846	3.3500	1.0000	0.0672	1.1800		0.1199	0.0793	2.2822	0.1309	1.1273	
	Elev	−0.0307	0.0035	77.5023	1.0000	0.0000	0.9700	*	−0.0173	0.0037	21.4214	0.0000	0.9829	*
	RdDen	1.0045	0.1085	85.7607	1.0000	0.0000	2.7300	*	0.1481	0.1638	0.8174	0.3660	1.1596	

(continued)

Table 5.8 (continued)

LULC TRANSITION[a]		B^{T1}	*Std. Error*^{T1}	*Wald*^{T1}	$df^{T1\ and\ T2}$	*Sig.*^{T1}	$Exp(B)^{T1}$		B^{T2}	*Std. Error*^{T2}	*Wald*^{T2}	*Sig.*^{T2}	$Exp(B)^{T2}$	
SECFOR TO PASAGR	Intercept	3.0463	0.5776	27.8140	1.0000	0.0000			1.8424	0.5393	11.6697	0.0006		
	CstDstTwn	−0.0001	0.0000	83.5395	1.0000	0.0000	0.9900	*	−0.0001	0.0000	5.3986	0.0202	0.9901	*
	DstMnRds	0.0001	0.0000	31.8281	1.0000	0.0000	1.0100	*	−0.0001	0.0000	0.0174	0.8952	0.9902	
	DstStrm	−0.0001	0.0001	0.5429	1.0000	0.4612	0.9900		−0.0002	0.0001	6.7040	0.0096	0.9998	*
	PopDen	−0.0028	0.0014	3.9700	1.0000	0.0463	0.9900	*	0.0019	0.0014	1.9272	0.1651	1.0019	
	TRI	−0.0820	0.0282	8.4370	1.0000	0.0037	0.9200	*	−0.0008	0.0029	0.0804	0.7767	0.9992	
	Slope	0.0034	0.0444	0.0059	1.0000	0.9386	1.0100		−0.0578	0.0148	15.1428	0.0001	0.9439	*
	Elev	−0.0101	0.0019	29.1004	1.0000	0.0000	0.9800	*	−0.0061	0.0017	12.4383	0.0004	0.9940	*
	RdDen	0.3900	0.0986	15.6356	1.0000	0.0001	1.4800	*	0.3424	0.0971	12.4228	0.0004	1.4083	*
	[a] *The reference category is: LINEV TO SECFOR.*				*T1 =*	*2013–2015*			*T2=*	*2015–2019*		** Significant at a 0.01 level*		

on the other hand, depict a positive relationship with the probability of change. For every one unit of change in these covariates, the likelihood increases by factors of 1.01 and 1.67 respectively. In period T2, the odd ratios of DstStrm and Elev show that as their values increase by one unit, the likelihood decreases by a factor of 0.99 for both covariates. PopDen and RdDen show a significant postive relationship with this type of transition. Their odds ratios indicate that for every one unit increase in their values the likelihood of change increases by factors of 1.01 and 1.51 respectively.

The likelihood of change from SECFOR to PALM in period T1 is significantly associated with CstDstTwn, DstMnRds, TRI, Elev, and RdDen. The odd ratios indicate that for every unit increase in these variables, the likelihood decreases by factors of 0.99 (CstDstTwn), 0.87 (TRI), and 0.97 (Elev), and increases by factors of 1.01 (DstMnRds) and 2.73 (RdDen). In period T2, CstDstTwn, DstMnRds, DstStrm, PopDen, and Elev were significant predictors. For one unit increase in all these variables, except PopDen, the likelihood decreases by a factor of 0.98, whereas, for PopDen, it increases by a factor of 1.01.

Finally, the likelihood of change from SECFOR to PASAGR in period T1 is explained by variation in CstDstTwn and DstMnRds, PopDen, TRI, Elev, and RdDen. The odd ratios indicate that for every one unit increase in CstDstTwn, PopDen, TRI, and Elev, the likelihood decreases by factors ranging between 0.92 and 0.99. The opposite is true for DstMnRds and RdDen: for every one unit increase in these two variables, the likelihood increases by factors of 1.01 and 1.48 respectively. In period T2, the likelihood of this type of land use transition is mostly explained by CstDstTwn, DstStrm, Slope, Elev, and RdDen. While the odd ratios show that for every one unit increase in the first four variables, the likelihood of change decreases by factors ranging between 0.94 and 0.99. However, for every one unit increase in RdDen, the likelihood of change increases by a factor of 1.41. In general, the spatial depiction of the MNLR results for periods T1 and T2 shows a clustered pattern of areas in the vicinity of towns and roads especially, where forest degradation is very likely (>50%) (Fig. 5.4).

4.3 *Multi-Temporal Water Quality Analysis*

We first present an overall summary of the water quality parameters measured in the communities of Pimampiro and Pacayacu (Table 5.9) (Fig. 5.5). Mean temperatures ranged between 24.94 ± 1.33 and 27.42 ± 1.20 °C, mean pH values between 5.49 ± 0.67 to 7.14 ± 0.48, mean DO from 1.72 ± 0.79 to 5.29 mg/L ± 2.97, and mean Cond from 78.65 ± 66.74 to 126.59 ± 59.16 μS/cm. Although there are no standardized guidelines for temperature of drinking water, some agencies have suggested that cold water (< 15 °C) is more preferable than warm because it reduces the risk of developing micro fungi that can grow inside the internal plumbing systems and protect bacteria from the effects of chlorination (Health Canada 2020). Our results showed that temperatures were warmer than such suggested guidelines

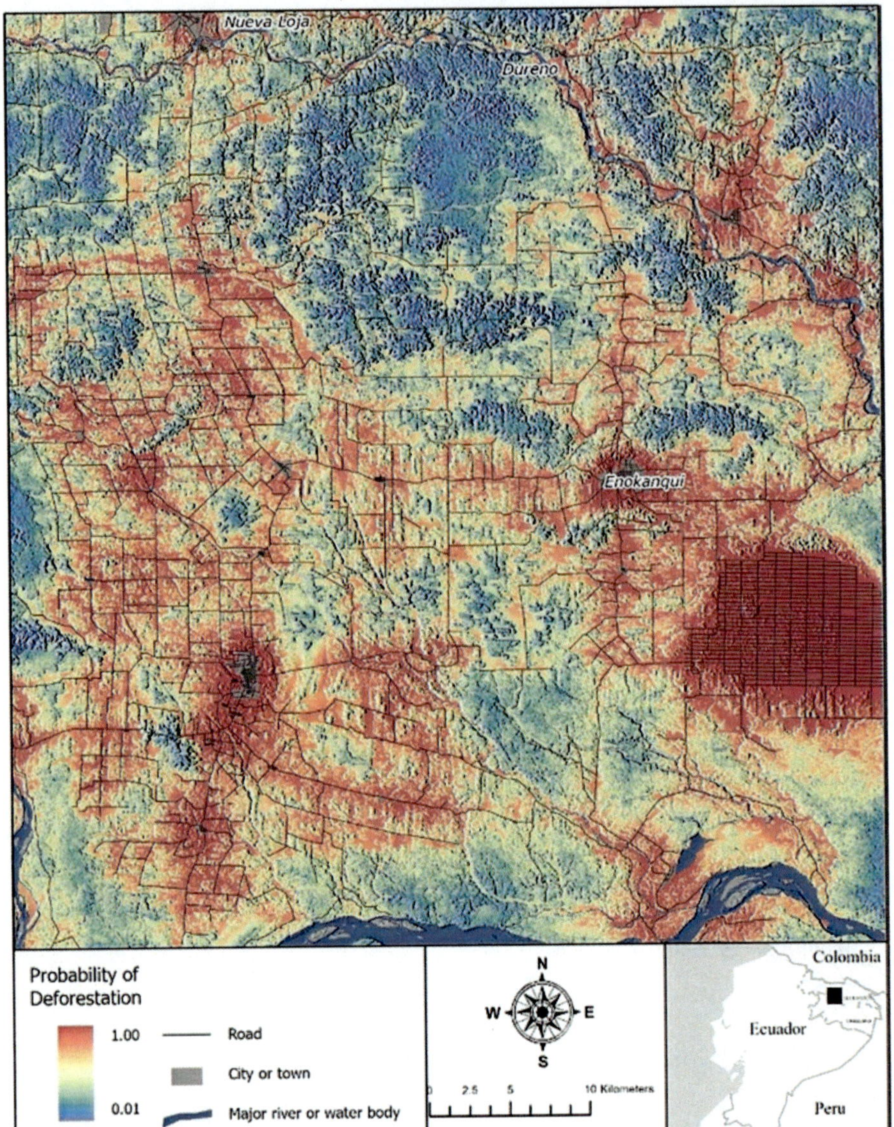

Fig. 5.4 Probability of deforestation based on the analysis of six major land cover transitions: LINEV to BARR, LINEV to PALM, LINEV to PASAGR, SECFOR to BARR, SECFOR to PALM, and SECFOR to PASAGR. (Modified from Lopez 2022)

(Mean Temp: 25.58 ± 1.81 °C in Pacayacu and 26.80 ± 1.39 °C in Pimampiro). Based on a comparison with different water quality guidelines (Table 5.1), our results showed that mean pH levels in 2011 for both Pacayacu

and Pimampiro were fairly lower than those recommended by Ecuadorian and international standards. Several water sources were found to be acidic with the lowest value (4.49) located in the community of Pacayacu. In addition, mean DO was in general lower than the levels recommended for drinking water (< 5.5 or 6.5 mg l^{-1}), with extremely low mean levels in 2011 (1.72 ± 0.79 mg l^{-1}) in Pacayacu, probably linked to high water temperature. The mean Cond range showed that most observations were lower than the maximum recommended value by the Council of the European Union (2500 μS cm^{-1}) (European Union Council Directive, 1998) and the World Health Organization (1500 μS cm^{-1}) (WHO 2018) and most values (99%) were very low (< 200 μS cm^{-1}). No recommendations are provided in the Ecuadorian regulations (TULSMA 2015). T-test analyses revealed that water quality

Table 5.9 Descriptive summary of water quality parameters in the communities of Pimampiro and Pacayacu in the Napo River basin

Parameter	*Year*	*N*	*Mean*	*Min*	*Max*	*Std Dev*	*Std Er*
Temp (°C)[a]	2011	29	27.42	25.90	30.20	1.20	0.22
	2013	17	26.05	24.70	29.00	1.31	0.32
	2017	8	26.19	24.00	27.80	1.25	0.44
pH (−log (H+))[a]	2011	29	5.49	4.10	6.90	0.67	0.13
	2013	17	6.56	5.76	7.76	0.63	0.15
	2017	8	6.47	6.03	6.79	0.24	0.08
DO (mg/L)[a]	2011	29	2.23	0.73	3.73	0.78	0.15
	2013	17	4.73	1.14	7.19	1.89	0.46
	2017	8	5.19	2.92	7.88	1.69	0.60
Cond (μS/cm)[a]	2011	29	103.06	57.80	163.16	32.66	6.07
	2013	17	109.25	16.70	189.60	47.91	11.62
	2017	8	126.59	59.70	214.90	59.16	20.92
Temp (°C)[b]	2011	27	25.72	22.40	28.87	1.96	0.38
	2013	22	24.94	20.90	26.60	1.33	0.28
	2017	9	26.72	25.00	31.30	1.88	0.63
pH (−log (H+))[b]	2011	27	5.97	4.60	8.20	0.95	0.18
	2013	22	6.02	4.49	7.53	0.69	0.15
	2017	9	7.14	6.43	8.04	0.48	0.16
DO (mg/L)[b]	2011	27	1.72	0.43	2.87	0.79	0.15
	2013	22	5.29	0.70	9.74	2.97	0.63
	2017	9	1.94	0.09	3.98	1.29	0.43
Cond (μS/cm)[b]	2011	27	104.50	7.72	551.00	129.69	24.96
	2013	22	78.65	6.90	272.90	66.74	14.23
	2017	9	84.08	26.00	170.00	42.79	14.26

[a]Pimampiro
[b]Pacayacu

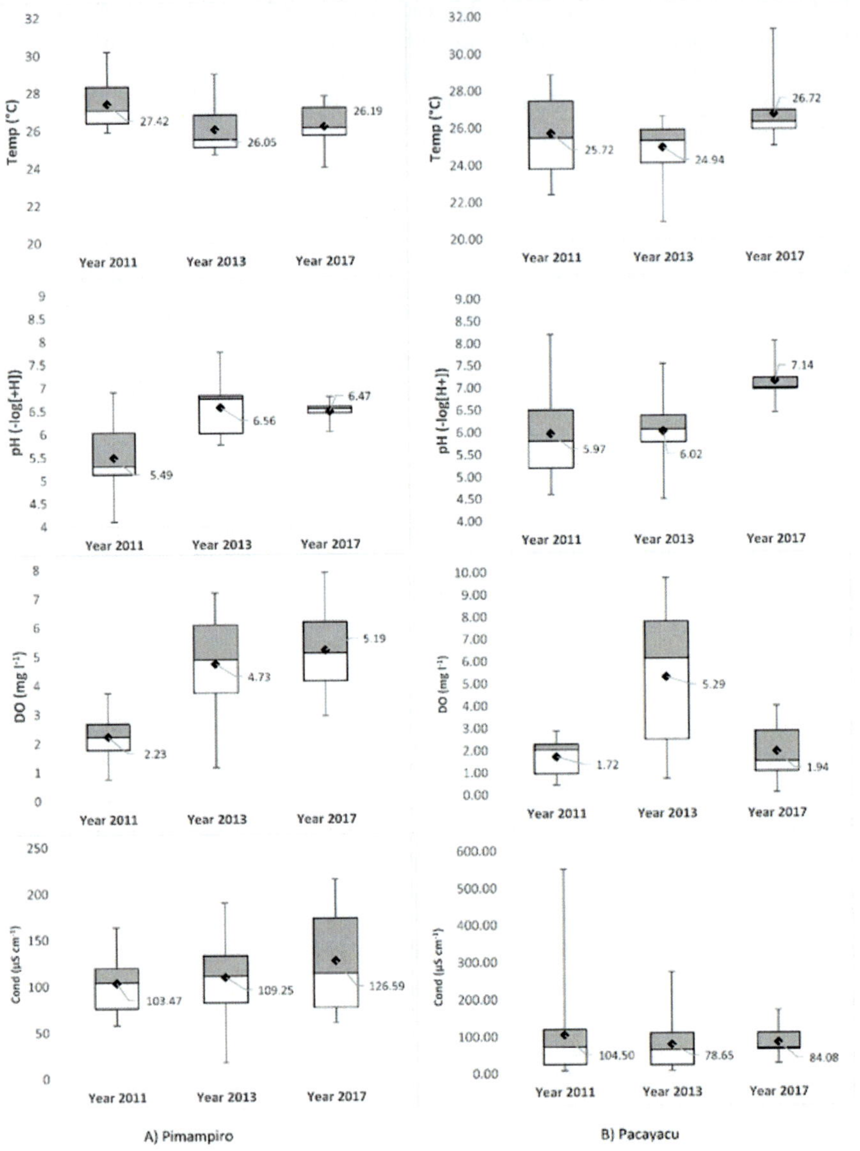

Fig. 5.5 Boxplots showing the spread and mean values of the water quality parameters measured in the communities of (**a**) Pimampiro and (**b**) Pacayacu in 2011, 2013, and 2017

significantly changed between 2011 and 2013, and between 2013 and 2017 (Table 5.10). Between 2011 and 2013, there were significant mean differences in Temp, pH, and DO in Pimampiro, and just DO in Pacayacu ($p < 0.01$). Cohen's d estimates show a positive change and big effect size for Temp ($d = -1.20$; $-1.84 < CI < -0.54$), negative differences and large effect sizes for pH ($d = -1.20$;

$-1.84 < CI < -0.54$) and DO ($d = -1.72$; $-2.38 < CI < -1.05$) in Pimampiro, and DO ($d = -1.72$; $-2.38 < CI < -2.38$) in Pacayacu. Between 2013 and 2017, we detected significant negative changes with strong effect sizes for Temp ($d = -1.19$; $-2.01 < CI < -0.34$) and pH ($d = -1.75$; $-2.64 < CI < -0.85$), and a significant positive difference with a strong effect size for DO ($d = 1.28$; $0.43 < CI < 2.11$).

5 Discussion

5.1 Land Use and Land Cover Change Dynamics

Our LULC transition analysis showed that forest cover extent, which includes both LINEV and SECFOR, decreased significantly from 2036 to 1661 km^2 at a rate of 3.33% $year^{-1}$ during the whole analysis period, which is consistent with most recent land use/land cover change studies (cf. Mena et al. 2006; Santos et al. 2019) if we consider the size of the study area and its location at the center of the area historically affected by intense deforestation. Out of the possible LULC transition pathways in the Napo region, the transition from forest to commercial plantations such as African palm deserves closer attention. According to Ecuador's National Institute of Census and Statistics (INEC), the provinces of Sucumbios and Orellana, where this study concentrates, are the third largest producers of oil palm in Ecuador (INEC 2010), with Ecuador in the top-10 list of the world's largest producers (Coral-Medina et al. 2019). The substitution of evergreen forest with African palm is more likely happening in areas closer to regional markets to minimize transportation costs, but not necessarily in the proximity of principal roads where land may be available but has marginal agricultural value. Forest areas with denser population concentrations and road infrastructure, and higher elevations (i.e., closer to the Andean foothills) are more likely to adopt this type of land use. This finding suggests that increased population pressure in areas with higher elevations, where the overall road network is denser and precipitation levels are high, are more likely to experience this type of transition. However, only a small percentage (2% – roughly 22 km^2) of evergreen forest transitioned to palm cultivation from 2013 to 2019. During this time period, most palm land use originated from the clearing of secondary forest (46 km^2), followed by the conversion of barren lands (40 km^2) and pastures (31 km^2) (Table 5.4). The transition probability from SECFOR to PALM is not only correlated with transportation costs and market proximity, but also terrain conditions such as ruggedness and elevation. Our results show that as ruggedness and elevation increase the transition probability from secondary forest to palm cultivation decreases, which suggests that this type of transition is occurring in areas of relatively smooth terrain conditions, farther away from the Andean foothills, probably due to drier climatic conditions. These findings indicate that although African palm encroachment into areas of old growth forest may not be happening as quickly or intensively as in other tropical regions (e.g., Indonesia or Malaysia), its impact on secondary and successional forests is an issue of concern.

Table 5.10 Results of the T test analyses showing significant differences between mean values of different water quality parameters obtained in 2011, 2013, and 2017

Parameter		N_{11}	N_{13}	N_{17}	*Mean*$_{11}$	*Mean*$_{13}$	*Mean*$_{17}$	*StdDev*$_{11}$	*StdDev*$_{13}$	*StdDev*$_{17}$	*t-test*	*df*	*t sig*		*Mn Diff*	*StdEr Diff*	*Cohen's d*	*d Low CI*	*d Up CI*	*Change Dir.*
Temp (°C)	a	29	17	–	27.42	26.05	–	1.20	1.31	–	3.61	44.00	0.0008	*	1.37	0.38	1.10	0.46	1.74	Decrease
pH (−log (H+))	a	29	17	–	5.49	6.56	–	0.66	0.63	–	−3.92	44.00	0.0003	*	−1.25	0.32	−1.20	−1.84	−0.54	Increase
DO (mg/L)	a	29	17	–	2.23	4.73	–	0.78	1.89	–	−5.19	19.28	0.0000	*	−2.50	0.48	−1.92	−2.63	−1.19	Increase
Cond (µS / cm)	a	29	17	–	103.06	109.25	–	32.66	47.91	–	−0.52	44.00	0.6052		−6.19	11.88	−0.16	−0.76	0.44	Increase
Temp (°C)	a	–	17	8	–	26.05	26.19	–	1.31	0.44	−0.25	23.00	0.8025		−0.14	0.55	−0.11	−0.95	0.73	Increase
pH (−log (H+))	a	–	17	8	–	6.56	6.47	–	0.63	0.08	0.49	22.38	0.6285		0.09	0.17	0.16	−0.69	1.00	Decrease
DO (mg/L)	a	–	17	8	–	4.73	5.19	–	1.89	0.60	−0.59	23.00	0.5640		−0.46	0.79	−0.25	−1.09	0.60	Increase
Cond (µS / cm)	a	–	17	8	–	109.25	126.59	–	47.91	20.92	−0.78	23.00	0.4411		−17.34	22.12	−0.34	−1.18	0.51	Increase
Temp (°C)	b	27	22	–	25.72	24.94	–	1.96	1.33	–	1.65	45.63	0.1066		0.78	0.47	0.45	−0.12	1.02	Decrease
pH (−log (H+))	b	27	22	–	5.97	6.02	–	0.95	0.69	–	−0.19	47.00	0.8473		−0.05	0.24	−0.06	−0.62	0.51	Increase
DO (mg/L)	b	27	22	–	1.72	5.29	–	0.79	2.97	–	−5.48	23.44	0.0000	*	−3.57	0.65	−1.72	−2.38	−1.05	Increase
Cond (µS / cm)	b	27	22	–	104.50	78.65	–	129.69	66.74	–	0.85	47.00	0.4012		25.86	30.52	0.24	−0.32	0.81	Decrease

Temp (°C)	b	–	22	9	–	24.94	26.72	–	1.33	1.88	−3.00	29.00	0.0055	*	−1.78	0.59	−1.19	−2.01	−0.34	Increase
pH (−log (H+))	b	–	22	9	–	6.02	7.14	–	0.69	0.48	−4.43	29.00	0.0001	*	−1.12	0.25	−1.75	−2.64	−0.85	Increase
DO (mg/L)	b	–	22	9	–	5.29	1.94	–	2.97	1.29	4.37	28.79	0.0001	*	3.35	0.77	1.28	0.43	2.11	Decrease
Cond (µS / cm)	b	–	22	9	–	78.65	84.08	–	66.74	42.79	−0.22	29.00	0.8237		−5.43	24.17	−0.09	−0.86	0.69	Increase

[a]Pimampiro
[b]Pacayacu
*Significant at the 0.01 level (t test)

Similarly, the conversion probability from LINEV to BARR and PASAGR increases with lower costs to markets, higher elevations, and higher population densities, but decreases in the vicinity of principal roads. About 99 km^2 (8%) and 47 km^2 (4%) of LINEV area in 2013 transitioned to pastures and agriculture, and barren lands respectively in 2019. Furthermore, our results indicate that areas with higher transportations costs, closer to waterways, higher population densities, smooth or flatter topography, lower elevations, farther away from main roads but higher overall road density are more likely to change from SECFOR to PASAGR and BARR. Between 2013 and 2019, about 229 km^2 (29%) and 97km^2 (12%) of SECFOR extent transitioned to PASAGR and BARR (Table 5.4). What these findings suggest is that spatial accessibility, market proximity, characteristics of land, and population conditions significantly contribute to the likelihood of forest loss. This is not surprising since several studies have found similar associations between deforestation and similar factors for this part of the Amazon and elsewhere (Busch and Ferreti-Gallon 2017; Rudel et al. 2009; Mena et al. 2006; Pfaff et al. 2013). However, our findings show that although the clearing of old growth forest for cattle raising and agriculture due to increased population pressure, better accessibility, and terrain conditions, is still an issue of concern, the transformation of secondary forests as reflected by their conversion to pastures and barren lands is of much greater magnitude (≈120% more). The importance of secondary forests for conservation cannot be understated. Similar to primary forests, secondary vegetation not only helps maintain critical environmental services such as carbon sequestration, pest control, and climate regulation (Martin et al. 2013; Guimberteau et al. 2017) but is also rich in biodiversity (Hernández-Ordoñez et al. 2015). This study also adds to this body of research by incorporating the spatial dimension to show where those land cover transitions are more likely to occur based on the combination of the abovementioned factors.

5.2 *Multi-Temporal Physico-Chemical Analysis of Water Quality*

Our water quality characterization showed that water temperature of our samples was generally high, with pH levels that increased since 2011 (Fig. 5.4) but were generally on the acidic side (< 7). The significant increase of pH between 2011 and 2013 in Pimampiro and between 2013 and 2017 in Pacayacu suggests a trend toward a more neutral water (pH values closer to 7 in both communities), however, on average, the acidity of the water remained relatively high when compared to the recommended national and international guidelines. In addition, DO and Cond levels remained very low throughout the entire period. DO levels may be influenced by the high temperature of the water and by the fact that most sample observations were obtained from shallow wells where water is stagnant in comparison to rivers and streams (Michaud 1991). Although low DO levels may not have a direct effect on public health, when considering the parameters altogether, our results indicate that water quality in these communities is poor. Particularly, acidic and low mineralized water (as indicated mainly by low pH and Cond measurements) may have important effects on both human and environmental health if not treated. Acidic water, for instance, can accelerate the leaching of toxic metals and other chemical compounds from distribution pipelines and other local water collection infrastructure and modify the taste (Maurice et al. 2019). In addition, drinking water with low mineral content has a negative effect on hormone secretion, kidney functions, and bone mineral density (Aptel et al. 1999; WHO 1980). It may also cause a reduction of calcium and magnesium intake and other essential elements. Furthermore, epidemiological studies have reported that soft water (low in minerals) is associated with increased morbidity and mortality from cardiovascular disease compared to hard water (Kozisek 2005).

6 Implications on Human-Environment Dynamics of the Napo Region

This chapter helped set the stage for some recommendations through an improved understanding of socio-environmental change in the Napo region. In the last 10 years, there have been major LULC alterations and water quality changes in the upper Napo River basin. These changes have been linked not just to the geographical features of the area but also to unsustainable resource management strategies that are negatively affecting Amazonian ecosystems and the communities therein.

From the standpoint of LULC change, socio-economic and environmental factors including transportation costs, road infrastructure and density, population dispersion and clustering, and land characteristics have traditionally been linked to land cover change in this area of the Amazon. However, this study provides fresh insights into lesser-known patterns of forest reduction. We demonstrate that, even though local deforestation rates have decreased or remained largely unchanged since the late 1990s and early 2000s, there are still serious concerns about the clearing of secondary and successional forests associated with the spread of African palm cultivation, pastures, and other types of agriculture. In tropical environments, secondary and successional forests play an important role in conservation by providing key environmental services like carbon sequestration (Martin et al. 2013; Hernández-Ordoñez et al. 2015) and habitat provision for critical biological species. Our results underscore the need to promote the conservation of this type of forests not only to maintain vital ecological services, but also to encourage the re-growth of forests situated in relatively marginal production lands (away from main highways), which could help attain agrobiodiversity and environmental sustainability goals. Some of these objectives could be achieved, for example, through environmental policy interventions. Important progress has been achieved by Ecuador‘s recent reactivation of its jurisdictional Roundtable on Sustainable Palm Oil (RSPO)

Certification plan in 2018 (RSPO 2021), which includes a commitment to zero deforestation. However, this type of commitment should be complemented with the promotion of sustainable agricultural practices. Maurice et al. (2019), for example, found that high concentrations of manganese (Mn) in well water in the region could be related to common agricultural practices including palm management, as Mn is one of the main components found in fertilizers, animal food, and pesticides. Other forest conservation such as the expansion of the protected area network using mixed-use frameworks (López et al., 2020) could also help to not only protect remnant secondary forest patches but provide income opportunities for local communities through, for example, community-based eco and agro-tourism.

From a hydrological standpoint, this study contributed to a better understanding of water quality in the region. In general, our results show that DO, pH, and Cond are outside the optimal ranges suggested by national and international guidelines for drinking water, which are generally based on maximum and/or minimum acceptable concentrations of inorganic and organic compounds. Health issues, commonly associated with water contamination, are usually attributed by local communities to oil extraction activities (Beristain et al. 2009). Although we cannot determine a direct correlation between oil contamination and water quality based on the physicochemical indicators we measured, some patterns can be discerned. The low mineralization and high acidification of water as indicated by our Cond and pH measurements suggest that common health issues reported by local people such as fatigue, stomachaches, dizziness, and muscle cramps could be linked to the high dehydration levels to which the population is exposed due to humidity and temperature and the low concentrations of minerals together with the leaching of metals from soils, metallic roofs (commonly used in family dwellings), wells, and distribution pipes. Maurice et al. (2019), for instance, suggested that the low levels of Ca, Mg, and F and high levels of Zn, Hg, and Mn found in water sources exacerbate the health impacts of overall poor water quality in the Napo region. These findings underscore the need for a detailed and systematic monitoring of water quality in areas historically affected by environmental degradation. We suggest a comprehensive plan to address all the different sources of pollution including oil and gas generation, and conventional agricultural practices. National policies and municipal ordinances should address these issues in a responsible manner by extending rural potable water supply and wastewater treatment systems to Amazonian communities, which are mostly inexistent in the Napo region.

Finally, we acknowledge that environmental change is complex and emerges from constantly changing interactions between people and the environment. Through this chapter, we contributed to an improved understanding of spatial heterogeneity as it pertains to environmental change in the Napo region by using geotechnologies and the analysis of a-spatial and spatially explicit environmental data. However, limited in-situ data collection efforts can make it difficult to conduct a thorough multi-temporal analysis of human-environment interactions. We offered a contextualized evaluation of environmental transformations in this biologically and culturally diverse region of the Amazon basin to get around some of these limitations. This work adds new insight into concerns of human-environmental dynamics that directly affect environmental sustainability and public health.

Acknowledgements The authors would like to acknowledge Muwen He's help in ground-truthing the classifications of satellite imagery and model development. Pacayacu and Pimampiro farmers from the provinces of Francisco de Orellana and Sucumbíos and members of local communities, as well as students from the University of Washington provided invaluable assistance and guidance during field campaigns. The comments from Becky Beard and colleagues from the University of Washington Bothell helped improve the overall accuracy and quality of this chapter.

References

Alves DS (2002) Space-time dynamics of deforestation in Brazilian Amazônia. Inter J Remote Sens 23:2903–2908

Aptel I, Cance-Rouzaud A, Grandjean H (1999) Association between calcium ingested from drinking water and femoral bone density in elderly women: evidence from the EPIDOS cohort. J Bone Miner Res 14(5):829–833

Asner GP, Broadbent EN, Oliveira PJ et al (2006) Condition and fate of logged forests in the Brazilian Amazon. Proc Natl Acad Sci 103:12947–12950

Azevedo-Santos VM, Garcia-Ayala JR, Fearnside PM et al (2016) Amazon aquatic biodiversity imperiled by oil spills. Biodivers Conserv 25(13):2831–2834

Barber CP, Cochrane MA, Souza CM et al (2014) Roads, deforestation, and the mitigating effect of protected areas iin the Amazon. Biol Conserv 17:203–209

Barbieri AF, Carr DL, Bilsborrow RE (2009) Migration within the frontier: the second generation colonization in the Ecuadorian Amazon. Popul Res Policy Rev 28(3):291–320

Barraza F, Maurice L, Uzu G, Becerra S et al (2018) Distribution, contents and health risk assessment of metal(loid)s in small-scale farms in the Ecuadorian Amazon: an insight into impacts of oil activities. Sci Total Environ 622–623:106–120

Belsley D (1991) Conditioning diagnostics: collinearity and weak data in regression. Wiley, New York

Beristain CM, Páez RD, Fernandez I (2009) Las palabras de la selva : estudio psicosocial del impacto de las explotaciones petroleras de Texaco en las comunidades amazónicas de Ecuador. Instituto de Estudios sobre Desarrollo y Cooperación Internacional, Bilbao

Bilsborrow RE (2003) Cambios demograficos y medio ambiente en la Región Amazónica de los paises andinos. In: Aramburu CE, Bedoya E (eds) Amazonia: Procesos demográficos y ambientales. Consorcio de Investigación Económica y Social, Lima, pp 53–86

Bilsborrow RE, Barbieri AF, Pan W (2004) Changes in population and land use over time in the. Acta Amazon 34(4):635–647

Brauman KA (2015) Hydrologic ecosystem services: linking ecohydrologic processes to human well-being in water research and watershed management. Wires Water 2(4):345–358

Breiman L (2001) Random forest. Mach Learn 45:5–32

Busch J, Ferretti-Gallon K (2017) What drives deforestation and what stops it? A meta-analysis. Rev Environ Econ Policy 11(1):3–23

Buytaert W, Bièvre BD (2012) Water for cities: the impact of climate change and demographic growth in the tropical Andes. Water Resour Res 48:1–13

Castellanos-Navarrete A, de Castro F, Pacheco P (2021) The impact of oil palm on rural livelihoods and tropical forest landscapes in Latin America. J Rural Stud 81:294–304

Castello L, Macedo MN (2016) Large-scale degradation of Amazonian freshwater ecosystems. Glob Change Biol 22:990–1007
Chomitz KM, Thomas TS (2001) Geographic patterns of land use and land intensity in the Brazilian Amazon. World Bank, Washington
Congalton RG, Green K (1999) Assessing the accuracy of remotely sensed data: principles and practices. CRC Press, Boca Raton
Coral-Medina JD, Magalhães AI, Zamora HD et al (2019) Oil palm cultivation and production in South America: status and perspectives. Biofuels Bioprod Biorefin 13:1202–1210
dell'Olio L, Ibeas A, Cecin P (2011) The quality of service desired by. Transp Policy 18(1):217–227
Dislich C, Keyel AC, Salecker J et al (2017) A review of the ecosystem functions in oil palm plantations, using forests as a reference system. Biol 92:1539–1569
Elmhagen B, Eriksson O, Lindborg R (2015) Implications of climate and land-use change for landscape processes, biodiversity, ecosystems services, and governance. Ambio 44:1–5
EPA (1986) Quality criteria for water "Green book". Office of Water Regulations and Standards, Washington
ESRI (2020) ArcGIS Pro (version 2.7). Environmental Research Institute, Redlands
European Union Council Directive (1998) 98/83/EC of 3 November 1998 on the quality of water intended for human consumption. Off J Eur Commun
Finer M, Jenkins CN (2012) Proliferation of hydroelectric dams in the Andean Amazon and implications for Andes-Amazon connectivity. PLoS ONE 7(4):e35126
Finer M, Jenkins CN, Pimm SL, Kean B, Ross C (2008) Oil and gas projects in the Western Amazon: threats to wilderness, biodiversity, and Indigenous peoples. PLoS ONE 3(8):e2932
Geist HJ, Lambin EF (2002) Proximate causes and underlying driving forces of tropical deforestation. BioSci 52(2):143–150
Gray C, Bilsborrow R (2013) Environmental influences on human migration in rural Ecuador. Demography 50:1217–1241
Guimberteau M, Ciais P, Ducharne A et al (2017) Impacts of future deforestation and climate change on the hydrology of the Amazon Basin: a multi-model analysis with a new set of land-cover change scenarios. Hydrol Earth Syst Sci 21:1455–1475
Health Canada (2020) Guidelines for Canadian drinking water quality summary table. Water and air quality bureau, healthy environments and consumer safety branch. Health Canada, Ottawa
Hernández-Ordoñez O, Urbina-Cardona N, Martínez-Ramos M (2015) Recovery of amphibian and reptile assemblages during old-field succession of tropical rain forests. Biotropica:1–12
Hilbert K, Soentgen J (2020) From the "Terra Preta de Indio" to the "Terra Preta do Gringo": a history of knowledge of the Amazonian Dark Earths. In: Mikkola HJ (ed) Ecosystem and biodiversity of Amazonia, IntechOpen. https://doi.org/10.5772/intechopen.91089
Hosmer D, Lemeshow S (2000) Applied logistic regression. Wiley, New York
INAMHI (2014) Base de datos meteorológica del Ecuador. INAMHI, Quito
INEC (2010) Censo de Población y Vivienda. http://redatam.inec.gob.ec/cgibin/RpWebEngine.exe/PortalAction?andMODE=MAINandBASE=CPV2010andMAIN=WebServerMain.inl Accessed 21 Jan 2021
Jocknick C, Normand R, Zaidi, S (1994) Estudio de exposición al petróleo y la salud" en Violaciones de derechos en la Amazonía ecuatoriana: Las consecuencias humanas del desarrollo petrolero. Centro para Derechos Económicos y Sociales, Quito, p 47–56
Joye SB, MacDonald IR, Leifer I et al (2011) Magnitude and oxidation potential of hydrocarbon gases released from the BP oil well blowout. Nat Geosci 4:160–164
Kozisek F (2005) Health risk from drinking demineralized water. In: Nutrients in drinking water. WHO Press, Geneva
Laurance WF, Albernaz AK, Schroth G et al (2002) Predictors of deforestation in the Brazilian Amazon. J Biogeogr 29(5–6):737–748
Laurence SG, Laurence WF, Nascimento HE et al (2009) Long-term variation in Amazon forest dynamics. J Veg Sci 20:323–333

Lessman J, Fajardo J, Muñoz J et al (2016) Large expansion of oil industry in the Ecuadorian Amazon: biodiversity. Ecol Evol 6(14):4997–5012

Liu J, Heiskanen J, Aynekulu E et al (2016) Land cover characterization in west Sudanian savannas using seasonal features from annual Landsat time series. Remote Sens 8(365):rs8050365

López S (2022) Deforestation, forest degradation, and land use dynamics in the Northeastern Ecuadorian Amazon. Appl Geogr 145:102749

López S, López-Sandoval MF (2016) The socio-ecological systems of southeastern Ecuador: processes and patterns of land use change. In: Young K, Postigo J (eds) Nature and society: socio-ecological perspectives on global changes in Latin America. Instituto de Estudios Peruanos, Lima, pp 67–108

López S, Sierra R (2010) Agricultural change in the Pastaza River basin: a spatially explicit model of native Amazonian cultivation. Appl Geogr 30:355–369

López S, López-Sandoval MF, Gerique A et al (2020) Landscape change in southern Ecuador: an indicator-based and multi-temporal evaluation of land use and land cover in a mixed-use protected area. Ecol Indic 115:106357

Lu F, Gray C, Bilsborrow RE et al (2010) Contrasting colonist and Indigenous impacts on Amazonian forests. Conserv Biol 24(3):881–885

Martin PA, Newton AC, Bullock JM (2013) Carbon pools recover more quickly than plant biodiversity in tropical secondary forests. Proc Royal Soc B 280:20132236

Maurice L, Lopez F, Becerra S et al (2019) Drinking water quality in areas impacted by oil activities in Ecuador: associated health risks and social perception of human exposure. Sci Total Environ 690:1203–1217

Mena CF, Bilsborrow RE, McClain ME (2006) Socioeconomic drivers of deforestation in the northern Ecuadorian Amazon. Environ Manag 37(6):802–815

Mena CF, Laso F, Martinez P et al (2017) Modeling road building, deforestation and carbon emissions due deforestation in the Ecuadorian Amazon: the potential impact of oil frontier growth. J Land Use Sci 12(6):477–492

Michaud JP (1991) A citizen's guide to understanding and monitoring lakes and streams. Washington State Departement of Ecology, Lacey

Millington JD, Perry GL, Romero-Calcerrada R (2007) Regression techniques for examining land use/land cover change: a case study of a Mediterranean landscape. Ecosystems 10:562–578

Morcote-Ríos G, Aceituno FJ, Iriarte J et al (2020) Colonisation and early peopling of the Colombian Amazon during the late Pleistocene and the early Holocene: new evidence from La Serranía La Lindosa. Quat Int. https://doi.org/10.1016/j.quaint.2020.04.026

Myers N (2003) Biodiversity hotspots revisited. Bioscience 53:916–917

Myers N, Mittermeier RA, Mittermeier CG et al (2000) Biodiversity hotspots for conservation priorities. Science 403:853–858

Nobre CA, Sampaio G, Borma LS et al (2016) Land-use and climate change risks in the Amazon and the need of a novel sustainable development paradigm. Proc Natl Acad Sci 113(39):10759–10768

Ochoa-Gaona S, Gonzales-Espinosa M (2000) Land use and deforestation in the highlands of Chiapas, Mexico. Appl Geogr 20:17–42

Olofsson P, Foody GM, Herold M et al (2014) Good practices for estimating area and assessing accuracy of land change. Remote Sens Environ 148:42–57

Oo TK, Arunrat N, Sereenonchai S et al (2022) Comparing four machine learning algorithms for land cover classification in gold mining: A case study of Kyaukpahto Gold Mine, Northern Myanmar. Sustainability 14:10754

Pan WK, Walsh SJ, Bilsborrow RE et al (2004) Farm-level models of spatial patterns of land use and land cover dynamics in the Ecuadorian Amazon. Agric Ecosyst Environ 101:117–134

Pfaff A, Amacher GS, Sills EO et al (2013) Deforestation and forest degradation: concerns, causes, policies, and their impacts. In: Shogren (ed) *Encyclopedia of energy, natural resource, and environmental economics* (Vol. 2). Elsevier, Amsterdam, pp 144–149

Pichón F (1997) Colonist land-allocation decisions, land use, and deforestation in the Ecuadorian Amazon frontier. Econ Dev Cult Change 45(4):707–744

Potes V (2010) Análisis de la aplicación del derecho ambiental en la Amazonía ecuatoriana y el rol de las fiscalías ambientales. CEDA

Rivera-Parra JL, Vizcarra C, Mora K et al (2020) Spatial distribution of oil spills in the northeastern Ecuadorian Amazon: a comprehensive review of possible threats. Biol Conserv 252:108820

RSPO (2021) RSPO roundtable on sustainable palm oil: https://rspo.org/certification Accessed 23 Jan 2021

Rudel TK, Bates D, Machinguiashi R (2002) Ecologically noble amerindians? Cattle ranching and cash cropping among Shuar and colonists in Ecuador. Lat Am Res Rev 37(1):144–159

Rudel TK, Defries R, Asner G et al (2009) Changing drivers of deforestation and new opportunities for conservation. Conserv Biol 23(6):1396–1405

Rusydi AF (2018) Correlation between conductivity and total dissolved solid in various types of water: a review. In: IOP conference series: earth and environmental science. IOP Publishing, 118, p. 012019

Santos F, Graw V, Bonilla S (2019) A geographically weighted random forest approach for evaluate forest change drivers in the northern Ecuadorian Amazon. PLoS ONE 14(12):1–37

Savenije HH, Hoekstra AY, van der Zaag P (2014) Evolving water science in the Anthropocene. Hydrol Earth Syst Sci 18:319–332

Schielen J, Börner J (2018) Recent transformations of land-use, and land-cover dynamics across different deforestation frontiers in the Brazilian Amazon. Land Use Policy 76:81–94

Sierra R (2000) Dynamics and patterns of deforestation in the western Amazon: the Napo deforestation front 1986-1996. Appl Geogr 20(1):1–16

Sierra R, Campos F, Chamberlin J (2002) Assessing biodiversity conservation priorities: ecosystem risk and representativeness in continental Ecuador. Landscape and urban planning and patterns of deforestation in the western Amazon: the Napo deforestation front 1986 to 1996. Appl Geogr 20:1016

Singh A (1989) Digital change detection techniques using remotely-sensed data. Inter J Remote Sens 10:989–1003

Sonter LJ, Herrera D, Barret DJ et al (2017) Mining drives extensive deforestation in the Brazilian Amazon. Nat Commun 8:1013

Swenson JJ, Carter CE, Domec JC et al (2011) Gold mining in the Peruvian Amazon: global prices, deforestation, and mercury imports. PLoS ONE 6(4):1–7

Thanh Noi P, Kappas M (2018) Comparison of random Forest, k-nearest neighbor, and support vector machine classifiers for land cover classification using Sentinel-2 imagery. Sensors 18(1):s18010018

TULSMA (2015) Texto unificado de legislación secundaria del ministerio del ambiente. Libro VI Anexo 1. Norma de Calidad Ambiental y de Descarga de Efluentes al Recurso Agua. Ministerio del Ambiente, Quito

USGS (2020) Earth explorer. https://earthexplorer.usgs.gov/ Accessed 15 Aug 2020

Vasco C, Bilsborrow RE, Torres B et al (2018) Agricultural land use among mestizo colonist and Indigenous populations: contrasting patterns in the Amazon. PLoS ONE 13(7):e0199518

Velázquez-Gómez M, Lacorte S (2020) Organic pollutants in indoor dust from Ecuadorian Amazonia areas affected by oil extractivism. Environ Res 186:109499. https://doi.org/10.1016/j.envres.2020.109499

Verissimo A, Souza C Jr, Stone S et al (1998) Zoning of timber extraction in the Brazilian Amazon. Conserv Biol 12:128–136

Vijay V, Reid CD, Finer M, Jenkins CN, Pimm SL (2018) Deforestation risks posed by oil palm expansion in the Peruvian Amazon. Environ Res Lett 3:114010

Walsh SJ, Shao Y, Mena CF et al (2008) Integration of hyperion satellite data and a household social survey to characterize the causes and consequences of reforestation patterns in the northern Ecuadorian Amazon. Photogramm Eng Remote Sens 74(6):725–735

Webb J (2005) Use of the ecosystem approach to population health: the case of mercury contamination in aquatic environments and riparian populations, Andean Amazon, Napo River valley, Ecuador. Can J Public Health 96(1):44–46

WHO (1980) Guidelines on health aspects of water desalination. WHO Press, Geneva

WHO (2008) Guidelines for drinking-water quality: incorporating first addendum, Vol 1. Recommendations. WHO Press, Geneva

WHO (2018) A global overview of national regulations and standards for drinking water quality. WHO Press, Geneva

Chapter 6
From Mapping to Guiding: An Emergent Framework for the Multiple Uses of Remote Sensing and GIScience in Socio-environmental Research in the Peruvian Andes

Julio C. Postigo, Javier A. Ñaupari, and Enrique R. Flores

Abstract The Andes as a mountain social-ecological system is the result of continued mutually transformative social and environmental processes operating over millennia. A significant portion of the transformations occurring in the last century has been documented and analyzed using remote sensing techniques. The four cases presented illustrate different uses of remote sensing and geographic information science (GIScience). The cases sequentially expand the spatial extent and aspects analyzed (e.g., ecology, economy, vulnerability). The first, mapped land cover change and ecological productivity in the upper watershed of the Santa River (Ancash, Peru). The second, evaluated rangeland conditions under different governance regimes in a basin. The next case assessed the vulnerability of a peasant community to the impacts of environmental and social change at the household and ecosystem levels. The fourth case evaluated the conservation status of two types of grassland to model future rangeland degradation of four departments on the Peruvian Central Andes. Based on these cases, we propose a multi-level and cross-scale framework to analyze human-environment interactions in Andean landscapes. The implementation of this framework will help bridge the gap between the conceptual work of researchers and the empirical work of practitioners who must apply the principles to the changing social and ecological situation of Andean landscapes. We argue that the identification of the proper scale of the processes at play and levels at which they happen allow us to improve the understanding of mountain social-ecological systems. The contribution to action goes beyond generating knowledge.

J. C. Postigo (✉)
Department of Geography, Indiana University, Bloomington, IN, USA
e-mail: jpostigo@iu.edu

J. A. Ñaupari · E. R. Flores
Laboratory of Ecology and Use of Rangelands, Universidad Nacional Agraria La Molina, Lima, Peru

S. López (ed.), *Socio-Environmental Research in Latin America*, The Latin American Studies Book Series, https://doi.org/10.1007/978-3-031-22680-9_6

It recognizes the multiple stakeholders in the Andes, as well as their needs and the effects of their practices on ecosystems' health and their conservation status.

Keywords Rangeland · Social-ecological system · Pastoralism · Multi-level · Vulnerability

1 Introduction

Remote sensing analysis of social and environmental aspects of the Andes goes back nearly 100 years (Vander Jagt et al. 2015; Veettil and Kamp 2017). The most analyzed aspect is probably changes in the Andean cryosphere (i.e., glaciers, snow, and permafrost) (e.g., Shaw et al. 2020, Vuille et al. 2018, Yarleque et al. 2018, Kozhikkodan Veettil and de Souza 2017, Rabatel et al. 2013, Vuille et al. 2008). The conspicuous cryospheric changes, however, have spurred research about multiple impacts downstream from glaciers (Wood et al. 2021; Motschmann et al. 2020; Cuesta et al. 2019; Anacona et al. 2018; Buytaert et al. 2017; Mark et al. 2017; De Bièvre et al. 2012; Postigo et al. 2012; Mark et al. 2010; Vergara et al. 2007; Carey 2005).

Below the lower end of glaciers, the Andean *puna* in the Peruvian Andes encompasses high Andean rangelands, wetlands, and grasslands. This chapter elaborates a framework to analyze human-environment interactions in these Andean landscapes. This elaboration builds upon four cases that use remote sensing to understand change in the *puna* ecosystem in Peru (Fig. 6.2). The cases sequentially expand the spatial extent and socio-environmental aspects (e.g., ecology, economy, vulnerability) analyzed. The first, maps land cover change and ecological productivity in the upper watershed of the Santa River (Ancash, Peru). The second evaluates rangeland conditions under different governance regimes in a basin on the western slopes of the Huascaran National Park—HNP (Ancash, Peru). Additionally, this case analyzes the effect of HNP regulation of livestock management on rangeland and proposes a compensation scheme to reduce cattle grazing. The next case assesses the vulnerability of a peasant community to the impacts of environmental and social change at the household and ecosystem levels in the department of Huancavelica. The fourth case examines the conservation status of two types of grassland ecosystems to model future rangeland degradation of four departments on the Peruvian Central Andes.

Understanding changes in the *puna* is relevant because they indicate both changes occurring upstream and their consequences downstream. The *puna* directly depends on the water available from precipitation, surface runoff, and groundwater, and is fundamental for supporting livestock and storing water and carbon (Postigo 2020; Cuesta et al. 2019; Ochoa-Tocachi et al. 2019; Polk 2016). Furthermore, impacts of climate change on the *puna* may compromise food security and livelihoods of livestock-based farming systems (Postigo 2020; Flores-Lopez et al. 2016; López-i-Gelats et al. 2015).

Changes in the *puna* ecosystem have been widely documented using remote sensing. The response of wetlands or mountain peatland pastures (locally called *bofedales*) to climate change mimics glacial runoff fluctuation, initially growing and then contracting. In Peru, the impacts on wetlands, range from degradation and desiccation to humidity reduction, to size and area contraction, fragmentation, and species composition change (Cuesta et al. 2019; Polk et al. 2017; Verzijl and Guerrero Quispe 2013), leading to negative consequences on herders' livelihoods (Yager et al. 2019; Dangles et al. 2017; Polk et al. 2017; López-i-Gelats et al. 2015; Postigo 2014). In Peru's Cordillera Blanca (the largest area of tropical glaciers), wetlands expanded from 1987 to 1999 but contracted in the first decade of 2000 (Young et al. 2017). The analysis of Peruvian grasslands >3000 m a s l shows a declining trajectory. Using Landsat TM and patch metrics for the period 1987–2007, Tovar and colleagues (2013) found annual losses of *puna*, and montane forest and shrubland of 1.5% and 2.8%, respectively.

These dramatic alterations represent a critical hazard for vulnerable pastoralists since highland wetlands are crucial for grazing alpaca (*Vicugna pacos*) herds, especially in the dry season when the rain-fed puna vegetation does not provide enough fodder for them (Yager et al. 2019; Polk et al. 2017; Rasmussen 2016; Verzijl and Guerrero Quispe 2013). Additionally, if these changes lead to variation in wetlands and pastures species composition, they could also alter the nutritional value of pastures and, potentially, their suitability as a food resource for livestock. Further, drying wetlands modify alpine plant communities, which has an impact on carbon storage, water regulation, and food provision for livestock.

It is increasingly important to link the changes in the *puna* with their impacts downstream on both natural and social systems. The consistent analysis of these linkages is fundamental to produce generalizable knowledge. An analytical framework is a crucial step toward research consistency and replicability. In this chapter, we propose a multi-level and cross-scale (Vanwey et al. 2005) framework to analyze human-environment interactions in Andean landscapes. Level and scale have multiple meanings and uses based on disciplinary idiosyncrasies. Following Cash et al. (2006), we define scale as the analytical or thematic dimensions (e.g., spatial, temporal, ecological, economic) studied in processes or phenomena (Fig. 6.1). Such study includes measurement and quantification of processes. For instance, the spatial dimension of land change in the *puna* involves quantifying the variation of land covers or biophysical characteristics. The ecological dimension may include monitoring changes in species composition and biotic/abiotic interactions. Other dimensions may comprise aspects of governance and social structures influencing policies that regulate land use. These dimensions can be analyzed at different levels. When doing so, levels are considered as the spatial or temporal units of analysis (e.g., plot, landscape, country, daily, seasonal) at a particular scale. For an example of land change in the *puna*, the level can be the whole region, the country, states, or other administrative units. The targeted land use policy dimension could be analyzed at the national, sub-national, or even global levels. The framework encompasses processes linking dimensions (i.e., cross-scale) such as local land zoning shaping land use, and events occurring across levels like local impacts of global warming.

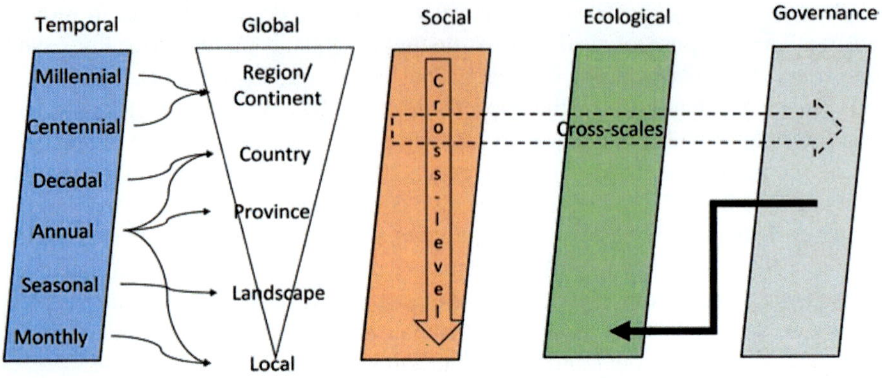

Fig. 6.1 Examples of scales and levels of emerging frameworks. The levels range from local to global. The temporal scale encompasses from millennial to monthly events. The interactions depicted include cross-level and cross-scale, and multiple temporal levels

Furthermore, the framework also includes linkages between dimensions and levels; for instance, global trade modifying national policies that impact local farmers (e.g., O'Brien et al. 2004). The following section presents the cases that support the emergent framework.

2 Case Studes in the Peruvian Andes

2.1 *Land-Cover and Rangeland Productivity of Andean Ecosystems: The Upper Watershed of the Santa River, Ancash, Peru*

The Santa River watershed is in the department of Ancash (Peru) (Fig. 6.2a). It extends from the lowlands on the coast of the Pacific Ocean to the summits. We used three Landsat 4-5 TM (level -1) imagery of the upper watershed of the Santa River to determine land cover types and biomass between 1987 and 2011. Images with a low cloud coverage (<5%) were selected. Landsat scenes were radiometrically corrected in ENVI and then classified using an unsupervised approach. The unsupervised classification involved: (i) identification of regions of interest based on thematic maps (life zones, ecological sites, and climate); (ii) analysis of spectral signatures of vegetation types, glaciers, water bodies, rocks, crops, and urban areas (McGraw and Tueller 1983). Cells with similar reflectance values were converted into regions of interest, whereas regions with significantly different reflectance values were clearly differentiated (Marchand and Golea 1993). Following an unsupervised method based on the Iterative Self-Organizing Data Analysis Technique (ISODATA), we refined the selection of regions of interest. With these areas, we performed a supervised classification of grasslands, shrubs, forests, crops,

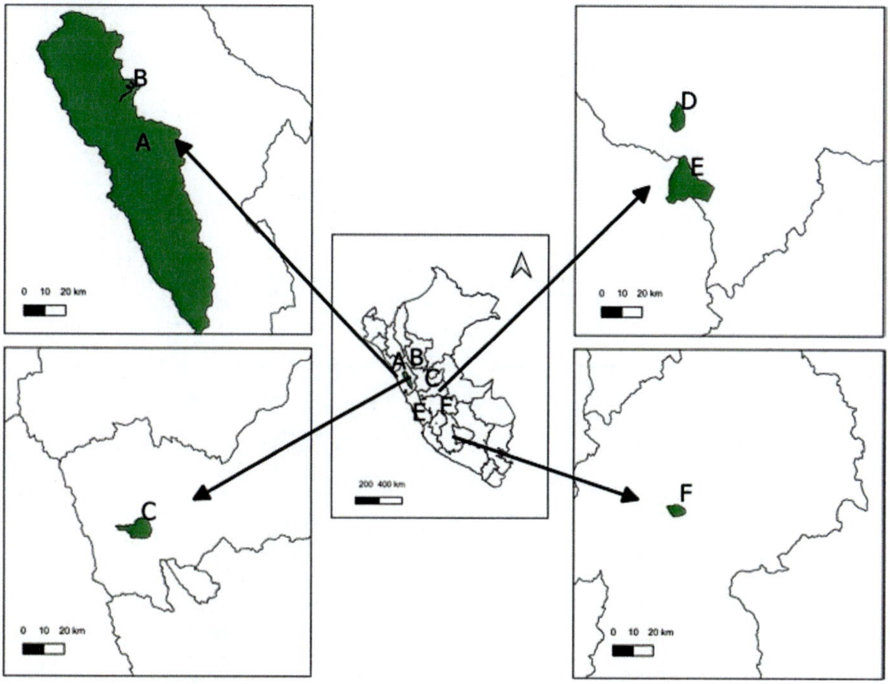

Fig. 6.2 Study areas. (**a**) Case Sta. River basin; (**b**) Case Ulta Valley; Case Peruvian Central Andes includes San Pedro de Racco (**c**), Canchayllo (**d**), Tomas (**e**); and (**f**) Case Santa Ana

agriculture, snow, and rocks based on maximum likelihood (Jensen 2014). The seven classes were defined at the beginning of the study based on previous studies developed by the Rangeland's Ecology and Use Laboratory (LEUP for its acronym in Spanish). However, the delimitation of mountain wetlands (*bofedales*) using reflectance values was not possible due to water reflectance noise. Bofedales were digitized from the national chart and refined with field information. The classified image was subsequently validated with field data and applying rigorous cartographic standards (Table 6.1 and Fig. 6.3). Table 6.1 shows the land cover change in the 1987–2011 period. Forests and farmlands increased by 2.1% and 4.2% respectively, while shrubland and snow had very slight changes (<0.5%), Grasslands extent reduced 14.4% and rocky areas increased by 7.8%.

We divided the ecological region central humid Andes (Josee et al. 2009) into ecologic sub-units or regional ecosystems defined by altitudinal, bioclimatic, and vegetation (e.g., humid high montane grassland) characteristics derived from thematic maps. A field inventory was carried out in key areas within each regional ecosystem. We used transects to collect botanical data and define vegetation community composition and structure.

The ecological productivity of land cover classes was estimated using Normalized Difference Vegetation Index (NDVI) values, auxiliary data, and in-situ analysis (Fig. 6.4). We conducted a preliminary evaluation of key ecological processes (i.e.,

Table 6.1 Area (ha) and percent per land cover class in the Santa River basin in the years 1987 and 2011

	1987		2011	
Physiognomic unit	ha	%	ha	%
Forest	137272.6	13.2	159582.6	15.4
Shrubland	110174.0	10.6	109694.7	10.6
Grassland	526229.2	50.6	376218.5	36.2
Farmland	125287.7	12.1	168796.4	16.2
Rocky areas	111032.3	10.7	192610.7	18.5
Snow	29372.8	2.8	32465.6	3.1

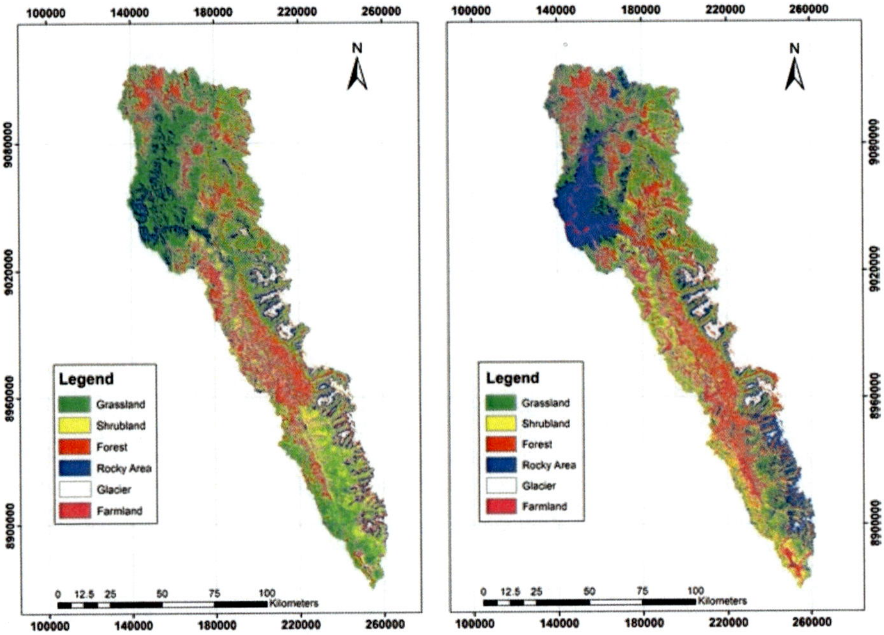

Fig. 6.3 Comparison of land cover in the Santa River basin (1:750,000) between 1987 (left) and 2011 (right) (Overall accuracy, Kappa = 0.87)

productivity and water function) considering vegetation type, climatic patterns, erosive potential, relative presence of functional taxonomic groups, and physical-geographical soil characteristics (Thurow 2000).

To generate a preliminary map of the ecological conditions of the ecosystems (ECE) we used the following equation:

$$ECE = a \quad \text{Slope class} + b \quad \text{Elevation class} + c \quad \text{NDVI class}.$$

The coefficient (a, b, and c) is the importance of the variables determined by a panel of the LEUP experts on rangelands. The range of values for each variable were

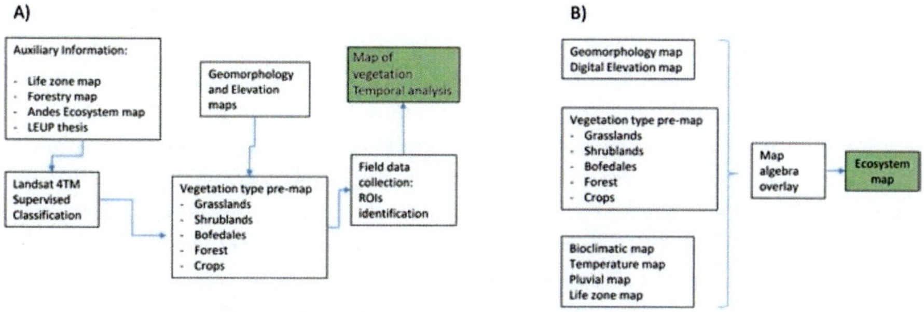

Fig. 6.4 (**a**) Process for elaborating a map of vegetation and conducting a temporal analysis for vegetation; (**b**) Map algebra overlay for elaborating an ecosystem map

Table 6.2 Ecological condition of ecosystems based on slope, elevation and NDVI

	Slope			Elevation			NDVI		
	Weight	Class	Range	Weight	Class	Range	Weight	Class	Range
Poor	2	1	45–75	0.25	1	>4000	1	1	<0.25
Fair	4	2	25–45	0.5	2	3000–4000	2	2	0.25–0.5
Good	6	3	<25	1	3	<3000	3	3	>0.5

reclassified as discrete classes. Cut-off values were calculated using the maximum values for each equation so that 'Good Condition' was set as 30, Fair condition as 20, and Poor condition as 7.

The values were then normalized from 0 to 1. The criteria for classifying the ecological condition were based on normalized values of NDVI after identifying slope and elevation classes (Table 6.2). We combined slope, elevation, and NDVI using a geographic information systems (GIS) framework. Field information from representative ecosystems, evaluated by rapid field surveys of vegetation (dominant species, coverage, litter) and soil conditions (erosion level, rockiness, slope), provided additional information about the threats and risks in these ecosystems.

We identified 41 ecosystems that differ in productivity, ecological condition, and spatial distribution based on the characteristics of plant communities (physiognomy) and abiotic factors, mainly climate and physiography. We identified and mapped two *bofedales* with a total area of 12,069 ha (1.6% of the area). The primary productivity of *bofedales* ranges from medium to high since humidity levels are, on average, higher and less variable than the rest of vegetation types. However, the ecological condition of these ecosystems was poor, suggesting high pressure from human use and grazing (Chimner et al. 2019).

We classified 13 types of tall grass ecosystem making a total of 482,093 ha, (63.9% of the area), with a low to medium productivity and some isolated patches with high productivity. Although the ecological condition varies from poor to fair, with very few sites in good condition, on average, they appear to be in better condition than the rest of the ecosystems we evaluated. Since the main use of these ecosystems is grazing, their integrity is primarily threatened by overgrazing and burning (Lipton 2014).

We detected 13 types of shrub ecosystem covering 116,968 ha, (15.5% of the area). Shrubs frequently appear interspersed within grasslands and forests, present low productivity in xerophytic regimes, and higher productivity in humid regimes. Most of these ecosystems are in poor condition, and very few reach the condition of fair or good. These ecosystems are used mainly for grazing, although they seem used with lower intensity than wetlands, possibly due to effective chemical anti-herbivore defense systems (Estell 2010).

The 13 types of forest ecosystem identified extend 143,191 ha (19% of the area). The species included non-native trees such as Eucalyptus (*Eucalyptus globulus*) and Pine (*Pinus sp.*), natural riverside forests, and montane forests. Forest ecosystem shows abundant biomass and leaf area index (between 5 and 6), though the ecological condition is mostly poor with isolated natural forests in regular condition. The main threats are grazing and illegal logging, followed by fire hazards.

2.2 Rangeland Condition in the Ulta Basin on the Western Slopes of the Cordillera Blanca in the National Park Huascaran, Peru

The Ulta basin is located on the western slope of the Huascaran National Park (HNP). It is surrounded by the mountains Southern Huascaran, Chopicalqui, and San Mateo (Fig. 6.2 B). The basin covers 405 ha. combining grasslands (37.9%), shrubs (33.3%), bofedales (17.1%), puna grassland (10.7%) and forest patches (1.0%). The Committee of Users (154 members with 165 Animal Units—AU) of the Ulta Basin Park (CUP Ulta acronym in Spanish) holds the right to graze livestock in the basin but under the norms of the HNP authority.

We assessed the ecological condition of the rangeland (ECR) of the CPU in Ulta basin using the following equation (Flórez and Malpartida 1987):

$$ECR = 0.5\ D + 0.2\ FI + 0.2\ Cov. + 0.2\ V$$

Where D is the desirable species, FI is forage index based on the desirable and less desirable species, Cov is basal plant coverage, and V is vigor based on plant height. The cut-off values for each class are every 20 points. The class 'very poor' has less than 20 points; the 'poor' class ranges between 21 and 40 points; the 'regular' class ranges between 41 and 60; 'good' between 61 and 80; and 'excellent' encompasses values over 81 points. The ecological condition of the rangeland reflects herbivores' preference for certain species; the most palatable species might become less abundant. The animal-dependent condition of the ecosystem is different from the ecological condition based on potential productivity estimated from NDVI values.

We overlaid geologic maps and satellite-derived slope conditions using ASTER GDEM data to generate sampling units throughout the basin. Vegetation classes were derived from RapidEye imagery 2017, at a spatial resolution of 5 m, and obtained through Google Earth and validated through ground truthing and GPS observations (n = 72). In the 29 sampling units, we evaluated the rangeland condition using Parker's three-step method (Ruyle and Dyess 2016). For example, Excellent condition means forage numbers over 80%, no observable erosion, abundant mulch, and clean water. Poor condition means less than 20% of forage species, significant erosion, low rates of soil infiltration, and water with sediment. Additionally, we estimated the condition of the vegetation units based on their floral composition, abundance of seedlings and young plants, presence of mulch, vigor of plants (based on the height of key species), andsoil surface conditions.

The results show that the Ulta basin is dominated by heterogeneous patches of vegetation, which respond differently to livestock preferences and grazing pressures. Grasslands areas dominated by *Calamagrostis sp.* and *Calamagrotis macrophylla* show a condition between good and regular. It is expected that this condition will continue or improve over time. Shrublands dominated by *Lupinus brachypremmon* are characterized by regular and poor conditions, which are expected to remain unchanged. Wetland areas, dominated by*Plantago rigida* and *P. tubulosa,* present a condition ranging from regular to poor, which is expected to remain unchanged or decline. The puna grasslands, dominated by *Calamagrostis brevifolia* and *Muhlenbergia fastigiata,* have a poor to regular condition, although it is expected to improve.

In general, grasslands were in 'regular' condition and expected to remain in such a state in the future. This suggests that the mixed governance system, the CUP-HNP combination, is regulating the number of livestock according to the basin's rangeland carrying capacity. This provides evidence that participatory governance (i.e., the processes and institutions through which decisions are made), policies, and incentives for rangelands conservation are successfully regulating land use and preventing degradation of rangeland ecosystems (Grünwaldt et al. 2016). However, the HNP lacks a robust program for monitoring rangeland conditions and change. This situation highlights the importance of capacity building of Park personnel to better assess rangeland conditions by using remote sensing and information technologies. This could include mobile applications such as EpiCollect or LandPKS for managing protected areas (Chimner et al. 2019; Herrick et al. 2017).

3 Multi-Level Vulnerability in the Peasant Community Sallca Santa Ana in Huancavelica, Peru

The peasant community Sallca Santa Ana (total area 61396 ha) is in the Castrovirreyna province of the Huancavelica region, at 4400 meters above sea level (Fig. 6.2f). Puna grasslands dominate the landscape. This case evaluates mountain pastoralist communities' vulnerability to change (VC) based on their sensitivity and adaptive capacity to environmental and social perturbations. We considered the ecosystem and household levels for further investigation.

At the ecosystem level, VC was calculated using climate variability (CV), potential rangeland degradation (PD), and a poverty index (PI) (Fig. 6.5). CV was obtained from raster images of average annual precipitation, and average annual temperature of Peru. We established climatic categories according to Josee et al. (2009). PD is a function of NDVI, land cover, and elevation. We obtained PI values from Peru's poverty map at the province level (INEI 2010). The model for the social-ecological system vulnerability (SESV) is: SESV = 2*CV + PD + 2*PI. Wei ghts were set by the LEUP expert panel. It is then normalized to allow the classification of ecosystems in high, medium, and low vulnerability (Table 6.3).

We selected a sample of 10% of the total number of households (n = 28). Household heads (men and women) were interviewed between 2012 and 2014. Collected socioeconomic data included livelihoods (livestock-related income, income fluctuations, remittances; Table 6.4) poverty level, and herd composition. The ecological data included vegetation type and rangeland condition. The variables for the indices of adaptive capacity and sensitivity to climate change have different weights in the model. Following Eakin and Bojórquez-Tapia (2008), an expert panel assigned weights and values to each of the variables (Table 6.3). The indices combine household and remotely sensed data and were computed using GIS (Table 6.4).

Households' main livelihood is herding mixed flocks of sheep (*Ovis aries*), alpaca and llama (*Lama glama*), and a few *criollo* cattle (*Box taurus*). Table 6.5 shows the most relevant socioeconomic characteristics of Sallca Santa Ana households. The most determining attributes of adaptive capacity, according to an expert panel and confirmed by a Mann–Whitney U test procedure, were age of the household's head (AHH), farm size (FS), access to credit (AC), technical assistance (TA), and amount of species of animals (LD); while sensitivity was mostly determined by rangeland's present and expected future condition (i.e., trend—T), and financial contributions of migrants (R). Table 6.6 shows some characteristics of the households of the community Sallca Santa Ana.

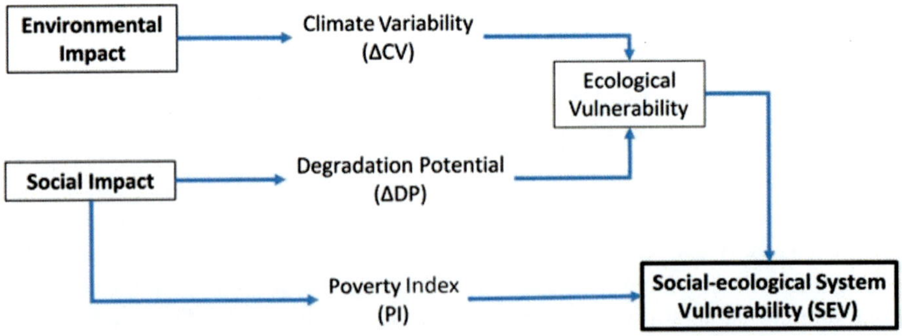

Fig. 6.5 Conceptual Model of Social-Ecological System Vulnerability (SESV)

Table 6.3 Social-Ecological System Vulnerability (SESV) based on values and weights of climate variability (CV), degradation potential (DP), and poverty index (PI).

CV (Weight 2)		DP (Weight 1)		PI (Weight 2)		SESV	
	Value		Value		Value	Class	Range
Xeric	9	High	9	Extreme	9	Extreme	35–45
Desert	6			High	7	High	25–34
Humid	3	Medium	5	Moderate	5	Medium	15–24
Perhumid	1	Low	1	Low	3	Low	5–14
				Very low	1		

Table 6.4 Indices of Adaptive Capacity and Sensitivity to Climate Change at the household level

Adaptive Capacity					Sensitivity	
Human Resources	Physical Resources	Financial Resources	Information Resources	Diversity Resources	Livelihood	Rangeland
Head of family size	**Farm size**	**Access to credit**	Organization level	Land use	Income from livestock	**Rangeland condition**
Adults in the family	Alpaca units	Health insurance	**Technical assistance**	**Animal species**	Income stability	**Rangeland trend**
Family size	Water sources	State support program		Income sources		Carrying capacity
Family education	Tenure	Private support program	Radio & TV access	Economic activities	**Immigrant contribution**	Mix grazing
Adults with primary education	Animal shelter					Vegetation diversity
	Machinery service					Extreme events

The most influential indicators of adaptive capacity and sensitivity, determined by an expert panel and confirmed by Mann–Whitney U test, are in bold

Our results showed that 19.8% of the puna ecosystem in Huancavelica has a low SESV, 61.3% medium, and 18.39% high. We found that 86.5% and 23.5% of the community's households have medium and high vulnerability, due to the combined effect of low adaptive capacity and high sensitivity.

4 Regional Assessment of Rangelands Dynamics in the Peruvian Central Andes (Cerro De Pasco, Junín, Lima, and Huancavelica Departments)

We evaluated rangeland degradation of the high Andean rangelands in the departments of Pasco, Junín, Lima, and Huancavelica (2,523,118.3 ha) (Fig. 6.2c–f) using Peru's Ministry of the Environment guidelines for evaluating rangeland conservation based upon ecosystems' functions and health (RM. N ° 183–2016 - MINAM)

Table 6.5 Weights, ranges, and equations of the variables in the indices of adaptive capacity and sensitivity.

Variable	Weight	Range of the variable	Equation
AHH	0.1497	X ≤ 19	1
		19 < X ≤ 25	−0.021X + 1.396
		25 < X ≤ 50	−0.035X + 1.75
		X > 50	0.01X − 0.5
S	0.0701	X ≤ 120	1
		120 < X ≤ 357	−0.001688X + 1.202616
		357 < X ≤ 714	−0.00168X + 1.19952
		X > 714	0
AC	0.0236	0, 1	
AT	0.0441	0, 1	
LD	0661	N/A	−0.25X + 1.25
R		X ≤ 375	1
	0.0156	375 < X ≤ 1500	−0.000889X + 1.3335
		X > 1500	0
T	0.1850	Negative	1
		Stable	0.5
		Positive	0

Table 6.6 Characteristics of the households of the community Sallca Santa Ana.

Average estate area (ha)	305.4
Average number of school years attained	7.2
Percent of households with stable income	70.5
Percent of households with limited access to credit	23.5

Source: Grünwaldt et al. (2016)

(Pyke et al. 2002). Rangeland conservation is estimated in the field based on 10 indicators: species richness, floristic composition, aerial cover, surface bare soil, loss of surface soil, surface horizon organic matter, canopy height of dominant plants, amount of aerial biomass, amount of litter, and invasive plants. The spatial model consisted of spatializing all possible variables to assess rangeland conservation status.

We elaborated an ecosystem map following LEUP's method, which overlays thematic layers of vegetation type, elevation ranges, and humidity regimes (Josee et al. 2009) and those maps used in Case 1 (*Land-cover and rangeland productivity of Andean ecosystems: the upper watershed of the Santa River, Ancash, Peru*) in this chapter.

The focus of our projection of rangeland degradation was short and tall grasslands, the two most extended plant communities in the Andean region (Flores 1991). We used the Google Earth Engine platform to classify Landsat 8 OLI in six classes including short and tall grasslands vegetation types. The overall accuracy of 0.96, a Kappa index of 0.95, which is considered excellent (Monserud and Leemans 1992). The spatial resolution of the images was 30 m, the minimum mapping area was 14 ha, although we included 5 ha patches in special cases (e.g., urban areas, wetlands, agricultural and reforested areas).

Initially, the model included 40 variables from previous LEUP studies such as rangeland condition and change. Moreover, soil variables from SoilGrids (On-line database for soil variables (Grunwald et al. 2011) were included. We reduced the number of variables from 40 to 5 after accounting for collinearity among variables using the Pearson method. The five variables were Erosion/Deposition, Maximum NDVI, Soil Adjusted Vegetation Index (SAVI) Change, vegetation cover, and soil organic carbon content (derived from SoilGrids) (Table 6.7).

The Revised Universal Soil Loss Equation (RUSLE) estimates erosion/deposition rate at the landscape scale. RUSLE uses the rainfall-runoff erosivity (R), soil erodibility, length and slope of the terrain, and management of vegetation cover (Renard et al. 2011). To obtain R, we used monthly precipitation (1981–2018) at a spatial resolution of 0.05° provided by the Climate Hazards Group InfraRed Precipitation with Station (CHIRPS), which utilizes the infrared spectrum of satellite imagery (Funk et al. 2015) to create rainfall time series. Soil erodibility (i.e., the amount of soil lost per unit of applied energy) was obtained from the SOILGRIDS database at a spatial resolution of 250 m (Hengl et al. 2014). The length and slope (LS) were estimated based on the ALOS PALSAR (Phased Array Type L-band Synthetic Aperture Radar) elevation model of the Japan Aerospace Exploration Agency. Management of vegetation cover—i.e., soil lost in types of vegetation with respect to uncovered areas—was estimated using the Fractional Vegetation Cover (FVC) from the NDVI of Landsat 8 OLI sensor bands (Durigon et al. 2014).

Maximum NDVI was estimated through a multi-temporal analysis of NDVI using the maximum composition of NDVI algorithm in Google Earth Engine for the period 2017–2018. SAVI was estimated from NDVI 16-day composites for the period 2000–2018, filtered by the Savitzky–Golay model (Chen et al. 2004), and then introduced into a simple linear regression model with time as an independent

Table 6.7 Processing method to measure five variables to predict rangeland degradation.

Layers	Resolution	Processing method	Source
Erosion/Deposition	30 m	RUSLE	Renard et al. (2011)
Maximum NDVI	30 m	Time series analysis	Landsat 8 OLI
SAVI Change	250 m	Time series trend analysis	MODIS: MOD13Q1.006/
Vegetation Cover	30 m	Image classification & NDVI	Landsat 8 OLI
Soil Organic Carbon Content	250 m	Kriging interpolation	SoilGrids

variable. The percent of vegetation cover was estimated through zonal statistical analysis by counting the pixels of a particular cover type within 25 ha-hexagonal units. Soil carbon content was derived from a soil organic carbon stock (OCSTHA) layer at 30 cm depth included in the SOILGRIDS database. We applied zonal statistics to determine the OCSTHA average per hexagonal unit.

We use the random forest approach to show a spatial pattern that better predicts the conservation status of rangeland. We used a dataset with estimates of ecosystem conservation (Pyke et al. 2002) to train the model. Because random forests are a set of individual decision trees, the Gini index (i.e., a measure of the total decrease in node impurities or IncNodePurity) was used to measure the relevance of the variables for the estimates. The most important variables—with the greatest average decrease in the Gini values—for estimating the conservation status were soil organic carbon content, erosion/deposition, and maximum NDVI. The least important variables were Change of SAVI index, and Vegetation Cover.

The values of the spatial model were ranked from 1 to 100 and then grouped, rendering the following categories of conservation status: Very poor (1–20); Poor (20–40); Fair (40–60); Good (60–80); and Very Good (80–100). The first two classes are considered degraded rangelands. Table 6.8 and Fig. 6.6 show the conservation status of short and tall grasslands of the Peruvian Central Andes. The North and East show better conservation status, whereas the western slopes of the Central Andes and Southwest concentrate most of the areas with very poor conservation status. It is worth mentioning that the model's precision and predictive power may be improved with more field data.

5 Emerging Framework for the Analysis of High Altitude Social-Ecological Systems

This section presents a framework for analyzing human-environment interactions in Andean landscapes. The implementation of this framework will help bridge the gap between researchers' conceptual work and practitioners' applied efforts on changing social and ecological Andean landscapes. Linking conceptual and empirical

Table 6.8 Conservation status of short and tall grasslands in the Peruvian Central Andes determined by NDVI and soil global models.

Status	Area (ha)	%
Very poor	259,614.5	10.3
Poor	1,199,089.0	47.5
Regular	828,930.9	32.8
Good	232,183.8	9.2
Very good	3,300.0	0.1
Total	2,523,118.3	100.0

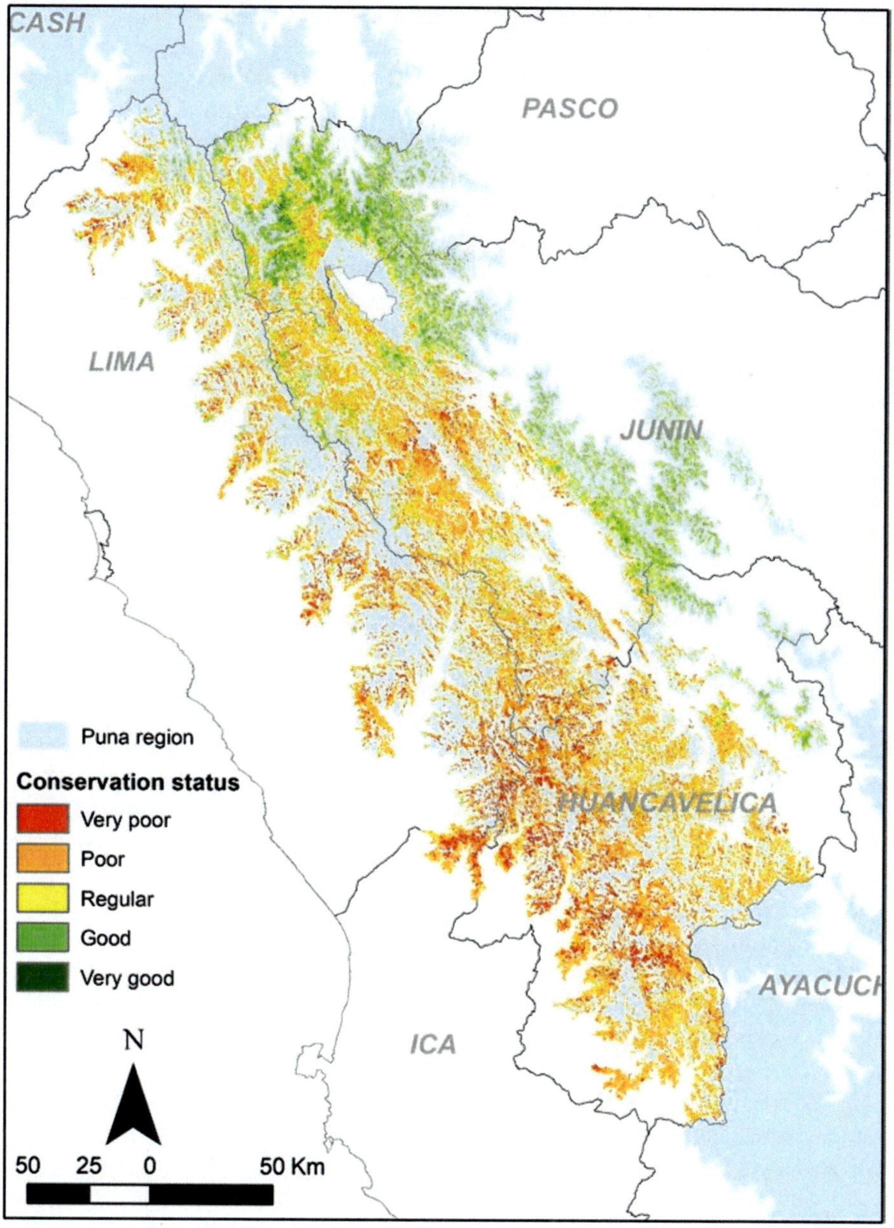

Fig. 6.6 Conservation Status of Rangeland in Peruvian Central Andes.

work bears the potential to generalizable formulations about mountain social-ecological systems.

We first evaluated change in land cover and productivity by merging thematic layers of topographic, climatic, and biological data. In doing so, the level of analysis was the upper watershed and the scale or dimension considered was, broadly speaking, environmental. The detection of change in both land cover and productivity assumed that alterations were related to ecological, climatic, and human factors. Then, we looked at the threats to this changing landscape, particularly those emerging from human use. This led us to include the influence of governance on managing the condition of Andean rangelands. The inclusion of the governance scale created a cross-scale relation (social to environmental) that aimed to improve our understanding of the influence of governance on rangeland conditions (see Fig. 6.1). Indeed, we analyzed a model of mix-governance that combined local users and authorities of the Peruvian National Service of Protected Areas. In this case, the scale of governance was multi-level from local regulation of resource use to national policies for protected areas. The level for the analysis of social-environmental dimensions also expanded to the basin.

Multi-level and cross-scale processes and interactions (see Fig. 6.1) are observed in the vulnerability analysis of the Sallca Santa Ana community. First, climatic change and variability occur at global and local levels respectively. To complicate things further, there are cross-level dynamics such that the physical manifestation of certain phenomena is not always consistent among levels. For instance, a global or regional climatic pattern may be absent or have an opposite signal at the local level. Second, the ecosystem-level analysis is cross-scale including the interactions and quantification of soil properties, vegetation types, hydrological aspects, and local perception and knowledge about rangeland. Third, vulnerability was analyzed at the household and ecosystem levels using factors from environmental (e.g., rangeland degradation) and social scales (e.g., poverty). In the social scale, cross-level interactions are observed in certain cases when, for example, households' poverty is largely shaped by meso- and macro processes such as access to credit.

The last case is a regional extrapolation of ecosystems' change; thus, it refers to environmental and temporal scales on the region level. This case, however, introduces a prospective analytic perspective ranging from past change to projected future ecosystem change.

In addition to the dynamics among and across levels and scales (see Fig. 6.1), the framework recognizes the multitude of users (e.g., farmers, practitioners) of mountain social-ecological systems. Further, these users have different—even contested—interests, and interact through unequal power and economic relations. These relations, on one hand, ought to be captured analytically for an accurate understanding of the system. On the other hand, they must be laid out and addressed for a sustainable and equitable system management. This applied aspect undergirded all the cases.

Though the framework shows the different dimensions and levels for conducting analyses on the Andean landscapes, it emphasizes the relations across levels (e.g., from global to local), across dimensions (e.g., environmental, governance, social), crossing from level to dimensions (e.g., climate to farm) and from dimension to level (e.g., added decisions of household changing water flow or nutrient load) (see Fig. 6.1). The framework aims to contribute to a sustainable and equitable mountain social-ecological system, considering that the users' legitimate participation and support are paramount for both management and knowledge co-production (Jack et al. 2020; Lam et al. 2020; Norström et al. 2020; López et al. 2017). This includes the involvement of disenfranchised groups (e.g., small farmers and pastoralists) not only when defining both research priorities and citizen-science data gathering protocols (e.g., key ecological variables, rangeland change indicators), but also discussing multi-level government support. At the community level, local research committees may monitor natural elements, and participate with professionals and policy-makers in designing sustainable rangeland management plans. Furthermore, actions implemented upon integrated data collection, analysis, and participatory policy-making, will be impactful and legitimate only if the requirements of communities and local users are included.

Some limitations of the framework relate to data in terms of availability, resolution, and quality. However, we believe that keeping posing questions about human-environment interactions in Andean landscapes will stimulate efforts to collect relevant data. Further, by putting the problems of the local communities front and center of knowledge co-production will lead to research questions that will trigger gathering relevant data for local people, practitioners, and policy makers. When data are available, though, a sound integration of diverse datasets requires careful examination of the differences involved in the scales and levels under analyses.

6 Concluding Remarks: Towards a Socio-environmental Research Agenda

There is nearly a centennial tradition of using remote sensing in the Andes. The analysis of changes in land use and land cover has been the most common application of satellite imagery. In addition to delineating different land uses and covers, remote sensing and GIS can be combined with other variables to build more realistic and comprehensive ecological units. In doing so, remote sensing and GIS are well suited for multi-cross scale and level analysis. For instance, the inclusion of climatic and soil conditions enables estimating relevant variables for rangelands such as productivity and conservation status. Further, the incorporation of key variables of ecosystems' structure and function, elicits understandings of the underlying processes of rangeland change (Briske et al. 2005).

The knowledge of these processes might be important for guiding actions to improve Andean rangeland conditions. Such information might be, for instance, particularly relevant for rangeland managers, from authorities of protected areas to government officials to pastoralists. For the latter, the condition of Andean rangeland is fundamental as this ecosystem is the main food source for livestock supporting people's wellbeing and way of life. In other words, the current ecological condition of rangeland greatly determines the vulnerability of pastoralists. Further, we show that vulnerability relates to social, economic, and environmental conditions (including climate) that influence both the ecosystem and household levels.

The impacts of interacting social and environmental processes on Andean social-ecological systems occur at multiple levels (e.g., from the plot to the landscape) and across scales and systems (e.g., ecologic, economic) (Postigo 2020; Mark et al. 2017). Based on the four case studies presented in this chapter, we elaborated a perspective that contributes to both an understanding of, and participatory action for an equitable sustainability of mountain social-ecological systems. The perspective argues for the relevance of identifying the scale(s) of the processes at play and the levels at which they happen to improve the understanding of mountain social-ecological systems. The contribution to action goes beyond mere generation of knowledge. It recognizes the multiple stakeholders in the Andes, as well as their needs and the effects of their practices on the ecosystems' health and conservation status.

At the regional level, spatial models help evaluate Andean ecosystems under intertwined climatic and biophysical change, and then identify the most important variables driving change. This information may be useful to guide decision-and policy makers' programs and actions to mitigate degrading processes and improve conditions of Andean social-ecological systems. Additionally, the knowledge generated and the recognition of multiple stakeholders at different levels, are two crucial elements for addressing rising pressures on mountain systems. For instance, the identification of these pressures and their levels of operation, offers insights in crafting targeted policies to improve adaptive capacity of households, and strengthen resilience of mountain social-ecological systems.

Using the results of analyses based on remote sensing data to inform decision-making (from farmers to policymakers, from the local to the country level) is yet to become widespread. Applications of remote sensing techniques play an important role in the monitoring and assessment of rangeland health so that managers can project and recognize threats to critical ecological processes. Satellite imagery offers both large spatial scale and high spatial and temporal resolution, iteratively gathering data at multiple wavelengths. Though field observations are important to calibrate data gathered remotely, their collection should be part of a participatory process too (Thorn et al. 2020). Community leaders should be involved in the monitoring and evaluation team. Training them to effectively gather data and share local perspectives about the value of ecological indicators, is crucial to increase the temporal and spatial resolutions of data and increase the legitimacy of research.

The co-production of a research agenda for mountain social-ecological systems in general, and for the Andes in particular, is paramount. Remote sensing and GIScience are important pieces of such co-production as they enable the integration of both perspectives and data from different stakeholders. Rising climatic and economic pressures in the Andes will deepen the challenges of sustainable development of this social-ecological system. Remote sensing and GIScience have improved the capability of monitoring land change and ecosystems health and condition. Yet, expectations about higher accuracy, and increased spatial and temporal resolutions to capture complex topographies remain as challenges. Some potential topics for further development are refining spatial models to assess land degradation, a system of reference sites as benchmark for monitoring ecosystems' health, and early warning systems for extreme weather events linked to household and livestock food security. Prospective use of remote sensing and GIScience may allow assessing the potential of rangeland ecosystems to provide environmental services, developing payment schemes for such services, and testing governance regimes for ecosystem services (Celleri 2010).

Finally, the centennial Andean connection to global circuits of capital has intensified in the last half century (Patel and Moore 2017). The impacts of such demands on the supplying landscapes are conspicuous (e.g., Rueda and Lambin 2013). Remote sensing and GIScience offer the possibility of monitoring the demand of markets and investors located far beyond mountain social-ecological systems, but who could have potentially substantial consequences on their sustainability.

References

Anacona PI, Kinney J, Schafer M et al (2018) Glacier protection laws: potential conflicts in managing glacial hazards and adapting to climate change. Ambio 47:835–845. https://doi.org/10.1007/s13280-018-1043-x

Briske DD, Fuhlendorf SD, Smeins F (2005) State-and-transition models, thresholds, and rangeland health: a synthesis of ecological concepts and perspectives. Rangel Ecol Manag 58:1–10

Buytaert W, Moulds S, Acosta L et al (2017) Glacier melt content of water use in the tropical Andes. Environ Res Lett 12:114014

Carey M (2005) Living and dying with glaciers: People's historical vulnerability to avalanches and outburst floods in Peru. Glob Planet Change 47:122–134

Cash DW, Adger WN, Berkes F et al (2006) Scale and cross-scale dynamics: governance and information in a multilevel world. Ecol Soc 11:8

Celleri R (2010) Estado del conocimiento técnico-científico sobre los servicios ambientales hidrológicos generados en los Andes. In: Quintero M (ed) Servicios ambientales hidrológicos en la región Andina. Estado del conocimiento, la acción y la política para asegurar su provisión mediante esquemas de pago por servicios ambientales. Agua y Sociedad, vol 12. IEP/CONDESAN, Lima, pp 25–45

Chen J, Jönsson P, Tamura M et al (2004) A simple method for reconstructing a high-quality NDVI time-series data set based on the Savitzky-Golay filter. Remote Sens Environ 91:332–344. https://doi.org/10.1016/j.rse.2004.03.014

Chimner RA, Bourgeau-Chavez LL, Grelik SL et al (2019) Mapping Mountain peatlands and wet meadows using multi-date, multi-sensor remote sensing in the cordillera Blanca, Peru. Wetlands 39:1057–1067. https://doi.org/10.1007/s13157-019-01134-1

Cuesta F, Llambí LD, Huggel C et al (2019) New land in the Neotropics: a review of biotic community, ecosystem, and landscape transformations in the face of climate and glacier change. Reg Environ Chang 19:1623–1642

Dangles O, Rabatel A, Kraemer M et al (2017) Ecosystem sentinels for climate change? Evidence of wetland cover changes over the last 30 years in the tropical Andes. Plos One 12:e0175814. https://doi.org/10.1371/journal.pone.0175814

De Bièvre B, Bustamante M, Buytaert W et al (2012) Síntesis de los impactos de los efectos del cambio climático en los recursos hídricos en los Andes Tropicales y las estraegias de adaptación desarolladas por los pobladores. In: Cuesta F, Bustamante M, Becerra MT, Postigo J, Peralvo M (eds) Panorama andino de cambio climático: Vulnerabilidad y adaptación en los Andes Tropicales. CONDESAN, SGCAN, Lima, pp 65–107

Durigon VL, Carvalho DF, Antunes MAH et al (2014) NDVI time series for monitoring RUSLE cover management factor in a tropical watershed. Int J Remote Sens 35:441–453. https://doi.org/10.1080/01431161.2013.871081

Eakin H, Bojórquez-Tapia LA (2008) Insights into the composition of household vulnerability from multicriteria decision analysis. Glob Environ Change 18:112–127. https://doi.org/10.1016/j.gloenvcha.2007.09.001

Estell RE (2010) Coping with shrub secondary metabolites by ruminants, vol 94. Elsevier. https://doi.org/10.1016/j.smallrumres.2010.09.012

Flores ER (1991) Manejo y utilización de pastizales. In: Fernández Baca S (ed) Avances y perspectivas del conocimiento de los camélidos sudamericanos. FAO, Santiago de Chile, pp 191–212

Flores-Lopez F, Galaitsi SE, Escobar M, Purkey D (2016) Modeling of Andean Paramo Ecosystems' hydrological response to environmental change. Water 8:18. https://doi.org/10.3390/w8030094

Flórez A, Malpartida E (1987) Manejo de Praderas Nativas y Pasturas en la Región Altoandina del Perú. Tomo II. Fondo del Libro del Banco Agrario, Lima, pp 116–140

Funk C, Peterson P, Landsfeld M et al (2015) The climate hazards infrared precipitation with stations – a new environmental record for monitoring extremes. Sci Data 2. https://doi.org/10.1038/sdata.2015.66

Grunwald S, Thompson JA, Boettinger JL (2011) Digital soil mapping and modeling at continental scales: finding solutions for global issues. Soil Sci Soc Am J 75:1201–1213. https://doi.org/10.2136/sssaj2011.0025

Grünwaldt JM, Castellaro G, Flores ER et al (2016) Pastoralismo en zonas áridas de Latinoamérica: Argentina, Chile, México y Perú. Revue Scientifique et Technique de l'OIE 35:543–560. https://doi.org/10.20506/rst.35.2.2526

Hengl T, Mendes de Jesus J, MacMillan RA et al (2014) SoilGrids1km – global soil information based on automated mapping. PlosOne 9(12):e114788. https://doi.org/10.1371/journal.pone.0105992

Herrick JE, Karl JW, McCord SE et al (2017) Two new Mobile apps for rangeland inventory and monitoring by landowners and land managers. Rangelands 39:46–55. https://doi.org/10.1016/j.rala.2016.12.003

INEI (2010) Mapa de Pobreza 2009:292-292

Jack JC, Gonet J, Mease A, Nowak K (2020) Traditional knowledge underlies one health. Science 369:1576–1576

Jensen JR (2014) Introductory digital image processing. Pearson, New York

Josee C, Cuesta F, Navarro G, Cabrera E (2009) Ecosistemas de los Andes del Norte y Centro. Bolivia, Colombia, Ecuador, Perú y Venezuela. Secretaría General de la Comunidad Andina, Programa Regional ECOBONA-Intercooperation, CONDESAN-Proyecto Páramo Andino, Programa BioAndes, EcoCiencia, NatureServe, IAvH, LTA-UNALM, ICAE-ULA, CDC-UNALM, RUMBOL SRL. Lima, Peru

Kozhikkodan Veettil B, de Souza SF (2017) Study of 40-year glacier retreat in the northern region of the cordillera Vilcanota, Peru, using satellite images: preliminary results. Remote Sens Lett 8:78–85. https://doi.org/10.1080/2150704X.2016.1235811

Lam DPM, Hinz E, Lang DJ et al (2020) Indigenous and local knowledge in sustainability transformations research: a literature review. Ecol Soc 25. https://doi.org/10.5751/es-11305-250103

Lipton JK (2014) Lasting legacies: conservation and communities at Huascaran National Park, Peru. Soc Nat Resour 27:820–833. https://doi.org/10.1080/08941920.2014.905888

López S, Jung J-K, López MF (2017) A hybrid-epistemological approach to climate change research: Linking scientific and smallholder knowledge systems in the Ecuadorian Andes. Anthropocene 17:30–45

López-i-Gelats F, Contreras Paco JL, Huilcas Huayra R et al (2015) Adaptation strategies of Andean pastoralist households to both climate and non-climate changes. Hum Ecol 43:267–282

Marchand M, Golea M (1993) On learning simple neural concepts: from halfspace intersections to neural decision lists. Netw Comput Neural Syst 4:67–85. https://doi.org/10.1088/0954-898X_4_1_005

Mark BG, Bury J, McKenzie JM et al (2010) Climate change and tropical Andean glacier recession: evaluating hydrologic changes and livelihood vulnerability in the cordillera Blanca, Peru. Ann Am Assoc Geogr 100:794–805

Mark BG, French A, Baraer M et al (2017) Glacier loss and hydro-social risks in the Peruvian Andes. Glob Planet Change 159:61–76

McGraw JF, Tueller PT (1983) Landsat computer-aided analysis techniques for range vegetation mapping. J Range Manag 36:627–627. https://doi.org/10.2307/3898357

Monserud RA, Leemans R (1992) Comparing global vegetation maps with the kappa statistic. Ecol Model 62:275–293. https://doi.org/10.1016/0304-3800(92)90003-W

Motschmann A, Huggel C, Carey M, Moulton H et al (2020) Losses and damages connected to glacier retreat in the cordillera Blanca, Peru. Clim Change:1–22. https://doi.org/10.1007/s10584-020-02770-x

Norström AV, Cvitanovic C, Löf MF et al (2020) Principles for knowledge co-production in sustainability research. Nat Sustain 3:182–190. https://doi.org/10.1038/s41893-019-0448-2

O'Brien K, Leichenko R, Kelkar U et al (2004) Mapping vulnerability to multiple stressors: climate change and globalization in India. Glob Environ Change 14:303–313

Ochoa-Tocachi BF, Bardales JD, Antiporta J et al (2019) Potential contributions of pre-Inca infiltration infrastructure to Andean water security. Nat Sustain 2:584–593

Patel R, Moore JW (2017) A history of the world in seven cheap things: a guide to capitalism, nature, and the future of the planet. University of California Press, Oakland

Polk MH (2016) "They are drying out": social-ecological consequences of glacier recession on mountain peatlands in Husacaran National Park, Peru. Dissertation, The University of Texas at Austin, Austin, TX, USA

Polk MH, Young KR, Baraer M et al (2017) Exploring hydrologic connections between tropical mountain wetlands and glacier recession in Peru's cordillera Blanca. Appl Geogr 78:94–103

Postigo JC (2014) Perception and resilience of Andean populations facing climate change. J Ethnobiol 34:383–400

Postigo JC (2020) The role of social institutions in Indigenous Andean pastoralists' adaptation to climate-related water hazards. Clim Dev. https://doi.org/10.1080/17565529.2020.1850409

Postigo JC, Peralvo M, López S et al (2012) Adaptación y vulnerabilidad en los sistemas productivos andinos. In: Cuesta F, Bustamante M, Becerra MT, Postigo J, Peralvo M (eds) Panorama andino de cambio climático: Vulnerabilidad y adaptación en los Andes Tropicales. CONDESAN, SGCAN, Lima, pp 147–177

Pyke DA, Herrick JE, Shaver P, Pellant M (2002) Rangeland health attributes and indicators for qualitative assessment. J Range Manag. https://doi.org/10.2307/4004002

Rabatel A, Francou B, Soruco A et al (2013) Current state of glacier in the tropical Andes: a multi-century perspective on glacier evolution and climate change. Cryosphere 7:81–102

Rasmussen MB (2016) Unsettling times: living with the changing horizons of the Peruvian Andes. Lat Am Perspect 43:73–86. https://doi.org/10.1177/0094582x16637867

Renard KG, Yoder DC, Lightle DT, Dabney SM (2011) Universal soil loss equation and revised universal soil loss equation. In: Mogan RP, Nearing MA (eds) Handbook of erosion modelling. Wiley, New Jersey, pp 135–167. https://doi.org/10.1002/9781444328455.ch8

Rueda X, Lambin EF (2013) Linking globalization to local land uses: how eco-consumers and gourmands are changing the Colombian coffee landscapes. World Dev 41:286–301. https://doi.org/10.1016/j.worlddev.2012.05.018

Ruyle G, Dyess J (2016) Rangeland monitoring and the Parker 3-step method: overview, perspectives and current applications rangeland monitoring and the Parker 3-step method: overview, perspectives and current applications. The University of Arizona, Tucson

Shaw TE, Gascoin S, Mendoza PA et al (2020) Snow depth patterns in a High Mountain Andean catchment from satellite optical Tristereoscopic remote sensing. Water Resour Res 56:e2019WR024880

Thorn JPR, Klein JA, Steger C et al (2020) A systematic review of participatory scenario planning to envision mountain social-ecological systems futures. Ecol Soc 25:1–55. https://doi.org/10.5751/ES-11608-250306

Thurow TL (2000) Hydrologic effects on rangeland degradation and restoration processes. Springer, pp 53–66. https://doi.org/10.1007/978-94-015-9602-2_5

Tovar C, Seijmonsbergen AC, Duivenvoorden JF (2013) Monitoring land use and land cover change in mountain regions: an example in the Jalca grasslands of the Peruvian Andes. Landsc Urban Plan 112(0):40–49. https://doi.org/10.1016/j.landurbplan.2012.12.003

Vander Jagt B, Lucieer A, Wallace L et al (2015) Snow depth retrieval with UAS using photogrammetric techniques. Geosci J 5:264–285

Vanwey LK, Ostrom E, Meretsky V (2005) Theories underlying the study of human-environment interactions. In: Moran EF, Ostrom E (eds) Seeing the Forest and the trees. Human-environment interactions in forest ecosystems. The MIT Press, Cambridge, MA/London, pp 23–56

Veettil BK, Kamp U (2017) Remote sensing of glaciers in the tropical Andes: a review. Int J Remote Sens 38:7101–7137

Vergara W, Deeb A, Valencia AM et al (2007) Economic impacts of rapid glacier retreat in the Andes. EOS Trans Am Geophys Union 88:261–263

Verzijl A, Guerrero Quispe S (2013) The system nobody sees: irrigated wetland management and alpaca herding in the Peruvian Andes. Mt Res Dev 33:280–293

Vuille M, Francou B, Wagnon P et al (2008) Climate change and tropical Andean glaciers: past, present and future. Earth-Sci Rev 89:79–96

Vuille M, Carey M, Huggel C et al (2018) Rapid decline of snow and ice in the tropical Andes–Impacts, uncertainties and challenges ahead. Earth-Sci Rev 176:195–213

Wood JL, Harrison S, Wilson R et al (2021) Contemporary glacial lakes in the Peruvian Andes. Glob Planet Change 204:103574. https://doi.org/10.1016/j.gloplacha.2021.103574

Yager K, Valdivia C, Slayback D et al (2019) Socio-ecological dimensions of Andean pastoral landscape change: bridging traditional ecological knowledge and satellite image analysis in Sajama National Park, Bolivia. Reg Environ Chang 19:1353–1369. https://doi.org/10.1007/s10113-019-01466-y

Yarleque C, Vuille M, Hardy DR et al (2018) Projections of the future disappearance of the Quelccaya ice cap in the Central Andes. Sci Rep 8:11. https://doi.org/10.1038/s41598-018-33698-z

Young KR, Ponette-Gonzalez AG, Polk MH, Lipton JK (2017) Snowlines and treelines in the tropical Andes. Ann Am Assoc Geogr 107:429–440. https://doi.org/10.1080/24694452.2016.1235479

Chapter 7
The Use of Remote Sensing in Air Pollution Control and Public Health

Cesar I. Alvarez-Mendoza

Abstract Due to the significant increase of environmental and health problems in recent years, remote sensing has become an alternative for finding new opportunities to minimize their impact. Thus, different options now exist to support public health interventions in Latin America, where resources to create prevention policies are limited compared to developed regions. Some developing countries do not have air pollution control measures because these are not considered the main focus of their current environmental policy in comparison to developed countries. However, it is well known that air pollution has significant negative impacts on respiratory health, which makes its control a priority. Remote sensing data are becoming a cost-effective alternative in Latin American countries for monitoring and controlling air pollution to improve public health. This chapter focuses on the use of remote sensing applied to air pollution control with an emphasis on public health. The chapter discusses examples of applied remote sensing methodologies in various Latin American cities and introduces an approach to analyzing the relationships between air quality and public health outcomes in Quito, Ecuador using land use regression models. The different remote sensing approaches discussed here could support Latin American public health authorities with the implementation of concrete public health interventions.

Keywords Remote sensing · Air pollution control · Public health · Latin America · LUR models

C. I. Alvarez-Mendoza (✉)
Environmental Research Group for Sustainable Development (GIADES), Environmental Engineering, Universidad Politécnica Salesiana, Quito, Ecuador
e-mail: calvarezm@ups.edu.ec

S. López (ed.), *Socio-Environmental Research in Latin America*, The Latin American Studies Book Series, https://doi.org/10.1007/978-3-031-22680-9_7

1 Remote Sensing in Environmental and Public Health Research

Climate change is currently the greatest global environmental threat (Hiraishi et al. 2013) that continues to exacerbate gradual processes of environmental degradation and the frequency and intensity of natural disasters, which requires the development of immediate mitigation actions to curb its negative impacts. Some of the effects of climate change are caused by the increment of heat-trapping gases concentrations globally. Changes in climate include significant increases in surface temperatures, intensification of precipitation events, prolonged droughts, enhanced seasonality, etc., that could lead to sea-level rise, changes in biogeographic patterns, dispersion of diseases with implications for public health, among other effects (NASA's Jet Propulsion Laboratory 2021). The impact of climate change on public health is problematic, considering the large number of people worldwide who suffer from health-related issues associated with degrading environmental conditions such as poor air quality exacerbated by increased fossil fuel emissions (IARC 2016; Fan et al. 2016; Kuo et al. 2019). However, skepticism about the severity of public health issues and their association with environmental problems remains high (Soutter and Mõttus 2020). A clear example is the issue of air pollution, its causes, and its impacts on human health. Air pollution is a complex physico-chemical process that involves the presence in or introduction into the atmosphere of a substance that has harmful effects on environmental or public health. The main air pollutants comprise particulate matter (PM) that consists of particles with aerodynamic diameters of less than 10 mm or 2.5 mm (PM10 and PM2.5), carbon monoxide (CO), nitrogen dioxide (NO_2), sulfur dioxide (SO_2), and ozone (O_3) (EPA 2017). Air pollution has important repercussions on respiratory health and significantly impacts global mortality (Agrawal et al. 2021). According to the World Health Organization (WHO), the health impact of air pollution and the prevalence of respiratory diseases represents approximately 6% of global annual deaths (WHO 2019). One of the factors exacerbating respiratory health is the worsening of air quality due to a significant increase of air pollutants caused by human activity such as transportation, industrial processes, food production, and others (Kumar et al. 2014).

Satellite remote sensing (RS) technologies for environmental monitoring expanded globally since the early 1960s. RS refers to the acquisition of information about an object through a sensor without touching it. The sensor can be onboard a crewed or unmanned aircraft or a satellite (Chuvieco 2020). Traditional RS environmental applications in Latin America include land cover change analysis based on supervised or unsupervised classifications and environmental feature analyses through the evaluation of multispectral indices (e.g., normalized difference vegetation index – NDVI, normalized soil moisture index – NSMI, normalized difference water index – NDWI, etc.) (Alvarez-Mendoza et. al. 2022). More recently, novel RS applications have emerged that incorporate public health approaches and environmental monitoring assessments, including the development of spatial models of environmental quality, monitoring and surveillance systems, and risk assessment tools to assess and predict environmental and public health spatial patterns (Viana et al. 2017), and air pollution concentrations (Xu et al. 2020; Omrani et al. 2020). The use of RS has increased in air pollution studies in recent years due to the availability of RS data and RS's regional coverage. In these studies, the goal is to assess air quality through the development of spatial models of air pollution that integrate pollutant concentration information and field data derived from air quality stations and other field sources. Spatially-explicit approaches of this type include land use regression (LUR) models. LUR methods are not new and have been extensively used to study the spatio-temporal patterns of air pollution associated with human activities. LUR models rely on geographical variables such as land use and land cover information, thermal surface conditions, and environmental indexes obtained from remotely sensed data to obtain a general understanding, usually in the form of a map, of air pollutants such as PM, O_3 or NO_2 concentrations (Alvarez-Mendoza et al. 2019a, b; Rivera-González et al. 2015). However, LUR models are of limited use because they are usually not fed by updated spatial data or simply rely on static geographic data such as roads, industrial site locations, and other pseudo-permanent features. Unfortunately, mapping agencies from many Latin American countries do not update their geospatial databases at the same rate of the changes observed in the field due to the lack of resources needed for data maintenance and quality control activities. Currently, nevertheless, new approaches have emerged that employ dynamic environmental variables obtained from RS data (e.g., land use and land cover data or multispectral index information of particular environmental features), and other mathematical techniques to compute LUR models incorporating artificial intelligence (AI) with machine learning (ML) techniques, which have filled important gaps in research and improved the prediction accuracy of the spatial models.

Currently, most public health RS applications focus on vector-borne disease modeling (i.e., malaria, dengue, and schistosomiasis), where the main contribution of RS is the prediction of tropical mosquito infestations in urban land scenarios (Lorentz et al. 2020) or the generation of multispectral indices to measure vegetation cover variability employing NDVI or NDWI indexes, to identify vectors' habitats (Hess et al. 2017). A less common RS application, on the other hand, is the study of cardiovascular and respiratory disease incidence (e.g., asthma prevalence). Ayres-Sampaio et al. (2014), for example, used remotely sensed data to predict asthma incidence. They showed that high NO_2 concentrations, low NDVI values, and high surface temperatures as measured by remote sensors, increase asthma hospitalizations.

Due to the relationship between environmental and public health factors, RS has been increasingly used in the extraction of surface conditions (e.g., vegetation types, water features, soil characteristics, land use, or surface temperatures) to develop spatially explicit models of disease dispersion (Dlamini et al. 2019). For example, it is well known that climate variation is highly associated with the incidence of vector-borne diseases (WHO 2022). Standard RS-derived variables include NDVI or land surface temperature (LST) (Wimberly et al. 2021), which are commonly used to accurately depict crucial environmental conditions that either facilitate or deter disease transmission. In the case of cardiovascular and respiratory disease prevalence, applications of RS concentrate on the prediction of air quality conditions based on environmental data such as ambient aerosol optical thickness (AOT), which measures aerosol concentration in the air (e.g., PM10, PM2.5, NO_2, or SO_2), and vegetation cover through the development of LUR models (Cilluffo et al. 2018). Air quality predictions from RS are then compared with epidemiological data (e.g., hospital discharge numbers or prevalence measurements). For example, some studies showed that green spaces and gardens are associated with reduced asthma hospitalizations when air pollutant concentrations are low (Alcock et al. 2017). Another study found that high NDVI values are correlated with a slightly increased risk of asthma in children (Andrusaityte et al. 2016).

Within this context, RS research on air pollution control and public health monitoring has been facilitated by the application of interdisciplinary approaches that combine environmental and public health data to predict public health outcomes. This integration is now possible due to the advancement of new modeling approaches such as ML techniques and the availability of remote sensors that specialize in measuring air pollutant concentrations. These new modeling frameworks allow the integration of different types of information, meteorological and social, to improve model estimation and prediction accuracy (Fig. 7.1). Given the growing interest in developing environmental monitoring strategies that adapt to the socio-environmental contexts where significant environmental degradation occurs, new opportunities have emerged to further develop and test RS strategies to improve planning and control of public health outcomes. This chapter focuses on how RS is currently applied in environmental and public health research in the Latin American context.

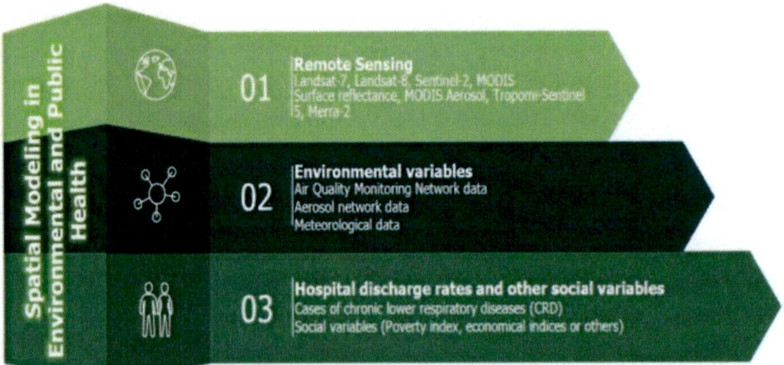

Fig. 7.1 Most frequent remote sensing (RS) products (01), environmental data emphasising air pollution control (02), and public health datasets (03) used to build LUR models of public health outcomes (Modified from Alvarez-Mendoza et al. 2020)

2 Remote Sensors Commonly Used in air Pollution Control And Public Health Research

Most studies about RS and air quality focus on areas with high levels of pollution, particularly China (Meng et al. 2015; Zhai et al. 2018; Li et al. 2016). These studies commonly rely on spatial modeling algorithms to explain air pollution spatialization with input from remotely sensed data. In most cases, potential predictors of air pollutants derived from RS comprise AOT, NDVI, LST, and others (Jethva et al. 2019; Heo and Bell 2019). The following sections describe primary RS sensing products obtained by sensors onboard Earth observation satellites (EOS) commonly utilized in public health studies and air pollution control. Two advantages of using this type of satellites are that they are publicly and globally available. Thus, Latin American researchers can take advantage of these sources of information to deepen the understanding of socio-environmental interactions mediated by changes in atmospheric composition.

2.1 The Landsat Program

The Landsat program is commonly viewed as a synonym of satellite RS. This system is the most extended continuous running EOS program in the world that started in the early 1970s. This program is a joint effort of the National Aeronautics and Space Administration (NASA) and the United States Geological Survey (USGS). The program started with the launching of the Earth Resources Technology Satellite, which was eventually renamed Landsat-1. The most recent satellites are Landsat-8, launched in February 2013, and Landsat-9, launched in September 2021. Landsat-8 is commonly used in air quality studies because of the type of sensors the satellite carries. This satellite has the Operational Land Imager (OLI) and the Thermal Infrared Sensor (TIRS) on board (Department of the Interior U.S. Geological Survey 2016). New air quality models rely on vegetation indexes and LST values derived from data captured by these sensors. The main advantages of current Landsat-8 data in comparison with other remotely sensed data are the spatial (30 m, adequate for regional studies), temporal (16-day repeat cycle) and spectral resolutions (11 bands, covering the visible, near, medium, and far infrared energy ranges of the electromagnetic spectrum). In comparison with Landsat-8, Landsat-9 presents some improvements, particularly regarding the spectral resolution of the TIRS sensor and minimization of stray light. Landsat-9 has two sensors: (1) a copy of OLI and OLI-2 sensor, and (2) the TIRS-2 sensor.

2.2 Moderate-Resolution Imaging Spectroradiometer (MODIS)

MODIS instruments are remote sensors on the Terra and Aqua satellites, launched by NASA in 1999 and 2002, respectively. The main advantages of MODIS are the spectral, spatial, and temporal resolutions. The sensors have 36 spectral bands and provide a view of the entire Earth's surface every 1 to 2 days with three different spatial resolutions (250, 500, and 1000 m). Thus, MODIS allows the detection of trends in air quality at a global level. One of the uses of MODIS in air pollution control is to obtain AOT information (Gupta et al. 2006; Liu et al. 2007; Chen et al. 2014). AOT can be acquired from MODIS Multi-Angle Implementation of Atmospheric Correction (MAIAC) (Just et al. 2018). This is an algorithm that allows to retrieve atmospheric aerosol concentrations at high spatial resolutions. AOT is similar to Aerosol Optical Depth (AOD) and obtained from ground stations of the AERONET network. They allow to obtain measurements of aerosol concentrations related to air pollutants (Jethva et al. 2019). Unfortunately, in most cases, Latin American countries do not have AOD AERONET stations, a limitation that has been lifted by the availability of AOT data derived from MODIS data in regional studies.

2.3 The TROPOspheric Monitoring Instrument (TROPOMI) Aboard Sentinel-5 Precursor (Sentinel 5-P) and the Future Sentinel 5

The Sentinel-5 Precursor (Sentinel-5P) mission is the first Copernicus mission developed by the European Space Agency (ESA) and designed to monitor the atmosphere. Launched in October 2017 onboard the Tropospheric Monitoring Instrument (TROPOMI), the sensor captures data at a spatial resolution of 0.01 arc degrees (aprox. 1 km) on a daily temporal resolution, which constitutes a significant advantage over other satellites. The TROPOMI provides concentration measurements of O_3, NO_2, SO_2, CO, CH_4, CH_2O, and aerosol properties (Veefkind et al. 2012), which is useful when measuring the impacts of these compounds on environmental and public health. In the near future, ESA will deploy a new generation of satellites like Sentinel-4 to monitor air quality across Europe, and Sentinel-5 with the Ultraviolet/Visible/Near Infrared/Short-Wave Infrared (UVNS) sensor, which will help improve air quality models with more accurate pollution concentration data than their precursors (Pérez Albiñana et al. 2017).

3 Applied Remote Sensing Research in Latin America on Air Pollution Control and Public Health: Opportunities and Limitations

Latin American regions are socially and environmentally heterogenous compared to developed areas in the Global North. Weather and topographic conditions, for instance, vary significantly on a daily basis and over small distances. Social conditions are another example of extreme heterogeneity, with nearby locations affected by significant poverty or depicting substantial wealth. This is a conspicuous characteristic of Latin American cities characterized by significant socio-economic inequities and disparities. Given Latin America's environmental and social conditions, it is a priority to develop geospatial approaches and tools to be able to monitor and control the impacts associated with uneven socio-economic conditions and the complex physical settings of urban locales. However, both national and local governments have other priorities and downplay the role that geospatial tools play in environmental, social, and land-use planning. Brazil and Argentina are perhaps two exceptions to this rule since they have created RS agencies and other smaller initiatives at universities or research institutes to support a more organized type of urban development and expansion that rely on environmental change assessments.

Recent research has highlighted the importance of geospatial approaches to quantify the extent of urban forests using multispectral indices, mainly NDVI. An example of such an application is the design of green corridors in Latin American cities as in the case in Temuco, Chile (Moreno et al. 2020). Green corridors are essential spaces that could potentially help improve the health, physical, and psychological conditions of urban dwellers. Moreover, urban green spaces contribute to improved air quality by minimizing the formation of harmful pollutants such as photochemical O_3. Therefore, by enhancing green urban growth policies, opportunities exist to improve the air quality of cities and the living conditions of their inhabitants.

New instruments such as the TROPOMI sensor onboard Sentinel-5P allow the collection of air pollutant concentration data, enabling their use in air quality and health studies. Sannigrahi et al. (2021), for example, studied air quality under COVID-19 conditions in Sao Paulo, Brazil and showed a positive impact of the pandemic on air quality due to the partial lockdown of cities, with temporary reductions of NO_2 and CO concentrations. Their study demonstrated that such effects could be monitored by comparing Sentinel-5P images after and before the lockdown. Using data from the same sensor, Toro et al. (2021) found a similar trend for Santiago de Chile also associated with mobility restrictions during the COVID-19 outbreak. Similarly, Rojas et al. (2021) showed a consistent reduction of NO_2 concentration during the lockdown using RS data in Lima, Peru. It is important to emphasize that NO_2 concentrations are significantly correlated with traffic emissions and other anthropogenic activities (Rojas et al. 2021). A study in Quito, Ecuador, for example, found a reduction in NO_2 (−68%), SO_2 (−48%), CO (−38%) and PM2.5 (−29%) concentrations based on data from TROPOMI

Sentinel-5P, thus demonstrating air quality improvement in this Latin America city during the COVID-19 pandemic (Zalakeviciute et al. 2020). Thus, the COVID-19 event provided an opportunity to evaluate the benefits of RS technologies in air pollution control and public health monitoring by allowing the characterization of air quality before and after lockdowns in different cities. RS technologies enabled the generation of synoptic views of the impacts that daily anthropogenic activities such as traffic emission and industries have on air quality and people's health (Fig. 7.2). In summary, these studies showed a clear trend: high air pollutant concentrations are strongly correlated with poor air quality and the deterioration of public health.

RS offers new opportunities to assist cities and municipalities in their air pollution control efforts, especially where data, and human and technological resources are limited. A common example of this type of limitation is the lack of air quality monitoring networks (AQMN) or the small number of AQMN stations in particular cities. These networks and stations allow monitoring air pollutant concentration by comparing good or bad air quality conditions at high temporal resolutions. It is well known that Latin American cities have considerable air quality and public health problems not only due to intense pollution caused by extreme vehicular traffic, use of old vehicles and technologies, or unsustainable industrial production, but also the lack of air quality policies and enforcement strategies that could help regulate pollution and reduce emission concentrations. In most cases, the absence of air quality regulation or weak policy enforcement is related to the absence of robust AQMNs.

Ecuador is a good example of a country with limited resources and disparities in air quality control. For instance, only the capital city, Quito, has an AQMN while the rest of major towns and cities in the country do not have air quality monitoring programs. Although having an AQMN is an important step toward improved systematic air quality monitoring, Quito's AQMN data are not trustworthy due to technical issues with the stations and poor quality control of the data, which calls into question the actual state of air quality and associated public health conditions at local levels. Thus, RS presents a novel cost-effective alternative to assist authorities with their air quality monitoring efforts, as several case studies in other Latin American cities such as Mexico City, Quito, Bogota, Lima, Santiago de Chile, etc. have demonstrated. These case studies suggest that RS allows us to not only monitor urban air pollution but link air quality data to respiratory and cardiovascular disease incidence.

One of the main advantages of RS is the availability of time-series data. This type of data allows the identification of spatio-temporal trends in air quality, which is relevant in long-term epidemiological studies. In addition, RS provides an opportunity for Latin American countries to keep up with more developed regions in their efforts to monitor atmospheric changes. When countries develop and use new technologies to obtain environmental data, they are better equipped to develop answers and solutions to complex environmental issues. Better air quality data means more opportunities to build reliable models and scenarios of atmospheric change, identify actual trends in air quality, derive accurate statistical characterizations, suggest

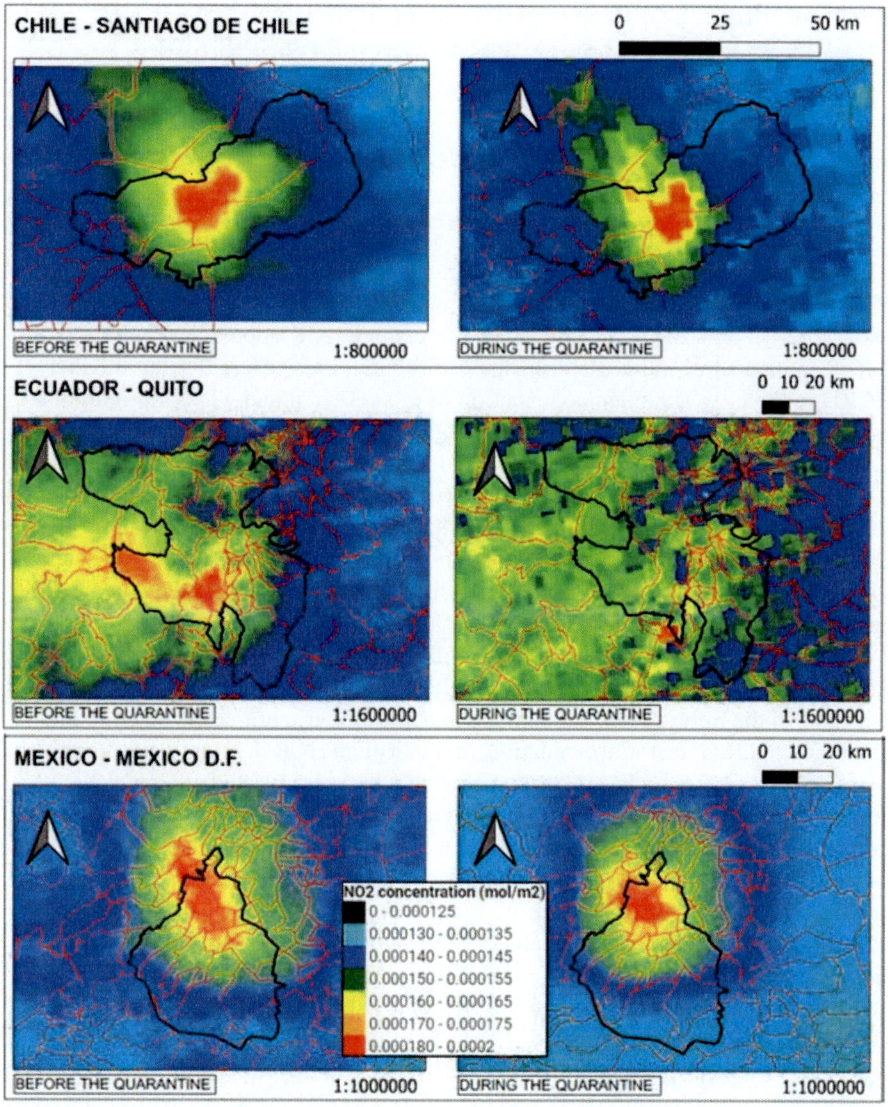

Fig. 7.2 NO_2 concentrations in mol/m^2 during Covid-19 lockdowns in different Latin American cities derived from TROPOMI Sentinel-5P data. The maps were generated using Google Earth Engine (GEE). Red tones depict high concentrations, and blue tones are low ones

better uses and applications of particular types of data, and plan for contextualized environmental or public health interventions. As we traverse the age of digital data, RS in general, and the data derived from RS technologies in particular, should be at the center of current data gathering efforts.

4 Modeling Chronic Respiratory Patient Discharge Based on Environmental Variables Extracted from Satellite Imagery in Quito, Ecuador: A Case Study

4.1 The Integration of Remote Sensing Data, Air Quality, and Respiratory Health

Alvarez-Mendoza et al. (2020) developed a novel RS application to estimate the number of hospital discharges of patients with chronic respiratory disease (HCRD) between 2013 and 2017 in Quito, Ecuador, using a LUR approach. They used different multispectral indices and hospital discharge data to predict the number of HCRD in the city. The main contribution of this research and what differentiated it from other studies is that they used RS data and spatially explicit techniques to predict HCRD and not only air quality conditions. The analysis was applied to a high-altitude, densely populated Andean city. Quito is in the core of the tropical Andean region and has different physical conditions in relation to low elevation cities. Quito's altitude (approximately 2800 m above sea level) influences the city's environmental and meteorological conditions, significantly impacting air quality and the epidemiology of respiratory diseases.

Moreover, Quito has a limited number of AQMN stations (Fig. 7.3), which makes RS data an ideal alternative to develop a better understanding of air quality in the city. Due to the influence of temperature and atmospheric pressure, the altitude of a city is an important factor in evaluating air quality and HCRD. Some Latin American cities, such as Quito, La Paz, Mexico City and others, are located at high elevations. At high elevations, the reduced oxygen content of the blood induces breathing instability with periods of deep and rapid breathing, alternating with sleep disorder episodes. Although this characteristic could have important effects on respiratory health, especially under conditions of poor air quality, high elevation conditions are usually underestimated in the evaluation of the possible respiratory health effects (Bravo Alvarez et al. 2013).

This study aimed to develop a LUR model to estimate HCRD based on dynamic geographical data derived from RS. The LUR models were built using four ML techniques: (1) multiple linear regression (MLR), (2) multilayer perceptron (MLP), (3) support vector regression (SVR), and (4) random forest regression (RFR). The conceptual LUR model is defined as:

$$HCRD = a + b_1x_1 + b_2x_2 + b_3x_3 \ldots b_nx_n$$

Where HCRD are the hospital discharge values and measured as the number of patients per 10,000 people who are admitted to the hospital with a chronic respiratory disease, a is the intercept, b_n are the regression coefficients, and x_n the independent variables. The model incorporates the spectral bands from Landsat imagery, multispectral indexes, and meteorological variables as predictors as they are

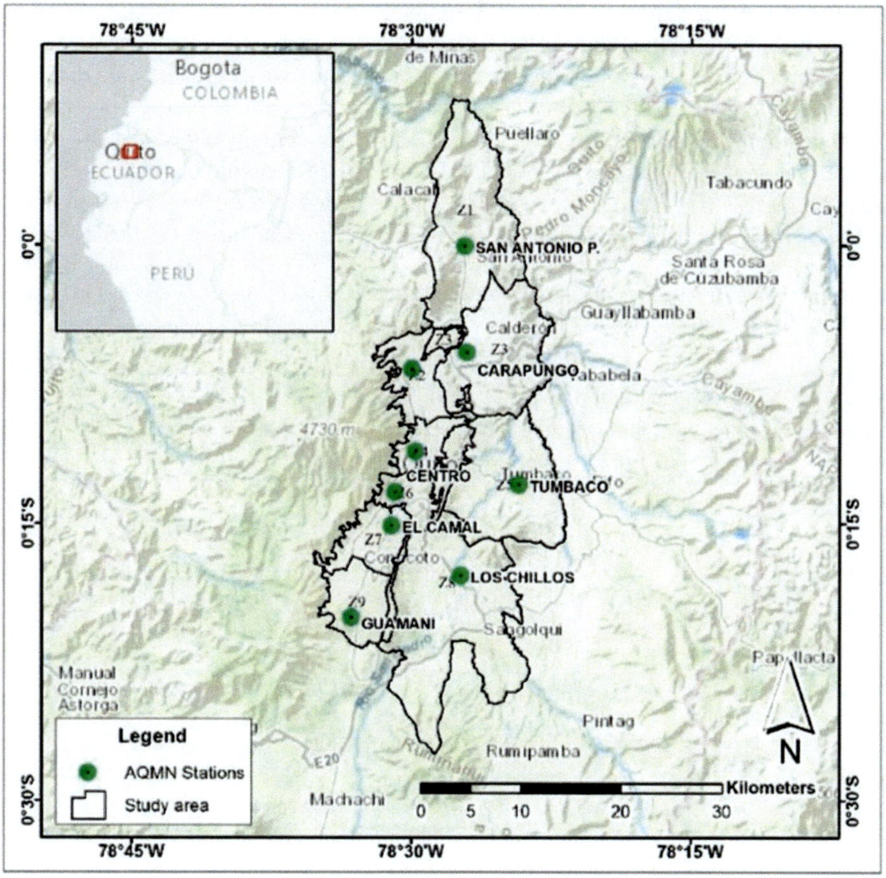

Fig. 7.3 Study area of Quito, Ecuador, a large urban metropolitan area. The green dots show the AQMN stations available in the city (Alvarez-Mendoza et al. 2020)

expected to play a significant role in the incidence of chronic respiratory diseases. Possible environmental factors include greenery as derived from NDVI, air quality characteristics, meteorological conditions, and epidemiological conditions.

4.2 Data Collection and LUR Models to Predict HCRD

The purpose of this research was to better understand the association between HCRD and the spatial context where hospital discharges occur. To reach this goal, the authors developed a LUR model to predict HCRD using a series of spatial covariates, assuming that environmental (vegetation indices from RS and air quality) and meteorological conditions are associated with HCRD variation as the revision of literature showed. The first step involved the collection of data to build a dataset to investigate the relationship between air quality and disease. The data collection process involved the operationalization of four variables (RS, air quality, meteorological, and HCRD), which resulted in a dataset with 23 predictors in addition to the dependent variable (Table 7.1). Each dataset contained the median yearly quarter value of the variables by AQMN stations. The RS predictors were obtained from Landsat-8 (OLI and TIRS sensors) and included several vegetation indices (NDVI, SAVI, and EVI), LST, and the visible and infrared surface reflectance bands.

Table 7.1 Input variables used in the construction of the LUR model to predict HCRD in Quito. (Modified from Alvarez-Mendoza et al. 2020)

No.	Type	Data	Variable	Units/scale
0	Dependent	Health	HCRD	Hospital discharges per 10,000 people with chronic respiratory disease
1	Independent	Remote sensing	Coastal aerosol band (B1)	Surface reflectance (0–1)
2			Blue band (B2)	Surface reflectance (0–1)
3			Green band (B3)	Surface reflectance (0–1)
4			Red band (B4)	Surface reflectance (0–1)
5			Near–infrared (NIR) (B5)	Surface reflectance (0–1)
6			Short-wave infrared 1 (B6)	Surface reflectance (0–1)
7			Short-wave infrared 2 (B7)	Surface reflectance (0–1)
8			Normalized Difference Vegetation Index (NDVI)	0–1
9			Soil-Adjusted Vegetation Index (SAVI)	0–1
10			Enhanced Vegetation Index (EVI)	0–1
11			Land Surface temperature (LST)	Degrees Celsius
12		Meteorological	Pressure (P)	Mb
13			Wind direction (WD)	Degrees
14			Relative humidity (RH)	Percentage (%)
15			Precipitation (PR)	Mm
16			Wind speed (WS)	m/s
17			Air temperature (AT)	Degrees Celsius
18			Solar irradiance (SR)	W/m^2
19		Environmental – air quality	CO	$\mu g/m^3$
20			NO_2	$\mu g/m^3$
21			O_3	$\mu g/m^3$
22			PM2.5	$\mu g/m^3$
23			SO_2	$\mu g/m^3$

Moreover, air pollutant concentrations (CO, NO_2, O_3, PM2.5 and SO_2) and meteorological variables (atmospheric pressure (P), wind direction (WD), relative humidity (RH), precipitation (PR), wind speed (WS), air temperature (AT), solar irradiance (SR)) were collected from the *Red Metropolitana de Monitoreo Atmosférico de Quito* (REMMAQ). Finally, HCRD data are available at the *Instituto Nacional de Estadísticas y Censos* (INEC) website.

The next step involved the selection of the most suitable empirical mathematical model with fewer independent variables to avoid statistical multicollinearity and overfitting. Using Bayesian information criteria (BIC), the predictors were reduced to 10. After reducing the number of independent variables, four ML techniques (MLR, MLP, SVR and RFR) were used to compute the LUR models. The model with the lowest root mean square error (RMSE) and highest coefficient of determination (R^2) was chosen for further interpretation. To train the models, the authors used 90% of the records, while the remaining 10% was used for validation and testing. A comparison between a simple-classical MLR approach with more robust non-linear MLP, SVR, and RFR algorithms allowed us to evaluate the robustness of the model and how it fitted the dataset. The power of these more robust algorithms helped construct a more accurate LUR model with fewer predictors than MLR. With the best fitting LUR model, HCRD was represented in yearly maps to assess public respiratory health in Quito. Figure 7.4 shows the general methodology used to generate the LUR models that includes the step about the selection of predictors. This is an approach that could allow researchers working on similar issues establish a workflow that could be relatively easy to implement to obtain a better understanding of respiratory public health in other Latin American cities.

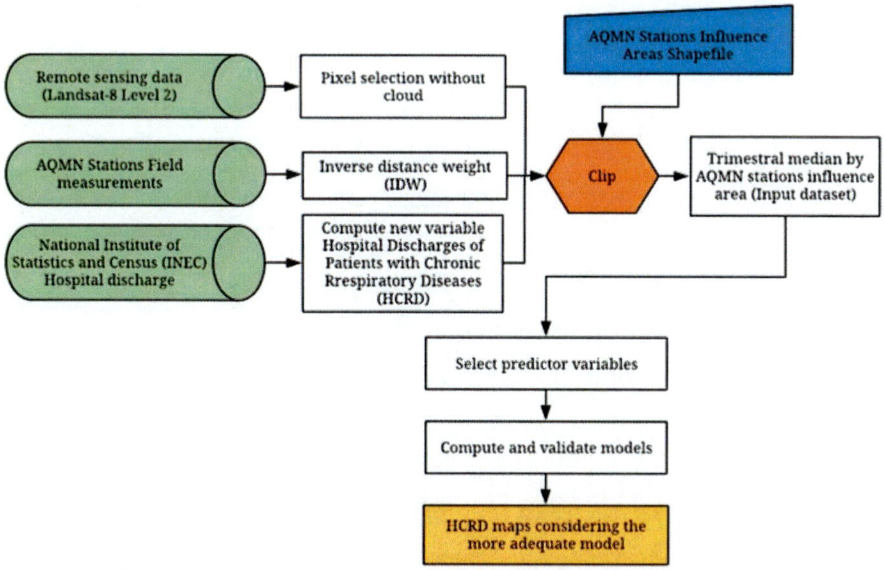

Fig. 7.4 Workflow methodology to build LUR models to predict HCRD in Latin American cities with a low quantity of AQMN stations. (Modified from Alvarez-Mendoza et al. 2020)

4.3 *The Usefulness of LUR Models in Understanding Air Quality and Public Health in Latin American Cities*

The fittest LUR model was constructed via SVR. The SVR approach generated the lowest RMSE (2.0439) and the highest R^2 (0.5066) in comparison to the other three ML algorithms (Table 7.2). However, the selection of the most adequate ML technique to construct the LUR depends on the quality and characteristics of the dataset. For example, lets assume we want to reproduce the methodology to assess public health outcomes in another Latin American city. The results could have suggested the selection of a different model and not necessarily the SVR outcome like in this project.

The most significant variables were air temperature and CO concentrations; both variables were collected from AQMN stations. Predictors derived from RS data like the EVI index appeared in third place (Fig. 7.5). It is reasonable to assume that greenness characteristics as depicted by vegetation indexes such as NDVI or EVI and air temperature will have an impact on atmospheric conditions (Wang et al. 2022), including air quality. The lack of greenspaces could also exacerbate health problems (Mozumder et al. 2013; Alvarez-Mendoza et al. 2019a). Therefore, policy

Table 7.2 LUR model results to predict HCRD (Alvarez-Mendoza et al. 2020)

Model	RMSE test data	R^2 test data
Multiple linear regression (MLR)	3.619	0.327
Multilayer perceptron (MLP)	3.052	0.254
Support vector regression (SVR)	2.043	0.506
Random Forest regression (RFR)	3.046	0.119

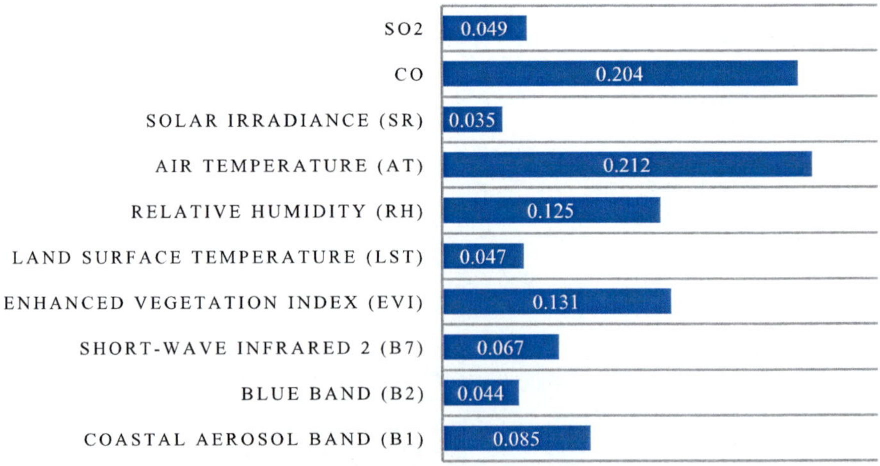

Fig. 7.5 Variable importance in the LUR models in the prediction of HCRD using SVR. (Modified from Alvarez-Mendoza et al. 2020)

to promote the expansion and preservation of green spaces in the city to decrease air pollution concentration could be developed to improve public health outcomes.

This type of approach allowed us to visualize HCRD values at particular locations within the city boundaries (Fig. 7.6). These maps helped identify HCRD variation using classed values on a color ramp. However, their main benefit is to help citizens track air pollution in their neighborhoods and better understand if they have or have not been exposed to respiratory diseases at their homes, which could increase citizen participation to address public health concerns at the community level. Since the maps have a spatial resolution of 30 m (Landsat spatial resolution), it is possible to accurately determine the probability of exposure to poor air quality at local levels using the georeferenced model outputs. The results from this type of spatially explicit LUR models can help local health authorities improve policy and decision-making to identify hot spots of respiratory disease incidence, in addition to the possible causes of health-related problems.

This research opened opportunities to replicate our methodology in other Latin American cities with similar resources and characteristics as Quito (i.e., similar environmental features, scarce air quality monitoring stations, and limited epidemiological data). Future work could include obtaining additional variables from different sources (RS, census institutes, environmental agencies) to be able to accurately model the prevalence of HCRD. For example, the inclusion of higher spatial

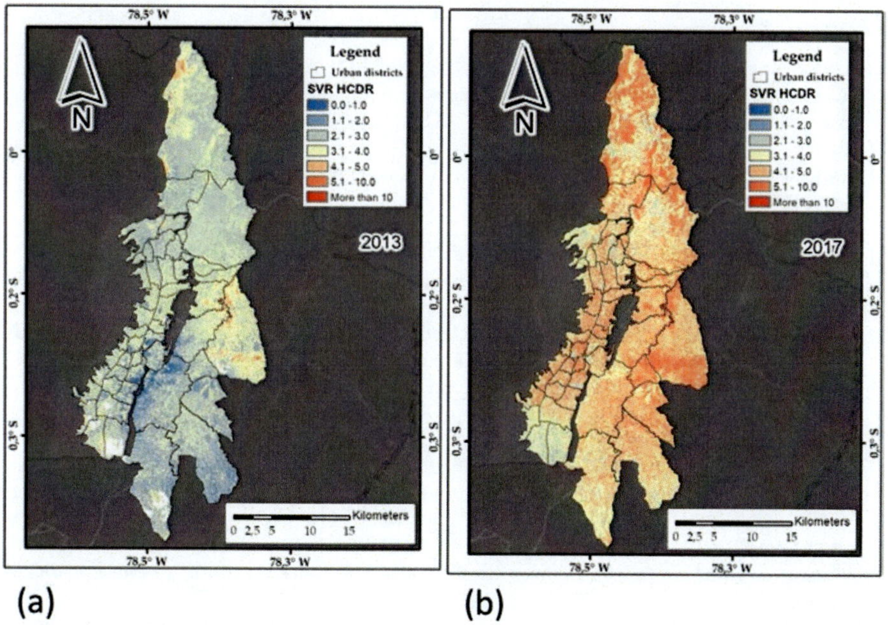

Fig. 7.6 Land use regression (LUR) model of hospital discharge of patients with chronic respiratory diseases (HCDR) in Quito, Ecuador using remote sensing (RS) and machine learning (ML) techniques during (**a**) 2013 and (**b**) 2017. HCDR values were computed through support vector regression (SVR). (Modified from Alvarez-Mendoza et al. 2020)

and temporal resolution imagery obtained from sensors such as Sentinel-2 or EOS developed for Latin American countries such as Amazonia-1 (Brazil), FASat-Charlie (Chile) or Perusat-1 (Peru) (Kopacz et al. 2020) could improve the quality of the LUR models. Other options could include the inclusion of air quality data obtained from air pollution remote sensors like Tropomi Sentinel-5P or AOT MODIS.

New opportunities to monitor air quality conditions via RS will be available in the near future as the Tropospheric Emissions: Monitoring of Pollution (TEMPO) program to monitor air quality in North America (including Mexico) (Zoogman et al. 2017) and the new Sentinel-5 system replacing Sentinel-5P are launched in the coming years. In addition, with the availability of accurate socio-economic and environmental information like poverty or income levels or greenness characterizations, new LUR models could be developed that accurately capture local socio-environmental conditions. Given the complex geographic characteristics of Latin American regions, data collection efforts should be directed toward improving our understanding of the socio-economic and environmental conditions of places, through the generation of accurate and robust spatial datasets. Additionally, with the widespread expansion of ML technologies, such as neural networks, increased accuracy in the spatialization of public health data in Latin American cities could be achieved.

References

Agrawal G, Mohan D, Rahman H (2021) Ambient air pollution in selected small cities in India: observed trends and future challenges. IATSS Res 45:19–30. https://doi.org/10.1016/j.iatssr.2021.03.004

Alcock I, White W, Cherrie M et al (2017) Land cover and air pollution are associated with asthma Hospitalisations: a cross-sectional study. Environ Int 109:29–41. https://doi.org/10.1016/J.ENVINT.2017.08.009

Alvarez-Mendoza C, Teodoro AC, Torres N, Vivanco V (2019a) Assessment of remote sensing data to model PM10 estimation in cities with a low number of air quality stations: a case of study in Quito, Ecuador. Environments 6(7):85. https://doi.org/10.3390/ENVIRONMENTS6070085

Alvarez-Mendoza C, Teodoro AC, Ramirez-Cando L (2019b) Spatial estimation of surface ozone concentrations in Quito Ecuador with remote sensing data, air pollution measurements and meteorological variables. Environ Monit Assess 191(3):155. https://doi.org/10.1007/s10661-019-7286-6

Alvarez-Mendoza C, Teodoro AC, Freitas A, Fonseca J (2020) Spatial estimation of chronic respiratory diseases based on machine learning procedures—an approach using remote sensing data and environmental variables in Quito, Ecuador. Appl Geogr 123:102273. https://doi.org/10.1016/j.apgeog.2020.102273

Alvarez-Mendoza CI, Guzman D, Casas J et al (2022) Predictive modeling of above-ground biomass in Brachiaria Pastures from satellite and UAV imagery using machine learning approaches. Remote Sens 14(22):5870. https://doi.org/10.3390/rs14225870

Andrusaityte S, Grazuleviciene R, Kudzyte J et al (2016) Associations between neighbourhood greenness and asthma in preschool children in Kaunas, Lithuania: a case-control study. BMJ Open 6(4):e010341. https://doi.org/10.1136/bmjopen-2015-010341

Ayres-Sampaio D, Teodoro AC, Sillero N et al (2014) An investigation of the environmental determinants of asthma hospitalizations: an applied spatial approach. Appl Geogr 47:10–19. https://doi.org/10.1016/j.apgeog.2013.11.011

Bravo Alvarez H, Sosa Echeverria R, Sanchez-Alvarez P, Krupa S (2013) Air quality standards for particulate matter (PM) at high altitude cities. Environ Pollut 173:255–256. https://doi.org/10.1016/J.ENVPOL.2012.09.025

Chen Y, Han W, Chen S, Tong L (2014) Estimating ground-level PM2.5 concentration using Landsat 8 in Chengdu, China. Proc SPIE 9259:925917-31. https://doi.org/10.1117/12.2068886

Chuvieco E (2020) Fundamentals of satellite remote sensing: an environmental approach, 3rd edn. CRC Press, Boca Raton. https://books.google.com.ec/books?id=MA3MDwAAQBAJ

Cilluffo G, Ferrante G, Fasola S et al (2018) Associations of greenness, greyness and air pollution exposure with children's health: a cross-sectional study in southern Italy. Environ Health 17(1):86. https://doi.org/10.1186/s12940-018-0430-x

Dlamini SN, Beloconi A, Mabaso S et al (2019) Review of remotely sensed data products for disease mapping and epidemiology. Remote Sens Appl Soc Environ 14:108–118. https://doi.org/10.1016/J.RSASE.2019.02.005

EPA (2017) Inventory of U.S. greenhouse gas emissions and sinks: 1990–2017. https://www.epa.gov/ghgemissions/inventory-us-greenhouse-gas-emissions-and-sinks

Fan J, Li S, Fan C, Bai Z, Yang K (2016) The impact of PM2.5 on asthma emergency department visits: a systematic review and meta-analysis. Environ Sci Pollut Res 23(1):843–850. https://doi.org/10.1007/s11356-015-5321-x

Gupta P, Christopher SA, Wang J et al (2006) Satellite remote sensing of particulate matter and air quality assessment over global cities. Atmos Environ 40(30):5880–5892. https://doi.org/10.1016/j.atmosenv.2006.03.016

Heo S, Bell ML (2019) The influence of green space on the short-term effects of particulate matter on hospitalization in the U.S. for 2000–2013. Environ Res 174:61–68. https://doi.org/10.1016/J.ENVRES.2019.04.019

Hess A, Davis JK, Lind BM, Wimberly MC (2017) Remote-sensing based risk mapping for mosquito-borne diseases: a spatial modelling study. Lancet 389:S10. https://doi.org/10.1016/s0140-6736(17)31122-4

Hiraishi T, Krug T, Tanabe K et al. (2013) Supplement to the 2006 IPCC guidelines for national greenhouse gas inventories: wetlands methodological guidance on lands with wet and drained soils, and constructed wetlands for wastewater treatment task force on national greenhouse gas inventories

IARC Working Group on the Evaluation of Carcinogenic Risks. International Agency for Research on Cancer (2016) Outdoor air pollution, IARC monographs. International Agency for Research on Cancer, World Health Organization. https://books.google.com.ec/books?id=bioFvgAACAAJ

Jethva H, Torres O, Yoshida Y (2019) Accuracy assessment of MODIS land aerosol optical thickness algorithms using AERONET measurements. Atmos Meas Tech Discuss:1–30. https://doi.org/10.5194/amt-2019-77

Just A, De Carli M, Shtein A et al (2018) Correcting measurement error in satellite aerosol optical depth with machine learning for modeling PM2.5 in the northeastern USA. Remote Sens 10(5):803. https://doi.org/10.3390/rs10050803

Kopacz JR, Herschitz R, Roney J (2020) Small satellites an overview and assessment. Acta Astronaut 170:93–105. https://doi.org/10.1016/j.actaastro.2020.01.034

Kumar P, Morawska L, Birmili W et al (2014) Ultrafine particles in cities. Environ Int 66:1–10. https://doi.org/10.1016/j.envint.2014.01.013

Kuo CY, Chan CK, Wu CY et al (2019) The short-term effects of ambient air pollutants on childhood asthma hospitalization in Taiwan: a national study. Int J Environ Res Public Health 16(2). https://doi.org/10.3390/ijerph16020203

Li Y, Wang J, Chen C et al (2016) Estimating PM2.5 in the Beijing-Tianjin-Hebei region using Modis Aod products from 2014 to 2015. Int Arch Photogramm Remote Sens Spatial Inf Sci ISPRS Arch 41(July):721–727. https://doi.org/10.5194/isprsarchives-XLI-B2-721-2016

Liu Y, Franklin M, Kahn R, Koutrakis P (2007) Using aerosol optical thickness to predict ground-level PM2.5 concentrations in the St. Louis area: a comparison between MISR and MODIS. Remote Sens Environ 107(1–2):33–44. https://doi.org/10.1016/j.rse.2006.05.022

Lorentz C, Chiaravalloti-Neto F, de Oliveira Lage M et al (2020) Remote sensing for risk mapping of aedes aegypti infestations: is this a practical task? Acta Trop 205:105398. https://doi.org/10.1016/j.actatropica.2020.105398

Meng X, Li C, Cai J et al (2015) A land use regression model for estimating the NO2 concentration in Shanghai, China. Environ Res 137(2):308–315. https://doi.org/10.1016/j.envres.2015.01.003

Moreno R, Ojeda N, Azócar J et al (2020) Application of NDVI for identify potentiality of the urban forest for the design of a green corridors system in intermediary cities of Latin America: case study, Temuco, Chile. Urban For Urban Green 55:126821. https://doi.org/10.1016/j.ufug.2020.126821

Mozumder C, Venkata Reddy K, Pratap D (2013) Air pollution modeling from remotely sensed data using regression techniques. J Indian Soc Remote Sens 41(2):269–277. https://doi.org/10.1007/S12524-012-0235-2/FIGURES/6

NASA's Jet Propulsion Laboratory (2021) Effects|facts – climate change: vital signs of the planet. https://climate.nasa.gov/effects/

Omrani H, Omrani B, Parmentier B, Helbich M (2020) Spatio-temporal data on the air pollutant nitrogen dioxide derived from sentinel satellite for France. Data Brief 28:105089. https://doi.org/10.1016/j.dib.2019.105089

Pérez Albiñana A, Martin D, Erdmann M et al (2017) Sentinel-5: the new generation European operational atmospheric chemistry mission in polar orbit. SPIE Intl Soc Opt Eng 23. https://doi.org/10.1117/12.2268875

Rivera-González LO, Zhang Z, Sánchez B et al (2015) An assessment of air pollutant exposure methods in Mexico City, Mexico. J Air Waste Manag Assoc 65(5):581–591. https://doi.org/10.1080/10962247.2015.1020974

Rojas JP, Urdanivia F, Garay RA et al (2021) Effects of COVID-19 pandemic control measures on air pollution in Lima metropolitan area, Peru in South America. Air Qual Atmos Health 14(6):925–933. https://doi.org/10.1007/s11869-021-00990-3

Sannigrahi S, Kumar P, Molter A et al (2021) Examining the status of improved air quality in world cities due to COVID-19 led temporary reduction in anthropogenic emissions. Environ Res 196:110927. https://doi.org/10.1016/j.envres.2021.110927

Soutter A, Mõttus R (2020) 'Global warming' versus 'climate change': a replication on the association between political self-identification, question wording, and environmental beliefs. J Environ Psychol 69:101413. https://doi.org/10.1016/j.jenvp.2020.101413

Toro A, Catalán F, Urdanivia FR et al (2021) Air pollution and COVID-19 lockdown in a large South American city: Santiago Metropolitan Area, Chile. Urban Clim 36:100803. https://doi.org/10.1016/j.uclim.2021.100803

USGS (2016) Landsat 8 (L8) data users handbook, vol 8. USGS, Reston

Veefkind JP, Aben I, McMullan K et al (2012) TROPOMI on the ESA sentinel-5 precursor: a GMES mission for global observations of the atmospheric composition for climate, air quality and ozone layer applications. Remote Sens Environ 120:70–83. https://doi.org/10.1016/j.rse.2011.09.027

Viana J, Vasco Santos J, Neiva RM et al (2017) Remote sensing in human health: a 10-year bibliometric analysis. Remote Sens 9(12):1225. https://doi.org/10.3390/rs9121225

Wang Y, Cai G, Yang L, Zhang N, Du M (2022) Monitoring of urban ecological environment including air quality using satellite imagery. PlosOne 17(8):e0266759. https://doi.org/10.1371/JOURNAL.PONE.0266759

WHO (2019) Chronic respiratory diseases (CRDs). https://www.who.int/respiratory/en/

WHO (2022) Vector-Borne diseases and climate change. https://tdr.who.int/activities/vector-borne-diseases-and-climate-change

Wimberly MC, de Beurs KM, Loboda TV, Pan WK (2021) Satellite observations and malaria: new opportunities for research and applications. Trends Parasitol 37(6):525–537. https://doi.org/10.1016/j.pt.2021.03.003

Xu X, Yang X, Zhu B et al (2020) Characteristics of MERRA-2 black carbon variation in East China during 2000–2016. Atmos Environ 222:117140. https://doi.org/10.1016/j.atmosenv.2019.117140

Zalakeviciute R, Vasquez R, Bayas D et al (2020) Drastic improvements in air quality in Ecuador during the COVID-19 outbreak. Aerosol Air Qual Res 20(8):1783–1792. https://doi.org/10.4209/aaqr.2020.05.0254

Zhai L, Li S, Zou B et al (2018) An improved geographically weighted regression model for PM2.5concentration estimation in large areas. Atmos Environ 181:145–154. https://doi.org/10.1016/j.atmosenv.2018.03.017

Zoogman P, Liu X, Suleiman RM et al (2017) Tropospheric emissions: monitoring of pollution (TEMPO). J Quant Spectrosc Radiat Transf 186:17–39. https://doi.org/10.1016/j.jqsrt.2016.05.008

Chapter 8
Human-Environmental Interactions and Their Impacts on Temperate Forests in the Exploradores Valley in Western Patagonia

Alejandro Salazar-Burrows, Jorge Olea-Peñaloza, Fernando Alfaro, Jorge Qüense, Didier Galop, and Francisca Flores-Galaz

Abstract The mid-latitude Patagonian ecosystems have great functional value: their high levels of endemism and biodiversity have the potential to mitigate global climate change impacts. Since the end of the nineteenth century, human occupation processes in western Patagonia have integrated, colonized, and exploited southern environments, anthropizing the Patagonian wilderness in a short period of time. The objective of this chapter is to analyze the processes underlying the transformation of land-cover in temperate rainforests in Patagonia, through the interpretation and quantification of environmental changes along with the identification of their main anthropic drivers. We use remote sensing and geographic information systems to perform a multi-temporal evaluation of the transformations of the Exploradores Valley, through the analysis of Landsat TM and OLI images from 1986, 2001, and 2018. The quantitative analysis was complemented by qualitative analysis of interviews and comparisons of historical photographs that allow the evaluation of forest change at different stages of the occupation of the valley. Although the rate of

A. Salazar-Burrows (✉)
Instituto de Geografía, Facultad de Historia, Geografía y Ciencia Política, Pontificia Universidad Católica de Chile, Santiago, Chile

Estación Patagonia de Investigaciones Interdisciplinarias UC, Pontificia Universidad Católica de Chile, Santiago, Chile

Observatoire homme-milieux International Patagonia-Bahia Exploradores (OHM-i), Labex DRIIHM (Programme Investissements d'avenir: ANR-11-LABX-0010), INEE-CNRS, Paris, France

Centro de Investigación en Ecosistemas de la Patagonia (CIEP), Proyecto PATSER ANID RF0F0002, Coyhaique, Chile
e-mail: asalazab@uc.cl

S. López (ed.), *Socio-Environmental Research in Latin America*, The Latin American Studies Book Series, https://doi.org/10.1007/978-3-031-22680-9_8

land cover change between 1986 and 2018 is high, the processes of intensification of land-use changes have a much longer trajectory. Our results suggest that the Exploradores Valley is a dynamic system where the impacts of human activities are imprinted in the spatial changes of vegetation cover.

Keywords Environmental geography · Environmental history · Human-environmental interactions · Remote sensing · Socio-ecological systems

---

J. Olea-Peñaloza
Estación Patagonia de Investigaciones Interdisciplinarias UC, Pontificia Universidad Católica de Chile, Santiago, Chile

Observatoire homme-milieux International Patagonia-Bahia Exploradores (OHM-i), Labex DRIIHM (Programme Investissements d'avenir: ANR-11-LABX-0010), INEE-CNRS, Paris, France

Departamento de Ciencias Ambientales, Facultad de Recursos Naturales, Universidad Católica de Temuco, Temuco, Chile

F. Alfaro
Estación Patagonia de Investigaciones Interdisciplinarias UC, Pontificia Universidad Católica de Chile, Santiago, Chile

GEMA Center for Genomics, Ecology & Environment, Universidad Mayor, Huechuraba, Chile

Instituto de Ecología & Biodiversidad (IEB), Santiago, Chile

J. Qüense · F. Flores-Galaz
Instituto de Geografía, Facultad de Historia, Geografía y Ciencia Política, Pontificia Universidad Católica de Chile, Santiago, Chile

Estación Patagonia de Investigaciones Interdisciplinarias UC, Pontificia Universidad Católica de Chile, Santiago, Chile

Observatoire homme-milieux International Patagonia-Bahia Exploradores (OHM-i), Labex DRIIHM (Programme Investissements d'avenir: ANR-11-LABX-0010), INEE-CNRS, Paris, France

D. Galop
Observatoire homme-milieux International Patagonia-Bahia Exploradores (OHM-i), Labex DRIIHM (Programme Investissements d'avenir: ANR-11-LABX-0010), INEE-CNRS, Paris, France

GEODE UMR 5602 CNRS, Université de Toulouse, Toulouse, France

Labex DRIIHM, (Programme Investissements d'avenir": ANR-11-LABX-0010). OHM Pyrénées, CNRS/INEE, Toulouse, France

1 Introduction

1.1 The Temperate Rainforests of Western Patagonia

Ecosystems located in mid-latitudes, such as in the south of Western Chilean Patagonia, are of global importance for biodiversity preservation and conservation (Inostroza et al. 2016; Dinerstein et al. 2020). These ecosystems, particularly temperate rainforests and their soils, support complex and dynamic wildlife systems with high endemism (Echeverría et al. 2007; Wilson et al. 2005; Armesto et al. 1998; Armesto et al. 1996). Likewise, the transitional areas between forests and Patagonian steppes are among the most dynamic ecosystems on the planet and are also highly vulnerable to climate change and human land use (Iglesias and Whitlock 2014). In this context, the following chapter presents the transformations of the Exploradores valley in recent decades, considering human action and its impact on the territory, through a mixed-methods approach using geographic information systems (GIS), remote sensing (RS), and qualitative research strategies.

Forest and soil systems act as important carbon sinks that contribute to partially mitigate and counteract the impacts of global climate change (Hoyos-Santillan et al. 2019; Leifeld and Menichetti 2018). Nevertheless, in the context of global warming, the increase of human activities has transformed forest ecosystems from carbon sinks into sources of carbon emissions (Leifeld and Menichetti 2018; Kurz et al. 2008). The forest ecosystems are relevant at both regional and global levels, as Western Patagonia has the most important temperate forest cover in South America and accounts for more than half of the total forested area in the Southern Hemisphere (Bizama et al. 2011; Donoso 1993), fulfilling hemispheric ecological and climatic functions (Dinerstein et al. 2020) that are still not fully understood.

Similar to global terrestrial ecosystems, in the last century, forested lands have been exposed to strong anthropic pressures, undergoing accelerated transformation processes that involve the loss of tree cover, degradation of soils, and habitat fragmentation, which impact local biodiversity (Bizama et al. 2011; Echeverría et al. 2006) and alter ecosystem functions (Pellegrini et al. 2021).

Climatic factors, closely interrelated with fire regimes, have been among the main drivers of change in southern forests in past and current history. Fire regimes and dynamics have been the major subject of several paleo-environmental studies in Western Patagonia (Holz and Veblen 2011; Moreno et al. 2010; Whitlock et al. 2001). Although it has been suggested that the influence of human actions on the occurrence of fires in prehistoric periods is difficult to separate from climatic influences, the appropriate climatic conditions for fires are a prerequisite for their spread and intensity (Huber et al. 2004). In this chapter, we evaluate the relative importance of fires on Patagonian forest's transformations, emphasizing the action of fire as a widespread tool used by settlers in the area to transform the landscape when they arrived in the valley, which translated into an accelerated transformation of Patagonian forests, and whose traces can still be seen today.

Whereas rainfall levels show a large decline, a progressive increase in temperatures in Patagonia since the nineteenth century has been detected, with a significant change in the twentieth century (Neukom et al. 2011, Neukom et al. 2010, Trenberth et al. 2007). Accordingly, recent research has determined the impact of increased temperatures and enhanced dry periods on the frequency and intensity of fires of human origin during the twentieth century (Veblen et al. 2011). For temperate rainforest biomes, it is anticipated that these climatic conditions, which are conducive to fires, will continue during the twenty-first century (Holz and Veblen 2011). Indeed, temperature increases of up to ~5 °C have already been observed in Western Patagonian forests in the last two decades (Olivares-Contreras et al. 2019).

In the recent history of Western Patagonia, the greatest changes in temperate rainforests have been detected between the end of the nineteenth century and throughout the twentieth century (Holz et al. 2012a). Since the early twentieth century, forest cover declined due to increased frequency and intensity of fires from human activities (Holz et al. 2012a), deforestation, and the expansion of the agricultural-livestock frontier (Jaksic and Fariña 2015; Bizama et al. 2011). All these factors are associated with the late colonization of Western Patagonia in the mid-nineteenth century (Urbina 2013; Martinic 2005). The geographic location and the severity of climatic conditions make mid-latitude transitional forest ecosystems such as those found between ice fields, Patagonian forests, and steppes very fragile (Iglesias and Whitlock 2014). Indeed, these ecosystems are expected to experience changes in their environmental conditions which will affect human-environmental interactions in the coming decades.

This study focuses on the basin formed by the Exploradores and Tranquilo rivers valleys (hereinafter the Exploradores Valley or study area). The main valley corresponds to one of the largest basins in Western Patagonia (Fig. 8.1). This periglacial area is unique due to its location in the northern edge of the "Northern Patagonian Ice Field", which is the second-largest ice sheet in the Chilean Patagonia, which makes it highly sensitive to increases in temperature.

Due to its relative geographic isolation and the intensity of recent anthropogenic interventions since the early twentieth century, the study area offers unique conditions for monitoring its socio-environmental systems and contemporary land cover transformations. The complex relationships are non-linear and capture a wide set of human-environmental interactions (Castree et al. 2016). We define this type of interaction as the set of relationships that characterize the influence of environments on the actions and functioning of human societies, and the effect of human activities and decisions on particular territories (Chenorkian and Robert 2014; Ostrom 2009).

The purpose of this chapter is to advance the understanding of land cover change in Western Patagonian temperate rainforests in the last four decades and the interplay between socio-environmental factors that not only contributed to these changes but also make up part of the region's environmental history. The emphasis is on the colonization processes that started in the late 20th and early 21st centuries.

We relied on RS and GIS to quantify changes in forest cover. Specifically, we analyzed Landsat TM and OLI images from 1986, 2001, and 2018 to quantify changes over time. With the support of GIS, we analyzed the spatial distribution and spatial patterns of changes in forest cover. At the same time, we compiled graphic and historical records of the area to complement and deepen the interpretation of the

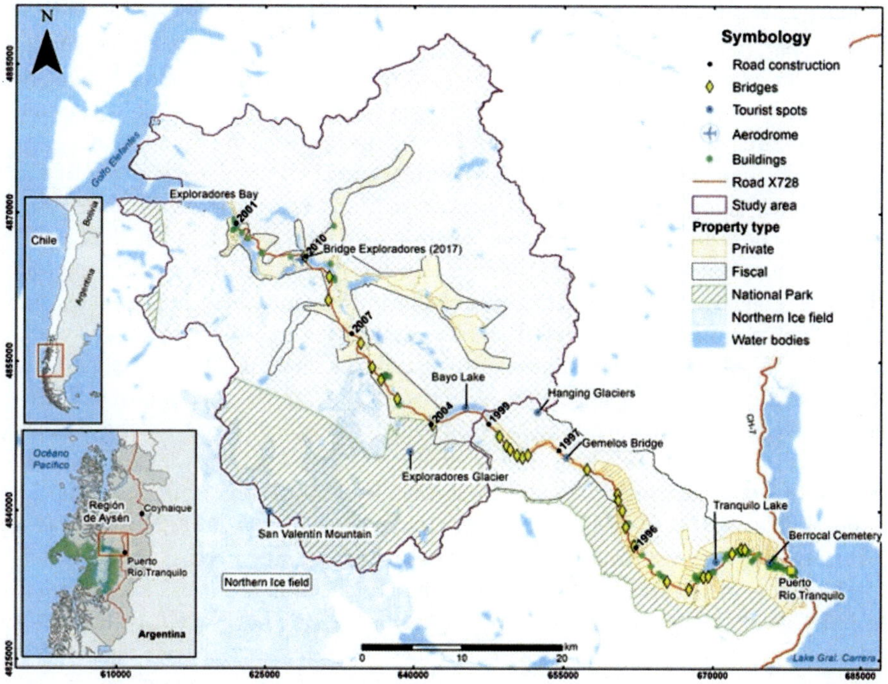

Fig. 8.1 Study area. Anthropization process in the Exploradores and Tranquilo valleys. The distribution of the main anthropic elements is shown (buildings, bridges, tourist spots, roads, and years of progress) (Cartography and design by the authors)

RS and GIS results. This goal was achieved through the use of a variety of sources, combining historical environmental analysis of the various stages of the valley's occupation, fieldwork, interviews with key informants that focused on present and past conditions, and the analysis of secondary sources and photographic records.

1.2 Human-Environmental Interactions in Patagonian Socio-environmental Systems

The traditional separation between those who focus on humans and those on nature, presents a significant barrier to deepening environmental knowledge in times of accelerated change. One way of understanding the synergies between global change and human activities is through the lens of a socio-ecological systems (SESs) framework (Maass et al. 2016; Lagadeuc and Chenorkian 2009). This integrating concept refers to complex and adaptive systems where multiple components (e.g., climatic, socio-political, cultural, ecological, economic) interact (Berkes and Folke 1998) with social and biophysical agents acting in the same environment at multiple spatial and temporal scales (Ostrom 2009; Janssen and Ostrom 2006).

The theoretical framework that accompanies SESs is not new since it answers to the need of generating a deeper understanding of the relationships and interactions between humans and nature at global-to-local scales (Farhad 2012) and a holistic or geographic perspective (Claval 2011). The conceptualization of SESs as integrated systems linking societies and nature aims to redefine ecosystems by explicitly considering all their components, thus integrating humans as active elements of the system under study (Lagadeuc and Chenorkian 2009; Liu et al. 2007a). This contrasts with studies that focus on human-environment interactions, in which the ecosystem conceptualization positions humans outside the natural system, limiting the ability to consider the complexity of existing interactions between humans and nature (Folke et al. 2007; Liu et al. 2007b; Redman et al. 2004). In this way, the dichotomy between nature and humans constitutes a challenge to fully appreciate SESs, making it necessary to deepen the understanding of the interactions and feedback between the components that cause vulnerability and those that strengthen resilience (Ostrom 2009; Folke 2006). In this context, it is relevant to carefully consider the particularities of each territory and its SESs (Bourgeron et al. 2009).

From this perspective, global environmental change can be understood as a transformative process, resulting from the interaction between and synergy of multiple components such as climate change, biological invasions, loss of biological diversity, and changes in land cover (Vitousek et al. 1997), as well as the socio-political, economic, and cultural elements that affect a certain territory. Thus, transformations of environmental conditions at a global scale could have differentiated impacts at regional and local scales (Rockström et al. 2009), which are determined by spatial and temporal variations in the relative contributions of the system's components (Mann et al. 1998). Thus, changes at a global scale (e.g., the generalized increase in temperature in the atmosphere) may be perceived geo-environmentally and historically at local levels with different intensities.

In summary, this chapter seeks to analyze the transformations that occurred in the Exploradores Valley using a mixed-methods approach combining GIS, RS, and qualitative methods. We specifically: (1) investigate the magnitude of changes in land cover, (2) identify the main anthropogenic drivers of landscape change and its spatial footprint; and (3) provide a methodological reflection on the value of combining the precision of GIS and RS with the depth and dynamism of qualitative methods, particularly regarding the use of interviews and historical photographs.

1.3 *Western Patagonia: Wild Frontiers of Socio-environmental Changes*

Archaeological records show a long-standing prehistoric occupation of Western Chilean Patagonia (Martinic 2005). These territories, however, present particular geographic characteristics in terms of climate, topography, as well as historical conditions, which hindered its occupation and exploration. The recent history of this region intersects with the search and exploitation of natural resources, as well as the exploration of natural spaces (wild or pristine). Thus, after the prehistoric (and pre-Hispanic) occupation of the region in the seventeenth century, there were processes of discovery, exploration, and exchanges with Europeans that occurred in Patagonia, that initiated its exploration, colonization, and exploitation during the 19th and 20th centuries (Urbina 2013; Araneda et al. 2007).

In Western Patagonia (and even later in the study area) there are important milestone-events that marked the transformation of its geographic space. Among them, the active occupation policies that promoted national and European settlers to "make a homeland" since the end of the nineteenth century, focusing on the extensive use of land for livestock (Urbina 2013), and the delivery of concessions to different livestock farming companies. These actions led to forest fires that affected Patagonia in the early and late nineteenth century, and beginning of the twentieth century. These territories witnessed late colonization initiatives due to their marginality in the territorial consolidation of the Chilean State.

2 Study Area in Western Patagonia

The study area includes environments in an elevation gradient that ranges between the edge of the Pacific Ocean and the General Carrera Lake sector in the so-called Central Patagonia, through the Exploradores and Tranquilo river valleys (Fig. 8.1). The area has a unique diversity of southern temperate forests, with evergreen and deciduous forests or Valdivian Forest according to the classifications of Luebert and Pliscoff (2005, 2006), and steppe, depending on the altitude and proximity to the ocean (Holz et al. 2012b).

A significant geographic characteristic is the highly isolated condition of an important part of the study area, which explains its paused colonization process. Thus, the area presents different levels of anthropization, probably due to a double and diachronic process of colonization and interactions within its socio-environmental system.

The first colonization phase began in the early twentieth century in the north-western sector of General Carrera Lake. In this period, the arrival of pioneers and settlers resulted in small settlements that incorporated sawmills and mining as well as productive activities linked to livestock raising and forestry. In the mid-1920s, a westward expansion from the settlement of Puerto Río Tranquilo led to the occupation and intervention of the Exploradores Valley, where major forest fires occurred at the time (year 1936). In the mid-1920s, in the same period, the State verified the potential productive value of Western Patagonia by hiring the explorer Augusto Grosse, who searched for a connecting pathway between General Carrera Lake and the Pacific Ocean in 1936 and completed the endeavor during the Exploradores Valley expedition in 1944 (Salazar et al. 2011; Grosse 1955).

The second colonization wave began in the 1950s, advancing from the Pacific Ocean to Exploradores Bay in the east, following the expeditions of the explorers

Grosse (1955) and Steffen (1902). After intense initial deforestation and successive livestock activity, together with the impact of extensive fires during the 1960s, 1970's and 1980's, the westernmost sector of the study area witnessed a 30-year pause in anthropogenic activity, which remained low in the last decades. These two colonization events of the Exploradores valley stayed disconnected until the beginning of the 2010s when the road connecting the valley with the rest of the region was completed (Fig. 8.1).

In the last 25 years, isolation conditions have changed due to the construction of a new road between 1996 and 2010 and a bridge in 2017, which finally connected Exploradores Bay with the town of Puerto Río Tranquilo. This development allowed a subsequent advance and intensification of new economic activities and territorial dynamics (i.e., their deployments, imprinting, patterns of dominance and occupation in a space or over the territory). Economic intensification was largely due to the change in the regional productive matrix, gradually promoted over the last three decades. This change in the production model is associated with nature tourism, global salmon farming activities, and energy initiatives, replacing the old distribution model between a valley dedicated to agriculture and cattle raising and an area close to the sea where cypress wood was exploited.

From a historical-geographical perspective, three major political-economic transitions territorially mark the socio-environmental system of the Exploradores valley in connection with the country's social and political history: (1) the Agrarian Reform (1960–1973) became a process of land redistribution that brought an end to the tenure system that had been in place since colonial times and granted access to land in Patagonia that was previously state owned. (2) The military dictatorship (1973–1989) involved a return to the previous process, which resulted in a new political economy based on a series of structural reforms that highlighted the privatization of water rights and the installation of a market-based economy that focused on foreign trade. After the return to democracy (1990) with the end of the military dictatorship, economic reforms that promoted the integration processes of the Aysén region into the global and national markets continued, with aquaculture and tourism as the main economic activities. In the first stage, the State played a crucial role in redistributing land to the inhabitants in frontier regions by providing "títulos de dominio" (titles of ownership)[1] (Ovalle 2011; Araya and Boris 2017), whereas in the second transition after the military coup, capital investment incentives for extractive production (i.e., forestry, aquaculture, and mining), and the construction of Route CH-7 (or Carretera Austral) were promoted. (3) Finally, with the consolidation of neoliberalism in Chilean democracy, the ethical revaluation of nature and as a niche market, and increased concerns about climate change, triggered the development of the "conservationist" tourism industry in Patagonia, a strategic economic activity in the region. Additionally, at a national and global scale, Patagonia's importance as a unique territory was recognized and consolidated, promoting

[1] Property or land tenure titles which were granted by the Chilean government, which granted ownership to settlers who effectively occupied the land.

research as if it were a laboratory for studies of global environmental change among different disciplines. Patagonia also became a renowned location for various environmental change research initiatives (Olea-Peñaloza et al. 2021; Moreira-Muñoz et al. 2014).

3 Methods and Data

3.1 Quantitative Remote Sensing and Geographic Information Analysis

Studies that focus on the analysis of the human-environmental interactions and impacts of changes in forest cover in the Chilean southern territories are relatively recent. Some of this literature analyzes the changes and evolution of temperate forests and geo-historically located spaces from different and interesting disciplinary perspectives, but that concentrate on other areas of Patagonia (Inostroza et al. 2016; Bizama et al. 2011; Solari et al. 2011; Torres-Gómez et al. 2009). Some of this research is based on the application of indexes for estimating human intervention levels for evaluating changes in landscape structure (Torres-Gómez et al. 2009), the degree, scope, and spatial distribution of anthropogenic alterations (Inostroza et al. 2016), or loss and fragmentation of native forest due to forest fires (Bizama et al. 2011). Most of this work uses an interdisciplinary approach to analyze, for example, distributions and development processes of human settlements and their economic activities based on various forms of space appropriation (Solari et al. 2011).

To determine the spatial changes in forest cover in the Exploradores valley, we used images from the Landsat platform to study two time periods. We relied on Landsat 5 TM and Landsat 8 OLI to map forest cover changes between 1986–2001 and 2001–2018. Since the study area is covered by clouds most of the year, we created a clear image mosaic with cloud-free sections of some scenes for each year. We processed images with a cloudiness index of less than 20% within the Google Earth Engine (GEE) platform.

Furthermore, we derived the following variables to support the supervised classification and improve the quality of classifications, especially in mountainous areas: Normalized Difference Vegetation Index (NDVI) (Tucker 1979); Normalized Difference Water Index (NDWI) (Gao 1996), slope, slope orientation, and elevation. The NDVI and NDWI indices were extracted from Landsat images and the topographic variables from a 30 m SRTM Digital Elevation Model (DEM). The results of this stage were added to the IDRISI Selva platform where, with the help of the Land Change Modeler module, we carried out a multi-temporal inter-comparison change analysis: 1986–2001 and 2001–2018.

We applied a supervised "random forest" classification on the resulting mosaics for each year, using control points to identify eight land use and land cover (LULC) classes based on the University of Maryland's Level 1 classification (Anderson 1976) and our knowledge of the area: (1) Forest, (2) Shrubland, (3) Pastures, (4)

Water, (5) Snow, (6) Ice, (7) Bare soil, and (8) Shadow. To interpret land cover changes in the study area, we selected four areas of interest according to the physical conditions of the different zones and the quality of the classifications, thereby reducing the amount of anomalous data (e.g., shadows, cloudiness, among others).

In addition to the RS and GIS analyses, we compiled other sources of qualitative information to critically assess the information derived from satellite data (Gregory and Geddes 2014). The combination of methods provides opportunities to develop more solid studies than single-method ones (Pole 2009), especially in data-scarce regions. Within this context, a mixed-methods approach may take different forms depending on the research team and the questions posed (Teddlie and Tashakkori 2003). In this study, we used four elements:

(a) First, to approximate the forest extent that existed in the early 1900s, we applied the methodology proposed by Bizama et al. (2011) to reclassify pastures and shrubland, agricultural areas, and human settlements from the National Forestry Corporation (CONAF) 2011 to forest cover. We used the AlosPalsar Digital Elevation Model with a 12.5 m resolution to create 250 m contour lines and establish the 1250 m maximum elevation limit denoting forest land boundaries. We used the oldest available image -year 1986- as the baseline to characterize changes trends within the Valley.
(b) Second, to validate the supervised classifications and rectify the results of the GEE procedure, we collected 20 random ground control points across the entire transect of the Exploradores valley. These data allowed us to characterize forest types and human occupation alterations at a larger scale through horizontal photographs taken in the field (Teixeira 2018) in 2020.
(c) Third, we relied on the interpretation of old photographs taken during Augusto Grosse's exploration (1955) in 1943 (from Puerto Río Tranquilo to Exploradores Bay) to perform a historical comparison of natural elements at the same scale and characterize historical forest cover conditions in the valley. The rereading of historical documents from an environmental perspective constitutes one of the main sources of information in this study. The tracking of historical information, can contribute to the reconstruction of the relationship between society and the natural environment (Solari et al. 2011). This procedure, together with local knowledge, allowed us to establish an initial forest cover baseline for the beginning of the twentieth century.
(d) Finally, interviews with key actors (Kvale 2012) allowed us to incorporate the accounts of the first settlers of the valley into the analysis. For this purpose, between 2018 and 2020, we interviewed five people who still live in the area. We asked them about their life histories and trajectories, how they had arrived in the valley, the main activities they had carried out, the fires, and what transformations they considered the most relevant, both past and present. This information facilitated a better understanding of the evolution of the ways of life and environmental changes over time by identifying relevant milestones, which helped integrate the different methodologies. This allowed the downscaling of the RS and GIS results and to perform the analysis chronologically.

4 Results and Discussion

4.1 Socio-environmental Transformations of the Valley

When characterizing the Patagonian territory, two key aspects must be considered: (1) the region is a predominantly forested space/ecosystem with steep slopes (Fig. 8.2); and (2) the most visible landscape transformations today occurred since the end of the nineteenth and during the twentieth century and these landscape change processes are characterized by a gradual modification of highly "pristine" environments to increasingly anthropized spaces. Since the first settlers' arrival, the area has been exposed to severe disturbances, mainly driven by extractive activities and subsistence needs. The initial need for pastures for grazing and livestock raising was met through the loss of forest cover caused by logging and fires, some common drivers of change in new colonization areas.

To obtain a comparative and longer-term visualization of the forest cover of the area of interest and partially resolve the technical and information limitations, we developed a potential estimate of the existing forest cover at the beginning of the twentieth century, based on the classification of pastures and shrublands areas provided by CONAF (2011) (Fig. 8.3).

As it was expected, the transformation of the valley's landscapes presents relevant nuances to consider. As Fig. 8.3 shows, the areas with the greatest loss of forest cover are located near the small town of Puerto Tranquilo. These reported losses have been produced mainly by the effects of large fires and human occupation activities. These areas are located especially in the lower sections of the slopes of the mountains and the river plains, which are the spaces dedicated to the development of grasslands for cattle and wood exploitation. In contrast, in the western end of the valley (Exploradores´ Bay) considerably minor losses of forest cover are observed.

Although this information and data suggests a decrease or loss of only 20% in forest cover in the period between 1900–2011 in the valley; fieldwork observations indicate that the surfaces with losses and regeneration or ecological succession are larger than the areas reported by RS. The burned areas documented during fieldwork, which presumably occurred prior to the last recorded fires (at the beginning and during the twentieth century), are identified as forest cover by the RS because they had a longer time to regenerate and a lower intensity of human exploitation in the last five decades. This means that the data collected during fieldwork provided critical and complementary information that RS alone was not able to show. Thus, the ecological situation is more complex at a finer temporal and spatial scales than what could interpreted from RS. Therefore, we identified four main areas where forest cover transformations were the greatest and where we applied complementary methods to delve deeper into the repercussions and evolution of change. These points are (A) the area surrounding Lake Tranquilo, which corresponds to the sector with the highest accumulated human intervention; (B) the beginning of the Exploradores Valley, where there is a critical transition between the presence/absence of forest; (C) Bayo Lake, which is one of the key areas in the transect, since it is not only the place where the Exploradores Glacier

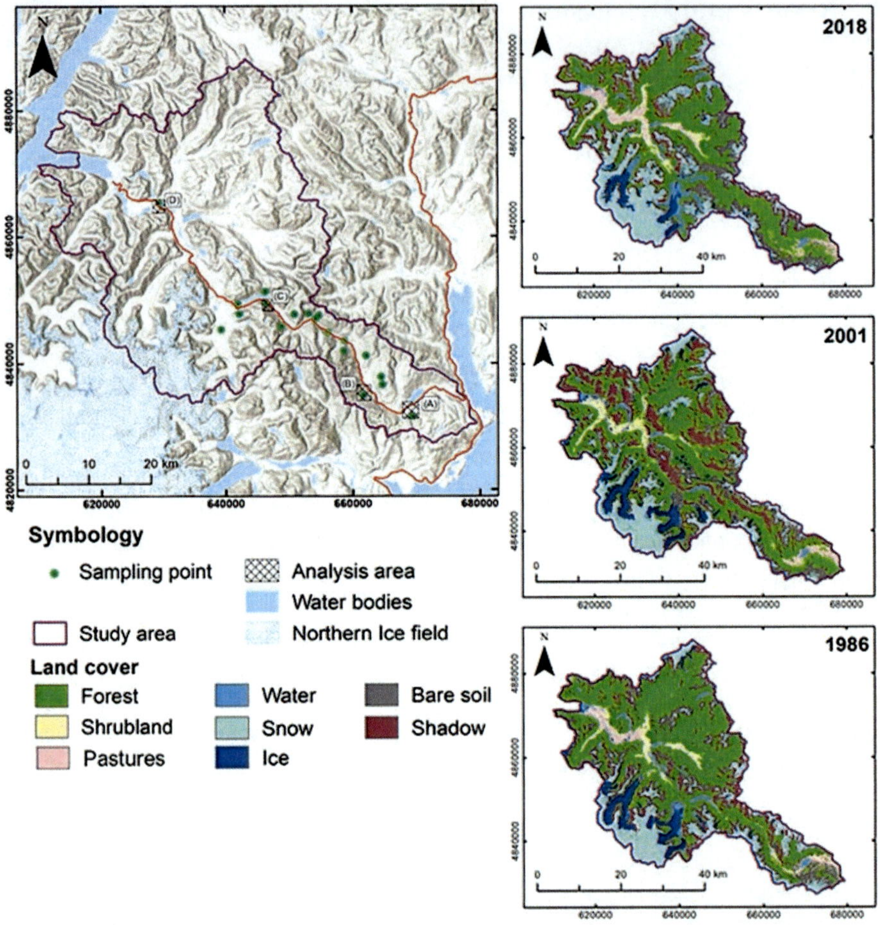

Fig. 8.2 Random forest supervised classification for the years 1986, 2001, and 2018 using Google Earth Engine, and the location of analysis areas in the Exploradores Valley. We validated the classification results through a confusion matrix and the calculation of overall, producer, and user accuracies, in addition to the Kappa (K) concordance index (Congalton 2004), with a target overall accuracy of 0.9 in the three dates analyzed (Cartography and design by the authors)

drains, but it is also a transitional area between the mountainous and coastal zones of the valley; and (D) the Teresa sector and Exploradores Bridge, which was built in 2017 (Fig. 8.3).

When analyzing LULC transformations in the four study areas for both periods (1986–2001) and (2001–2018), we observed that the overall Kappa index improved for areas (A) and (B) between 2001 and 2018. This is because the disturbances in the first period (or earlier) were more evident than in the second period, with greater spectral differences between land cover types (Table 8.1). On the other hand, the

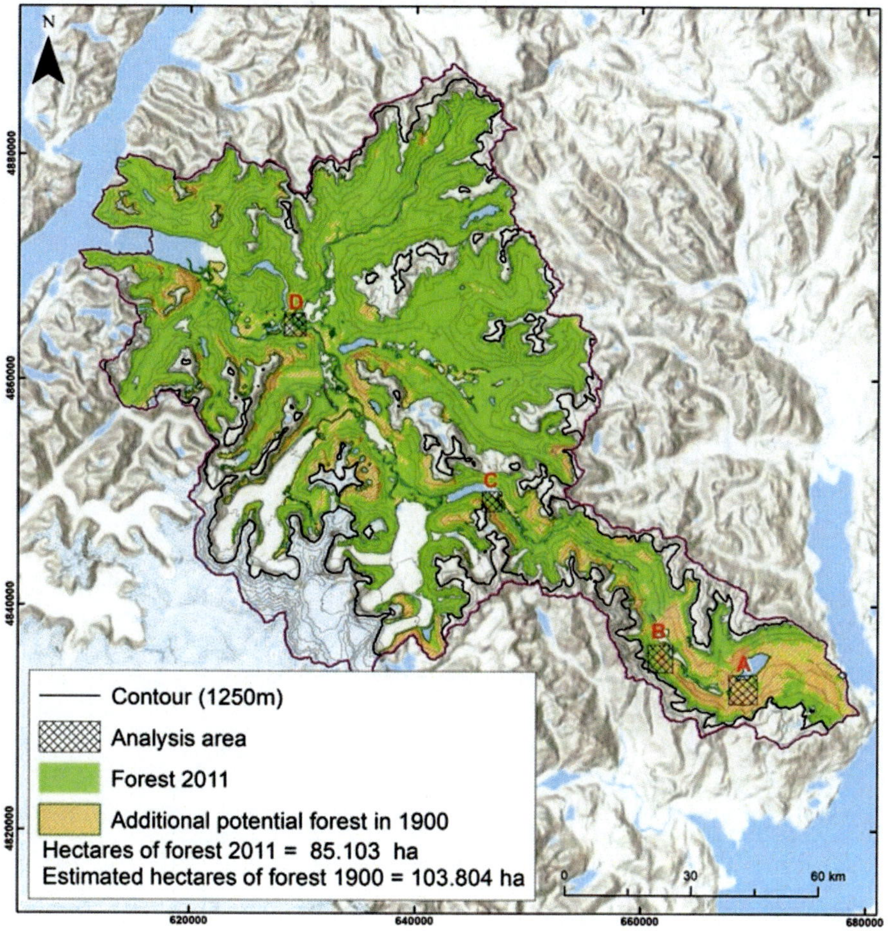

Fig. 8.3 Potential temperate forest cover between 1900 and 2011. Analysis areas within the study area: (**a**) Area surrounding Lake Tranquilo; (**b**) Beginning of the Exploradores Valley; (**c**) Bayo Lake; (**d**) Teresa sector and Exploradores Bridge. Classification based on CONAF (2011) for the reconstruction of the forest cover (Cartography and design by the authors)

recovery of vegetation stands out in analysis area (A), which predominantly changed from bare soil to forest and pastures. In the second period, land cover transitions are mainly characterized by changes from shrubland to forest, and from forest to pastures. In the case of analysis areas (B) and (C), there was an increase in forest cover between 1986 and 2001. These areas changed to bare soil and pastures between 2001 and 2018. Likewise, in analysis area (D), there was a significant change from forest to shrubland during the first period. These areas became mainly pasture in period two (Figs. 8.4 and 8.5).

Table 8.1 Land cover changes in the Exploradores Valley using the software package IDRISI Selva

	Cover change 1986–2001					Cover change 2001–2018				
Analysis area	N°	Cover 1986	Cover 2001	Cover change (ha)	Overall Kappa Index	N°	Cover 2001	Cover 2018	Cover change (ha)	Overall Kappa Index
A	1	Bare soil	Forest	204.84	0.63	1	Shrubland	Forest	64.53	0.77
	2	Bare soil	Pastures	117.18		2	Forest	Pastures	62.55	
	3	Shrubland	Pastures	57.06		3	Pastures	Forest	47.79	
	4	Bare soil	Shrubland	34.02		4	Pastures	Shrubland	43.92	
B	1	Pastures	Forest	36.4	0.89	1	Forest	Bare soil	58.95	0.9
	2	Bare soil	Forest	31		2	Pastures	Forest	15.21	
	3	Shrubland	Forest	17.5		3	Forest	Pastures	14.04	
	4	Shrubland	Pastures	8		4	Shrubland	Forest	11.61	
C	1	Bare soil	Forest	24.57	0.88	1	Forest	Bare soil	40.41	0.88
	2	Pastures	Forest	13.32		2	Water	Forest	14.67	
	3	Water	Forest	10.71		3	Bare soil	Forest	8.19	
	4	Water	Bare soil	10.53		4	Bare soil	Water	3.06	
D	1	Pastures	Shrubland	249.57	0.61	1	Shrubland	Pastures	149.85	0.75
	2	Pastures	Forest	16.47		2	Forest	Pastures	13.5	
	3	Forest	Shrubland	9.63		3	Forest	Shrubland	7.2	
	4	Water	Shrubland	4.77		4	Water	Pastures	4.95	

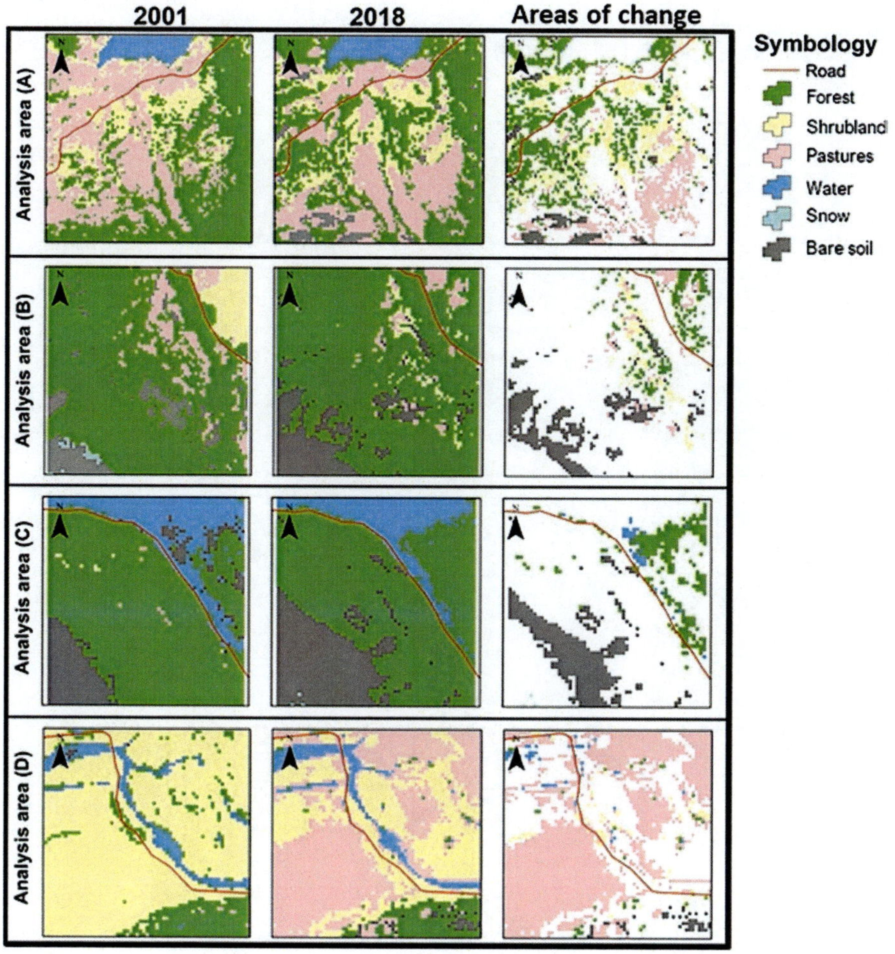

Fig. 8.4 Land cover changes between 1986 and 2001 in the four study areas: (**a**) Area surrounding Lake Tranquilo; (**b**) Beginning of the Exploradores Valley; (**c**) Bayo Lake; (**d**) Teresa sector and Exploradores Bridge (Cartography and design by the authors)

4.2 The Area Surrounding Lake Tranquilo (A)

This sector corresponds to the area where the town of Puerto Río Tranquilo is located (Fig. 8.6a,b,c) and registers one of the highest rates of change due to human activity during the historical occupation of the valley. A significant change was detected in this area in the two analysis periods. Between 1986 and 2001 the main land cover change was from bare soil to forest and pastures (205 ha and 117 ha respectively), while between 2001 and 2018 the transition from shrubland to the forest (65 ha) was prominent, showing an eventual succession from bare lands to forest or vegetation recovery. However, this transition must be further assessed

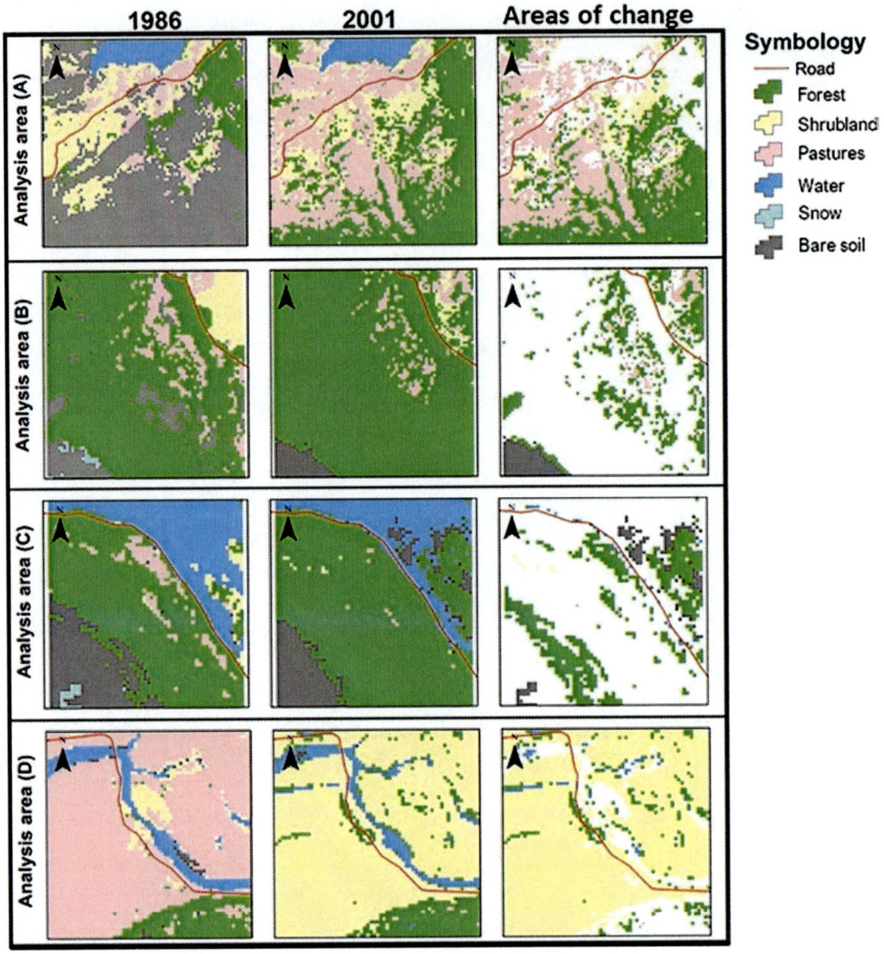

Fig. 8.5 Land cover transformations between 2001 and 2018 in the four study areas: (**a**) Area surrounding Lake Tranquilo; (**b**) Beginning of the Exploradores Valley; (**c**) Bayo Lake; (**d**) Teresa sector and Exploradores Bridge (Cartography and design by the authors)

because vegetation recovery is mostly associated with shrub invasion and the abandonment of human activities, which implies a type of ecological succession with lower ecosystem functionality than a transition to native forest. In addition, part of the recovery is attributable to reforestation efforts with introduced species such as *Pinus radiata*, *Pinus ponderosa*, and eucalyptus (Figs. 8.6b and 8.7). This type of transition occurred precisely during the same period as a livestock crisis that affected the whole region, which, according to our interviewees, gave way to other economic activities such as tourism.

Within Sector A, new anthropic elements, incorporated since the mid-1990s, play a significant role in environmental change. The emergence of nature-based

Fig. 8.6 (**a**) View Puerto Río Tranquilo in 1965 with visible burned surfaces in the background. (Source: D. Gruenwoldt 1965. Puerto Río Tranquilo Public Library Collection. Modified by authors); (**b**) Comparative view of Puerto Río Tranquilo in 2021 with visible afforestation in the background. (Source: Fieldwork by A. Salazar 2021); (**c**) Puerto Río Tranquilo in 1965. Surroundings include burned forests with active sheep farming. (Source: D. Gruenwoldt 1965. Puerto Río Tranquilo Public Library Collection)

tourist attractions that draw thousands of season tourists to the area is one of the most relevant factors of change today. Since the 1990s, the number of visitors to the region increased from 9959 in 1990 to 580,046 in 2017 (SERNATUR 2017). This increase together with changes in land tenure explain the intensification and densification of tourism-centered urbanization in the last two decades. The rapid densification of Puerto Río Tranquilo, which currently hosts a large floating population particularly during summers, represents a change in economic activities from livestock farming and forestry to tourism in the region. Likewise, after road and land connectivity improvements occurred in the area, a renewed valorization of wild spaces and the promotion of tourism resurfaced the notion of the importance of land ownership as a mechanism of socio-territorial transformation. This change in the way that people valued the land resulted in the gradual tesellation of the territory and contributed to land speculation during the last two decades.

In this sense, both field observations and interviews allowed us to go back in time and trace the socio-environmental history of these territories. We learned that forest cover was devastated by anthropogenic fires, both voluntarily and involuntarily, during the early twentieth century until ca. 1937 (Figs. 8.6c and 8.7). As Grosse must

Fig. 8.7 Sector A. Slopes of Tranquilo valley with traces of old fires, secondary forest, and reforested areas with non-native species. (Source: Fieldwork by A. Salazar 2015)

Fig. 8.8 Lake Tranquilo in 1944. Sector A. Lakeshores and valley slopes with extensive areas devoid of vegetation due to fires. (Source: Grosse 1955)

have observed in his expedition (Fig. 8.8a, b), and as it is remembered by the first inhabitants of the valley, the initial human footprint was much greater than what is possible to see today. The vestiges of that first occupation are still perceptible today, and they remind us that human-environment interactions must be observed at different spatial and temporal scales to fully understand them. RS, for example, shows extensive vegetation recovery in the Exploradores valley, however, this geotechnical approach is just a first step in the research process and must be accompanied by other local-level in-situ methods to better understand the intensity of these changes and the geographical peculiarities of such transformations.

Today, when you walk through the valley, you can see the paradox of being in a place with almost no human presence, but with deep traces of human activity. Going through the memories of its inhabitants allows us to once again see that the 30 hectares of change reflect processes invisible to GIS and the human eye.

4.3 Start of Exploradores Valley (B)

Secondary forest predominates in this area, with traces of fires in high elevation areas and the presence of pine plantations for erosion control, principally *Pinus radiata* and *Pinus ponderosa* (Fig. 8.9). Between 1986 and 2001, land cover changed from shrubland, bare ground, and pastures to the forest (36, 31, and 18 ha, respectively). On the contrary, between 2001 and 2018 we found evidence that forest cover was decreasing since land cover changed predominantly from forest to bare soil, shrubland, and pastures (59, 14, and 12 ha, respectively) (Figs. 8.4 and 8.5). This sector has been a space of constant human intervention and one of the areas with the greatest impacts of fires.

Fig. 8.9 Start of Exploradores valley (B). Traces of fires, second forest, and forest plantations. (Source: Fieldwork by J. Olea 2021)

Fig. 8.10 Start of Exploradores valley (B). Secondary forest and scrub vegetation along the slopes and in wetland areas. (Source: Fieldwork by A. Salazar 2015)

Fig. 8.11 Start of Bayo Lake (C). (Source: Fieldwork by A. Salazar 2015)

Similarly, the analysis area (B) (Figs. 8.9 and 8.10) is characterized by the presence of forest with patches of bare ground because of past burning. There are also areas of pastures and shrubland, which are linked to anthropogenic activity in the sector. The topographic conditions of the area, such as a wider valley floor, together with its proximity to the town of Puerto Río Tranquilo, allowed an earlier human occupation. This town had a connection with the mining settlements on the opposite shore of General Carrera Lake since 1931, which contributed to commercial exchanges that energized the economic activity in the area until 1992 (Fig. 8.6c).

Fig. 8.12 Start of Bayo Lake (C). (Source: Fieldwork by G. Mendoza 2015)

4.4 Bayo Lake (C)

The Bayo Lake is an intermediate point in the valley, where mature and taller forest stands are observed. However, traces of large fires can still be observed on the mountain slopes and lower parts of the basing, including some meandering areas of the river (Figs. 8.11 and 8.12). The area has the lowest number of hectares undergoing forest cover change. Between 1986 and 2001, this area was characterized by the transition from bare soil (25 ha) and shrubland (13 ha) to forest. Conversely, between 2001 and 2018, the recovered forests changed back to bare soil (44 ha) (Figs. 8.4 and 8.5).

Sector (C) Bayo Lake fundamentally serves as a site for comparison between past and present forest cover conditions. This area presents greater complexity due to the ecological evolution of forest cover after exposure to extensive fires and the apparent decrease of anthropogenic activity. Thus, the change from bare soil to forest is partly explained by the abandonment of livestock and/or forestry activities, with cleared areas that were later covered by forest. Likewise, glacial retreat in the sector (Alfaro et al. 2020; Mardones et al. 2011) gives way to bare soil. The slow but sustained recovery (succession) of native forest in this area could be described by the following characteristics: (1) the greatest changes occurred before satellite data became available; (2) sucession was uneven compared to the rest of the valley and occurred without human-aided reforestation; (3) recovery occurred in a period of increased local temperatures (Araneda et al. 2007), with less intense anthropogenic activity. In this sense, the changes and transformations evident in satellite images are limited by the availability and resolution of the images. Delving further back into the history of the occupation of the valley requires the use of other approaches.

Fig. 8.13 (**a**) The Exploradores River valley, near Bayo Lake, year 1955. (Source: Grosse 1955); (**b**) view of burned forests in the same area, year 2018. (Source: Fieldwork by D. Galop 2018); and (**c**) view of burned forests in the same area, year 2015. (Source: Fieldwork by A. Salazar 2015)

Thus, the photographs taken during the Grosse expedition continue to be an invaluable resource (Fig. 8.13a–c).

In the Bayo Lake area (C) is the Exploradores glacier, one of the main tourist attractions in the Aysén region. In the journey documented by Grosse, this place was one of the turning points in the search for a route to the interior. In his story, he describes both the lake fauna and the characteristics of the lake, but it is his photography that is the most relevant as it recorded the intensity and scope of the fires, which at the time of his exploration had already occurred (Fig. 8.14a, b).

4.5 *Exploradores Bridge (D)*

The fourth analysis area is characterized by the presence of one of the most recent infrastructure projects, the Exploradores Bridge, which crosses the junction of the Exploradores, Oscuro, and Teresa rivers. This area has changed over the years, with pastures changing to shrubland in a 250-ha area during the period 1986 to 2001, and conversely, 150 ha of shrubland changing to pastures between 2001 to 2019. During this last period, shrubland cover remained constant. These changes were due to the

Fig. 8.14 (**a**) Exploradores River valley, near Bayo Lake, year 1955. (Source: Grosse 1955), and (**b**) recovered burned forests in sectors B and C, year 2015. (Source: Fieldwork by A. Salazar 2015)

progressive and large-scale abandonment of livestock activities in the 1990s. As in other places, the decline in animal husbandry was marked by lower demand from the mining sectors on the other side of General Carrera Lake (since the late 1980s), which meant a moment of crisis for the inhabitants of the valley. Part of the population chose to migrate and seek new opportunities in other territories, while those who remained in the area switched to other activities, mainly tourism. One of the inhabitants of the sector recounted how he sold a large part of his animals to buy a boat to transport people to the area before the installation of the bridge, and later, as a means for tourists to practice sport fishing (Personal conversation with T.B.) 50 years ago inhabitant of the valley).

The bridge signified an important change in the way the valley was inhabited and used since it opened a link between General Carrera Lake and the town of Puerto Tranquilo, which has a jetty that leads to the ocean. The construction of this bridge meant the development of large scale tourism that included tours to the San Rafael lagoon since the bridge made it possible for people to come from Coyhaique, the regional capital of Aysén. Likewise, it allowed the movement of local workers to salmon processing operations. As it can be seen in Figs. 8.15 and 8.16, when Augusto Grosse explored the valley in search of maritime access to General Carrera Lake in the 1940s, the initial occupation of the valley depended on the available local resources. The first inhabitants had to clear densely forested areas with difficult topography. According to local testimonies, the tools to do this kind of work at the time were inadequate and scarce, so almost everything depended on the use of fire as a way to enter the territory (Fig. 8.16).

Fig. 8.15 The first settler's house was found in the Exploradores river valley by the 1944 Grosse expedition. Sector (D). Visible logging of trees for construction, energy sources, and grazing pastures. (Source: Grosse 1955)

Fig. 8.16 Aerial view of sector D. Visible traces of fires, logging for construction, and grazing pastures around the settlements of the first settlers. (Source: Fieldwork by A. Salazar 2019)

Fig. 8.17 Sector D. Intersection of the Teresa, Oscuro, and Exploradores rivers. At the top of the mountain are traces of past fires and growing forests. (Source: Fieldwork by A. Salazar 2015)

Figure 8.17 shows an isolated area with extensive traces of fires within the valley, where there was no record of human occupation at that time of Grosse's expedition. This sector, where the Teresa, Oscuro, and Exploradores rivers intersect, was the last section to be connected to the road, with the construction of the Exploradores Bridge (Fig. 8.18), allowing the most extreme section of the valley to connect to the jetty in the area. This incorporation of road infrastructure changed the forms of transportation. Before the construction of the road, the valley could only be crossed on horseback, as Grosse did and as some residents mentioned. Then, horses were replaced by boats (Fig. 8.19) until the road was finished, leading to significant transitions in the ways of life and occupation of the Exploradores valley.

5 Conclusions

As we stated, the purpose of this chapter was to analyze the main transformations and forces that shaped the changes in temperate rainforests in the Chilean Patagonia. We provided an account of the intensities and most important consequences of these transformations, and how they couldbe approached from a mixed methods perspective. The combination of GIS and RS with qualitative methods allowed us to provide an in-depth analysis of territorial metamorphosis.

Fig. 8.18 Exploradores Bridge (D), the crossing of the Teresa, Oscuro and Exploradores rivers. (Source: Fieldwork by D. Galop 2017)

Fig. 8.19 Sector D. Before the Exploradores Bridge, which provided land access to the settlers' jetty and a tourist route to Laguna San Rafael. (Source: Fieldwork by A. Salazar 2015)

By assessing the spatial transformations of the southern biomes in the Chilean

Patagonia, we provided evidence of the impact of anthropogenic disturbances, both past, and present, on forest cover and ecosystem integrity. This first assessment, which has been approached from the perspective that humans can be removed from nature, is only the starting point in the evaluation. We recognize the existence of other, more complex processes that permeate both spheres. Human-environmental interactions are constantly recreated and transformed, and a multi-temporal evaluation of their intensity can help identify trajectories of change in the territories as a whole.

The study area analyzed in this work extends from the edge of General Carrera Lake to the coast of the Pacific Ocean in Exploradores Bay. This area represents a transect for exceptional interpretation and experimentation, which allows us to better understand the dynamics of this complex socio-environmental system. The effective occupation of the area during the twentieth century is highlighted by evidence of differentiated human impacts, both during the period of occupation and by the means and ends with which it has been occupied. Therefore, the Exploradores area in the Chilean Patagonia is an ideal space to carry out retrospective and comparative analyses of the various interactions between the people who have inhabited it, the changes that the environment experienced, and the elements that function as poles of attraction today.

Although human activities evolved differently in intensity and over time, we can observe their footprint in the territory, as well as the diverse forms of these impacts. Each stage of occupation, the exploitation of cypress in the coastal marine area, as well as the advance of agribusiness at the lake's edge, left evident landscape marks on these fragile southern ecosystems. However, a record also exists of the withdrawal of human activities, which is directly reflected in the valley's ecological dynamics, where, depending on the geographical characteristics of the sectors, the marks or signs of human intervention are increasingly difficult to detect. Limitations with RS –a common approach to measure landscape transformations within a territory–, makes the use of other geospatial data sources, such as paleo-climatological or geo-historical field observations among others, urgent and necessary. In this way, information derived from historical photographs, interviews, and accounts of the inhabitants of the valley, complement the results obtained from the cartographic work.

In this sense, the changes and transformations that could be evident in satellite images are limited by their availability and resolution. To explore the reconfiguration of the history of the occupation of the valley, we needed to rely on other tools. The photographs taken during Augusto Grosse's expeditions continue to be a fundamental source of information since the landscape record in the expedition journal allows us to have additional years of data for comparison, as well as the possibility of identifying the activities that led to the occupation of the valley.

From the perspective of RS and GIS technologies, we suggest that the land use and land cover change characteristics and the pressure of human activity to settle in the valley are the most evident processes leading to environmental changes in recent times. However, RS and GIS approaches are not enough to understand the different dynamics and intensities of use, transformation, and recovery of vegetation cover, ecosystems, and the socio-environmental interactions. The detection of anthropogenic and wilderness areas with RS and GIS depends on the scale, techniques, and individual interpretations (Inostroza et al. 2016). For example, the temperate forest in its respective successions shows an early appearance of secondary forest, which requires, at least, more detailed work with RS to be better characterized. This is where qualitative methods allowed us to identify areas that were affected but not easily detectable. For instance, there are several stages of human interventions in the territory, which appear homogenous through the use of satellite images, but the qualitative descriptions by the first settlers gave us the opportunity to understand territorial transformations in a more nuanced manner.

Likewise, although the approximation of the sensors is useful to represent and understand the specific dimensions that we assign to nature, the measurements and categories we derive from RS must be used critically. In this case, the definition of forest cover had to be complemented by alternative methods and resources that contributed to a better interpretation of human-environmental interactions and their impact on land.

The Exploradores valley is a good example that reveals the complexity and depth of territorial transformations driven by long-lasting human-environmental interactions. Traditional approaches focusing on cause-effect relationships can be highly schematic and do not allow a nuanced and contextualized assessment of environmental change. The case study of the Exploradores valley provides an opportunity for deepening our understanding of human-environmental relationships if they are analyzed and interpreted from an interdisciplinary and territorialized retrospective, containing various scales and temporalities to address the particular history of interactions.

This alternative approach allows us to analyze the human-environmental system by adding a reflection on the transformation of dominant ideas and economic structures in tandem with discourses about nature. The social and cultural constructs that emerge from late capitalism, touristifying natural areas, and emphasizing the pristine and green as values themselves, have hidden the territorial transformations that have occurred. The imaginaries constructed about Patagonia are based on the dominant idea of a low human impact, an issue that we discussed in this chapter. The SES perspective offers a framework in which the social dimension is considered in relation to the natural system under study. This framework seeks to fully integrate human beings into the evolution of nature and not just as circumstantial data. Such coupling makes it possible to measure the speed and impact of interactions with nature, adding an explanatory narrative that is difficult to pursue via satellite images. The notion of human adaptation to the environment, through the lens of cultural geography, sheds new light on the structures and forces that influence the use and modification of geographic spaces (Harden 2012). This notion becomes even more relevant if we think of it in terms of global change since human action on the environment must be considered at both planetary and local scales.

Currently, several alternatives are being developed to study spatial transformations and human-environmental interactions. Attempts to link various methodologies and sources of information not only enrich the debate around the impacts of humans on the environment but also reveal the role that nature plays in the process of adaptation to new conditions. The resilience capacity of the forest is also analyzed as a counterpart of the impact that humans have on the environment. Both spheres are necessary to contemplate a science that aims not only to measure physical transformation but also create ways of valuing and preserving natural environments.

In summary, digital satellite data and GIS tools allow us to observe with relative clarity the most relevant aspects of territorial transformation. However, it is essential to apply alternative methods of obtaining, processing, and analyzing data at different temporal and spatial scales. An interdisciplinary methodological framework could lead to new questions and provide interrelated diverse perspectives on human-environmental issues. The Exploradores valley is an exciting laboratory that is worth continuing investigating along these lines of inquiry. The transformations that occurred in recent decades in almost every corner of the planet justify the need of a better understanding of the intensity and speed of changes, not only because of their relevance but also for their potential to help us reflect on the required actions that can still be implemented to reverse detrimental consequences on the socio-environmental system. The case of the Exploradores valley provides a relevant but limited spatio-temporal context in which we can observe a process of land cover anthropogenization, but at the same time explore ways of mitigating the impacts associated with this process that could contribute to a better understanding of human-environmental adaptation to global change dynamics.

Acknowledgments This work was funded by the Chilean National Agency for Research and Development (Agencia Nacional de Investigación y Desarrollo de Chile, ANID), Regular Fondecyt #1191865, and the OHM-I Patagonia-Bahía Exploradores, Labex DRIIHM (Programme Investissements d'avenir": ANR-11-LABX-0010), INEE-CNRS, Paris, Francia. We thank Felipe Jorquera for his fieldwork and initial management of this document, the Public Library of Puerto Río Tranquilo and Luis Pardo for the photographic materials provided. We also want to thank Adrian Fernández, Ian Farmer, Rosa Vera and Jaime Schienfeldt for his support during our stays in the Exploradores Valley, and to all the research assistants and students who supported us and made this work possible.

References

Alfaro D, Salazar-Burrows A, Bañales-Seguel C et al (2020) Soil microbial abundance and activity across Forefield glacier Chronosequence in the northern Patagonian ice field, Chile. Arct Antarct Alp Res 52(1):553–562. https://doi.org/10.1080/15230430.2020.1820124

Anderson JR (1976) A land use and land cover classification system for use with remote sensor data, 964. US Government Printing Office

Araneda A, Torrejón F, Aguayo M et al (2007) Historical Records of san Rafael Glacier Advances (north Patagonian icefield): another clue to 'Little ice Age' timing in southern Chile? The Holocene 17(7):987–998. https://doi.org/10.1177/0959683607082414

Araya V, Boris A (2017) Los orígenes de la Construcción Discursiva del Territorio de Aisén por parte del Estado de Chile (1818–1929). Magallania (Punta Arenas) 45(1):47–73. https://doi.org/10.4067/s0718-22442017000100047

Armesto J, León P, Kalin-Arroyo MT (1996) Los Bosques Templados Del Sur de Chile y Argentina: Una Isla Biogeográfica. In: Armesto J, Villagrán C, Kalin-Arroyo MT (eds) Ecología de Los Bosques Nativos de Chile, Vicerrectoría. Académica-Universidad de Chile, Santiago, pp 23–28

Armesto J, Rozzi R, Smith-Ramirez C, Kalin-Arroyo MT (1998) Conservation targets in south American temperate forests. Science 282(5392):1271–1272. https://doi.org/10.1126/science.282.5392.1271

Berkes F, Folke C (1998) Linking social and ecological systems for resilience and sustainability. In: Berkes F, Folke C (eds) Linking social and ecological systems: management practices and social mechanisms for building resilience. Cambridge University Press, Cambridge, pp 1–26

Bizama G, Torrejón F, Aguayo M, Muñoz MD, Echeverría C, Urrutia R (2011) Pérdida y fragmentación del bosque nativo en la cuenca del río Aysén (Patagonia-Chile) durante el siglo XX. Revista de geografía Norte Grande 49:125–138. https://doi.org/10.4067/S0718-34022011000200008

Bourgeron PS, Humphries HC, Riboli-Sasco L (2009) Regional analysis of social-ecological systems. Nat Sci Soc 17(2):185–193

Castree N, Demeritt D, Liverman D, Rhoads B (eds) (2016) A companion to environmental geography. Wiley, Chichester, UK

Chenorkian R, Robert S (eds) (2014) Les interactions Hommes-Milieux. Questions et pratiques de la recherche en environnement. Éditions Quae, Collection Indisciplines, Versailles

Claval P (2011) Epistemología da Geografía. UFSC, Florianopolis

Congalton RG (2004) Putting the map back in map accuracy assessment. In: Lunetta R, Lyon J (eds) Remote sensing and GIS accuracy assessment, 1st edn. CRC Press, Boca Raton, pp 1–11

Dinerstein E, Joshi AR, Vynne C et al (2020) A "global safety net" to reverse biodiversity loss and stabilize Earth's climate. Sci Adv 6(36):eabb2824. https://doi.org/10.1126/sciadv.abb2824

Donoso C (1993) Bosques templados de Chile y Argentina: variación, estructura y dinámica. Editorial Universitaria, Santiago

Echeverría C, Coomes D, Salas J et al (2006) Rapid deforestation and fragmentation of Chilean temperate forests. Biol Conserv 130(4):481–494. https://doi.org/10.1016/j.biocon.2006.01.017

Echeverría C, Newton AC, Lara A et al (2007) Impacts of forest fragmentation on species composition and forest structure in the temperate landscape of southern Chile. Glob Ecol Biogeogr 16(4):426–439. https://doi.org/10.1111/j.1466-8238.2007.00311.X

Farhad S (2012) Los sistemas socio-ecológicos. Una aproximación conceptual y metodológica. XIII Jornada de Economía Crítica, Sevilla

Folke C (2006) Resilience: the emergence of a perspective for social–ecological systems analyses. Glob Environ Change 16(3):253–267. https://doi.org/10.1016/j.gloenvcha.2006.04.002

Folke C, Pritchard L Jr, Berkes F et al (2007) The problem of fit between ecosystems and institutions: ten years later. Ecol Soc 12(1)

Gao BC (1996) NDWI—A normalized difference water index for remote sensing of vegetation liquid water from space. Remote Sens Environ 58(3):257–266. https://doi.org/10.1016/S0034-4257(96)00067-3

Gregory I, Geddes A (2014) From historical GIS to spatial humanities: deepening scholarship and broadening technology. In: Towards spatial humanities. Spatial Humanities. Indiana University Press, Bloomington, pp ix–xix. ISBN 9780253011862

Grosse A (1955) Visión de Aisén

Harden C (2012) Framing and reframing questions of human–environment interactions. Ann Am Assoc Geogr 102(4):737–747. https://doi.org/10.1080/00045608.2012.678035

Holz A, Veblen TT (2011) Variability in the southern annular mode determines wildfire activity in Patagonia. Geophy Res Lett 38(14). https://doi.org/10.1029/2011GL047674

Holz A, Haberle S, Veblen TT et al (2012a) Fire history in western Patagonia from paired tree-ring fire-scar and charcoal records. Clim Past 8:451–466. https://doi.org/10.5194/cp-8-451-2012

Holz A, Kitzberger T, Paritsis J, Veblen TT (2012b) Ecological and climatic controls of modern wildfire activity patterns across southwestern South America. Ecosphere 3(11):1–25. https://doi.org/10.1890/ES12-00234.1

Hoyos-Santillan J, Miranda A, Lara A, Rojas M, Sepúlveda-Jauregui A (2019) Protecting Patagonian peatlands in Chile. Science 366:1207–1208. https://doi.org/10.1126/science.aaz9244

Huber UM, Markgraf V, Schäbitz F (2004) Geographical and temporal trends in late quaternary fire histories of Fuego-Patagonia, South America. Quat Sci Rev 23:1079–1097. https://doi.org/10.1016/j.quascirev.2003.11.002

Iglesias V, Whitlock C (2014) Fire responses to postglacial climate change and human impact in northern Patagonia (41-43°S). Proc Natl Acad Sci 111(51):E5545–EE555. pnas.1410443111

Inostroza L, Zasada I, König HJ (2016) Last of the wild revisited: assessing spatial patterns of human impact on landscapes in Southern Patagonia, Chile. Reg Environ Change 16(7):2071–2085. https://doi.org/10.1007/s10113-016-0935-1

Jaksic FM, Fariña JM (2015) Incendios, sucesión y restauración ecológica en contexto. Anales del Instituto de la Patagonia 43(1):23–34. https://doi.org/10.4067/S0718-686X2015000100003

Janssen MA, Ostrom E (2006) Governing social-ecological systems. In: Tesfatsion L, Judd KL (eds) Handbook of Computational Economics, 2. Elsevier, pp. 1465–1509. https://doi.org/10.1016/S1574-0021(05)02030-7

Kurz WA, Stinson G, Rampley G (2008) Could increased boreal forest ecosystem productivity offset carbon losses from increased disturbances? Philos Trans R Soc Lond Ser B Biol Sci 363(1501):2259–2268. https://doi.org/10.1098/rstb.2007.2198

Kvale S (2012) Las entrevistas en investigación cualitativa, vol 2. Ediciones Morata, Madrid

Lagadeuc Y, Chenorkian R (2009) Les systèmes socio-écologiques: vers une approche spatiale et temporelle. Nat Sci Soc 17(2):194–196

Leifeld J, Menichetti L (2018) The underappreciated potential of peatlands in global climate change mitigation strategies. Nat Commun 9:1071. https://doi.org/10.1038/s41467-018-03406-6

Liu J, Dietz T, Carpenter SR et al (2007a) Complexity of coupled human and natural systems. Science 317(5844):1513–1516. https://doi.org/10.1126/science.1144004

Liu J, Dietz T, Carpenter SR et al (2007b) Coupled human and natural systems. Ambio 36(8):639–649. https://doi.org/10.1579/0044-7447(2007)36[639:CHANS]2.0.CO;2

Luebert F, Pliscoff P (2005) Sobre los límites del bosque valdiviano en Chile. Chloris Chilensis 8(1):8–10

Luebert F, Pliscoff P (2006) Sinopsis bioclimática y vegetacional de Chile. Editorial Universitaria, Santiago

Maass M, Balvanera P, Bourgeron P et al (2016) Changes in biodiversity and trade-offs among ecosystem services, stakeholders, and components of Well-being: the contribution of the international long-term ecological research network (ILTER) to Programme on ecosystem change and society (PECS). Ecol Soc 21(3):31. https://doi.org/10.5751/ES-08587-210331

Mann ME, Bradley RS, Hughes MK (1998) Global-scale temperature patterns and climate forcing over the past six centuries. Nature 392(6678):779–787. https://doi.org/10.1038/33859

Mardones M, Gonzalez L, King R, Campos E (2011) Holocene glacial variations in Central Patagonia, Aisen, Chile: geomorphological evidences. Andean Geol 38(2):371–392

Martinic MB (2005) De la Trapananda al Aysén. Pehuén Editores, Santiago

Moreira-Muñoz A, García JL, Sagredo E (2014) Reserva de la Biosfera Laguna San Rafael: sitio de importancia global para la investigación del cambio climático. In: Moreira-Muñoz A, Borsdorf A (eds) Reservas de la Biosfera de Chile: Laboratorios para la Sustentabilidad. Serie Geolibros 17, Academia de Ciencias Austriaca, Ponticia Universidad Católica de Chile, Instituto de Geografía, pp. 210–227

Moreno PI, Kitzberger T, Iglesias V, Holz A (2010) Paleofires in southern South America since the last glacial maximum. Pages News 18:75–77. https://doi.org/10.22498/pages.18.2.75

National Forestry Corporation (CONAF) (2011) Catastro de Uso del Suelo y Vegetación. Monitoreo y Actualización en la XI Región de Aisén del General Carlos Ibáñez del Campo., SIT CONAF

Neukom R, Luterbacher J, Villalba R et al (2010) Multi-centennial summer and winter precipitation variability in southern South America. Geophy Res Lett 37(14). https://doi.org/10.1029/2010gl043680

Neukom R, Luterbacher J, Villalba R et al (2011) Multiproxy summer and winter surface air temperature field reconstructions for southern South America covering the past centuries. Clim Dyn 37(1–2):35–51. https://doi.org/10.1007/s00382-010-0793-3

Olea-Peñaloza J, Salazar-Burrows A, Jorquera-Guajardo F (2021) La Patagonia como frontera científica: exploraciones contemporáneas desde una ciencia global. Diálogo Andino 66:95–105. https://doi.org/10.4067/S0719-26812021000300095

Olivares-Contreras VA, Mattar C, Gutiérrez AG, Jiménez JC (2019) Warming trends in Patagonian subantartic forest. Int J Appl Earth Obs Geoinf 76:51–65. https://doi.org/10.1016/j.jag.2018.10.015

Ostrom E (2009) A general framework for analyzing sustainability of social-ecological systems. Science 325:419–422. https://doi.org/10.1126/science.1172133

Ovalle L (2011) Reseña general de los procesos de colonización. In: Ovalle L (ed) Ocupación y desarrollo de la Provincia de Aisén. Ediciones Ñire Negro, Coyhaique, pp 89–96

Pellegrini AF, Refsland T, Averill C et al (2021) Decadal changes in fire frequencies shift tree communities and functional traits. Nat Ecol Evol 5:504–512. https://doi.org/10.1038/s41559-021-01401-7

Pole K (2009) Diseño de metodologías mixtas. Una revisión de las estrategias para combinar metodologías cuantitativas y cualitativas. Renglones (60):37–42

Redman CL, Grove JM, Kuby LH (2004) Integrating social science into the long-term ecological research (LTER) network: social dimensions of ecological change and ecological dimensions of social change. Ecosystems 7(2):161–171. https://doi.org/10.1007/s10021-003-0215-z

Rockström J, Steffen W, Noone K et al (2009) A safe operating space for humanity. Nature 461(7263):472–475. https://doi.org/10.1038/461472a

Salazar A, Moreira A, Arenas F, Osses P (2011) Bitácora de una expedición Geografía UC: Río y Bahía Exploradores. Región de Aysén, 2011. In: Arenas F, Salazar A, Núñez A (eds) El aislamiento Geográfico: ¿problema u oportunidad? Experiencias, interpretaciones y políticas públicas. Instituto de Geografía UC, Santiago, pp 103–117

Servicio Nacional de Turismo (SERNATUR) (2017) Barómetro de Turismo 2016. Ministerios de Economía, Fomento y Turismo, Gobierno de Chile

Solari ME, Cueto C, Hernández F et al (2011) Procesos territoriales y bosques en la cuenca del río Valdivia (siglos XVI-XIX). Revista de Geografía Norte Grande 49:45–62. https://doi.org/10.4067/S0718-34022011000200004

Steffen H (1902) Viajes de esploracion i estudio en la Patagonia Occidental: 1892–1902. Santiago de Chile: Impr. Cervantes, 1909-1910. 2 v

Teddlie C, Tashakkori A (2003) Major issues and controversies in the use of mixed methods in the social and behavioral sciences. In: Tashakkori A, Teddlie C (eds) Handbook of mixed methods in social and behavioral research. SAGE, pp 13–50

Teixeira S (2018) Qualitative geographic information systems (GIS): an untapped research approach for social work. Qual Soc Work 17(1):9–23. https://doi.org/10.1177/1473325016655203

Torres-Gómez M, Delgado LE, Marín VH, Bustamante RO (2009) Estructura del paisaje a lo largo de gradientes urbano-rurales en la cuenca del río Aisén (Región de Aisén, Chile). Rev Chil Hist Nat 82(1):73–82. https://doi.org/10.4067/S0716-078X2009000100005

Trenberth KE, Jones PD, Ambenje P, Bojariu R et al. (eds) Climate Change 2007: The physical science basis. Contribution of working group I to the fourth assessment report of the intergovernmental panel on climate change. Cambridge University Press, Cambrigde

Tucker CJ (1979) Red and photographic infrared linear combinations for monitoring vegetation. Remote Sens Environ 8(2):127–150

Urbina X (2013) Expediciones a la Costa de la Patagonia Occidental en el Periodo Colonial. Magallania 41(2):51–84. https://doi.org/10.4067/S0718-22442013000200002

Veblen TT, Holz A, Paritsis J et al (2011) Adapting to global environmental change in Patagonia: what role for disturbance ecology? Austral Ecol 36(8):891–903. https://doi.org/10.1111/j.1442-9993.2010.02236.x

Vitousek PM, Mooney HA, Lubchenco J, Melillo JM (1997) Human domination of Earth's ecosystems. Science 277(5325):494–499. https://doi.org/10.1126/science.277.5325.494

Whitlock C, Bartlein PJ, Markgraf V, Ashworth AC (2001) The mid-latitudes of north and South America during the last glacial maximum and early Holocene: similar paleoclimatic sequences despite differing large-scale controls. In: Markgraf V (ed) Interhemispheric climate linkages: present and past interhemispheric climate linkages in the Americas and their societal effects. Academic, Cambridge, pp 391–416

Wilson RJ, Gutiérrez D, Gutiérrez J et al (2005) Changes to the elevational limits and extent of species ranges associated with climate change. Ecol Lett 8(11):1138–1146. https://doi.org/10.1111/j.1461-0248.2005.00824.x

Chapter 9
El Chaltén, Argentine Patagonia: A Successful Combination of Conservation and Tourism?

Andrés Gerique Zipfel and Kim André Vanselow

Abstract El Chaltén is a village created in 1985 within Los Glaciares National Park, in Argentine Patagonia. Since 1985, this urban center has experienced rapid development due to immigration and nature tourism. We use the concept of naturbanization to investigate why and how El Chaltén was created, who are its inhabitants and visitors, and above all, what are the socio-environmental dynamics and impacts of the resulting urbanization process within a national park. To answer these questions, we use a mixed methodology that includes fieldwork, quantitative and qualitative interviews, and remotely sensed data time series based on the Normalized Difference Vegetation Index. Although our research finds limited negative environmental impacts caused by the creation of a village inside a national park, further urbanization can endanger the fragile balance between a growing population and tourism demand on the one side and effective conservation on the other.

Keywords Nature-based tourism · Protected areas · Naturbanization · Remote sensing · El Chaltén · Argentina · Patagonia

1 Introduction

As terrestrial biodiversity is tied to specific sites, protected areas remain the cornerstone of *in situ* conservation (Watson et al. 2018; Geldmann et al. 2013). During the last decade, the role of protected areas has been reinforced by two key decisions. The first one was the adoption of the Strategic Plan for Biodiversity 2011–2020 by

A. G. Zipfel (✉) · K. A. Vanselow
Institute of Geography, Friedrich-Alexander-Universität Erlangen-Nürnberg, Erlangen, Germany
e-mail: andres.gerique-zipfel@fau.de

S. López (ed.), *Socio-Environmental Research in Latin America*, The Latin American Studies Book Series, https://doi.org/10.1007/978-3-031-22680-9_9

almost all nations, which envisaged for 2020 that at least 17% of terrestrial and inland water areas will be conserved through protected areas (CBD 2010). The second decisive event was the adoption of the United Nations 2030 Agenda for Sustainable Development (DSDG 2015) in 2015, which confirmed the recognition of the value of protected areas to safeguard biodiversity. To date, around 15.4% of global terrestrial lands are protected areas (UNEP-WCMC and IUCN 2021); in Latin America and the Caribbean, this coverage reaches 24.2% (UNEP-WCMC 2021). The post 2020 global biodiversity framework has certainly raised this percentage considerably. The Kunming-Montreal Global Diversity Framework, which has been adopted by the 15th United Nations Conference of the Parties to the Convention of Biological Diversity in December 2022, aims to achieve the effective conservation and management of at least 30% of the world's terrestrial and marine areas (CBD 2022). Thus, it is necessary to analyze how to successfully combine conservation and land use, especially in inhabited areas. In this chapter, we analyze the challenges and opportunities for conservation in a protected area in Argentine Patagonia that has witnessed increased population growth in recent decades.

The establishment of protected areas can lead to conflicts. In the past, especially dramatic cases were those in which, following the path of "people-free parks" (Terborgh and Peres 2002), Indigenous and local communities were expelled from their lands (Neumann 1997). On other occasions, protected area creation has limited local people's access to vital ecosystem services such as food, fuel, or medicines (Soliku and Schraml 2018; Mathelvet et al. 2016).

More recently, conflicts have emerged from rapid urban expansion, which has brought large cities close to national parks (Elmqvist et al. 2013). Another event that threatens the viability of natural reserves is their attractiveness, as it can stimulate migration of urban populations to the vicinity of remote and relatively inaccessible protected areas. Berry (1980) and Champion (1989), called the process counterurbanization, a demographic de-concentration that brings urban migrants into rural spaces. Moss (2009) coined the term amenity migration or amenity-led migration, which refers to the study of the movement of people to places, permanently or part-time, principally because of their perceived higher environmental quality and cultural differentiation. These studies focus on the motivations behind these migrations to destinations that signify an improved quality of life or a favorable lifestyle choice, and/or on the variety of movements or mobilities (cf. Urry 2000) associated with the process of voluntary relocation (Benson and O'Reilly 2016; McIntyre 2009). Whereas amenity-led movements have been historically circumscribed to domestic migration to rural areas in mountain regions of the Global North (Perlik 2011; Loeffler and Steinicke 2007), in recent decades amenity migration has reached the mountainous areas in Latin America and the coasts of the Caribbean as an intranational phenomenon (Medina 2017; Borsdorf and Hidalgo 2009; González et al. 2009) or as a transnational event, where (mainly) U.S. citizens relocate to rural communities (Kordel and Pohle 2018; van Noorloos and Steel 2015; Janoschka 2009). Nevertheless, the spatial consequences of these migration processes on protected areas have seldom been the main objective of these studies (cf. Bender et al. 2017).

Remote sensing can assist with this challenge (Bailey et al. 2015; Nagendra et al. 2013; Fraser et al. 2009; Wang et al. 2009). Remote sensing applications range from rather simple objectives, like quantifying the amount and direction trend of built-up areas (de la Fuente et al. 2020) to complex issues, such as mapping habitat types and their floristic variability (Feilhauer et al. 2014). Particularly, time series of the Normalized Difference Vegetation Index (NDVI) have been used extensively to quantify trends of land and vegetation cover (Forkel et al. 2013; Verbesselt et al. 2010) and hence support classical fieldwork-based monitoring methods, e.g., in detecting and quantifying degradation or recovery driven by particular processes or events (Nagendra et al. 2013). However, although remote sensing is ideal to unravel the spatial impacts of human-environmental dynamics, it needs to be accompanied by other methods of analysis to understand how such dynamics are generated and operate. In this regard, Prados (2009) presented the concept of "naturbanization" to explain how the presence of protected areas stimulates immigration, urbanization, and land use change and describes the spatial consequences. The concept of naturbanization expands studies on amenity migration with land use/land cover change and environmental impact research, combining demographic data, digital mapping, and remote sensing techniques. While this concept has been widely used to analyze protected areas in Europe (del Valle and Prados Velasco 2019, Tulla et al. 2017, Calvache et al. 2016, Giusti and Prados Velasco 2013, Prados 2009) and literature therein, its utilization has been very limited elsewhere (for Argentina, see Medina 2017 for San Carlos de Bariloche and the Nahuel Huapi National Park).

2 Aims and Structure of this Study

In this chapter we use the naturbanization concept to study the case of El Chaltén (ECh), a village established during the late 1980s inside the Los Glaciares National Park (LGNP) in Argentina. El Chaltén has become a famous destination with population and tourist numbers increasing rapidly (INDEC 2010). The study first focuses on the causes that led to the creation of this village, then it investigates who are the inhabitants and who are the visitors, and finally it analyzes the consequences in terms of environmental impact and land cover change. From a methodological perspective, we describe the benefits of the combination of the fieldwork and remote sensing and reveal their limitations. Finally, we briefly discuss whether ECh represents a model of urban development applicable to other regions from a conservation point of view.

Our research questions are:

- Why was ECh created inside a national park?
- Who are the residents and who are the tourists?
- What are the major land cover changes since the foundation of ECh?
- Does ECh represent a model of coexistence between urbanization, tourism, and conservation applicable to other protected areas?

3 Methods

To analyze the creation of ECh and the colonization process as well as the type of migrants and visitors we first reviewed scientific publications, grey literature, institutional documents, and statistics. Missing information was found using different interview techniques in ECh. We conducted a problem-centered interview with key actors and administrators, namely the head of the park rangers in the northern area of the LGNP, the mayor of ECh, and the Secretary of Tourism of ECh. We also interviewed one of the few people who already lived in the area before the creation of ECh and who had witnessed its development. In March 2016, with the support of four students, we conducted 151 quantitative interviews with residents and 200 with tourists (CI = 90%, est. margin error = 7%). We asked about the respondents' origin, family status, age, education level, and inquiries about their experiences with the national park and tourism. In March 2019, we completed the data collection with 54 quantitative tourist interviews and an expert interview with the person who conceived and partially designed ECh in 1984. Fruitful conversation took place with two local entrepreneurs. We processed the interview data using Microsoft Excel. To calculate the total number of visitors per year, we used data from the local Secretary of Tourism and the visitor's registry system of the national park, which contains information from those arriving by public transportation during open hours. Following the advice of local experts, we added 35% to the registered number of visitors to include tourists arriving by night bus or by other means of transportation.

To unravel the environmental impact caused by the naturbanization process on the national park, we conducted participant observations of the behavior of inhabitants and tourists along the main trekking routes and documented impacts of touristic activity and other landscape alterations. To check the effectiveness of wastewater infrastructure, we analyzed organic pollution in water samples in Las Vueltas and Fitz Roy rivers with a portable Horiba NO_3-Meter (LAQUAtwin B-74X).

We used Landsat imagery to analyze land cover change during the last 35 years inside and around the LGNP in the ECh area. Landsat products provide a sufficient spatial resolution of 30 m and a relatively long time series. Altogether, we obtained 69 preprocessed and cloud-masked January scenes from 1985 to 2020 of the sensors Landsat 5 TM, Landsat 7 ETM+, and Landsat 8 OLI from the USGS ESPA service (https://espa.cr.usgs.gov/), which we used to create annual time series rasters. Unfortunately, there were no data available between 1987 and 1996. Then, we applied a trend estimation function using generalized additive mixed models (GAMMs; Wood 2017), with 'NDVI' as the response and 'YEAR' as the explanatory variable. In doing so, we were able to account for typical challenges, such as non-linearity in the trends, an irregular data distribution due to missing data, and temporal autocorrelation (Simpson 2018). Furthermore, to account for potential differences in the spectral characteristics between the three sensors, we introduced the variable 'SENSOR' as a random effect. After masking out pixel time series with too few valid observations to calculate reliable trends, we applied the GAMM model to the remaining 654,224-pixel time series, covering an area of 589 km^2 and derived the significance of the trend as well as the amount of change in the NDVI time series between 1985 and 2020.

4 The Study Site

4.1 Los Glaciares National Park in Argentinian Patagonia

The Glaciares National Park (LGNP) is in the Department of Lago Argentino, in the western part of Santa Cruz Province, southern Patagonia (Fig. 9.1). With 726,927 hectares, the LGNP is the largest protected area in Argentina (APN 2019) and one of the most popular glacial tourism destinations worldwide (Welling et al. 2015). The name refers to the ice fields and glaciers that cover roughly half of the park's area. In 1981, UNESCO declared the LGNP a World Heritage Site (UNESCO 2021). The Park is bounded by National Highway 40 to the west; borders the Lago del Desierto Provincial Reserve (RPLD) and private properties to the north, while the southern edge aligns with the Chilean border, private *estancias* (ranches), and the city of El Calafate. The LGNP borders Chile to the east. However, in the southern ice field area, the exact path of the international border and therefore the national park border remain pending.

Founded in 1927 by governmental decree as a place to supply ranchers and as the institutional headquarters near Argentino Lake, El Calafate is nowadays the third largest urban center in Santa Cruz Province and the most important in the Department (Ampuero et al. 2015). According to the last (outdated) census of 2010, El Calafate has 16,000 inhabitants (INDEC 2010). The town's airport is the LGNP's entry point for most national and international tourists (Ampuero et al. 2015). The main attraction of LGNP, the renowned Perito Moreno Glacier, is about one hour away from El Calafate (Fig. 9.2 Top). The second largest urban center in the Department is ECh, a destination that attracts nature and glacier tourism (Fig. 9.2 Bottom). ECh sits in the northern part of the LGNP northwest of Lake Viedma (Fig. 9.1), 215 km away from El Calafate and its airport.

Regarding human presence in the region, studies by Belardi et al. (2020) show first occupations around 9700 BP in the basins of the Lakes Argentino and Tar-San Martín, while archaeological records near ECh suggest an occasional low intensity occupation in prehistoric times. During subsequent periods, the area corresponds to the territory of the Anoikénk, the southernmost group of the Tehuelche culture (Pérez 2014). Europeans reached this region in 1782, when the Spaniard Francisco de Viedma went up the Santa Cruz River. However, the area remained almost unexplored until the expeditions by the Argentine explorer Perito Moreno during the last decades of the nineteenth century (Halvorsen 1999). Between 1878 and 1885, military campaigns known as the "Conquest of the Desert" extended Argentine domination within Patagonia, massacred most of the Indigenous people, provided land to

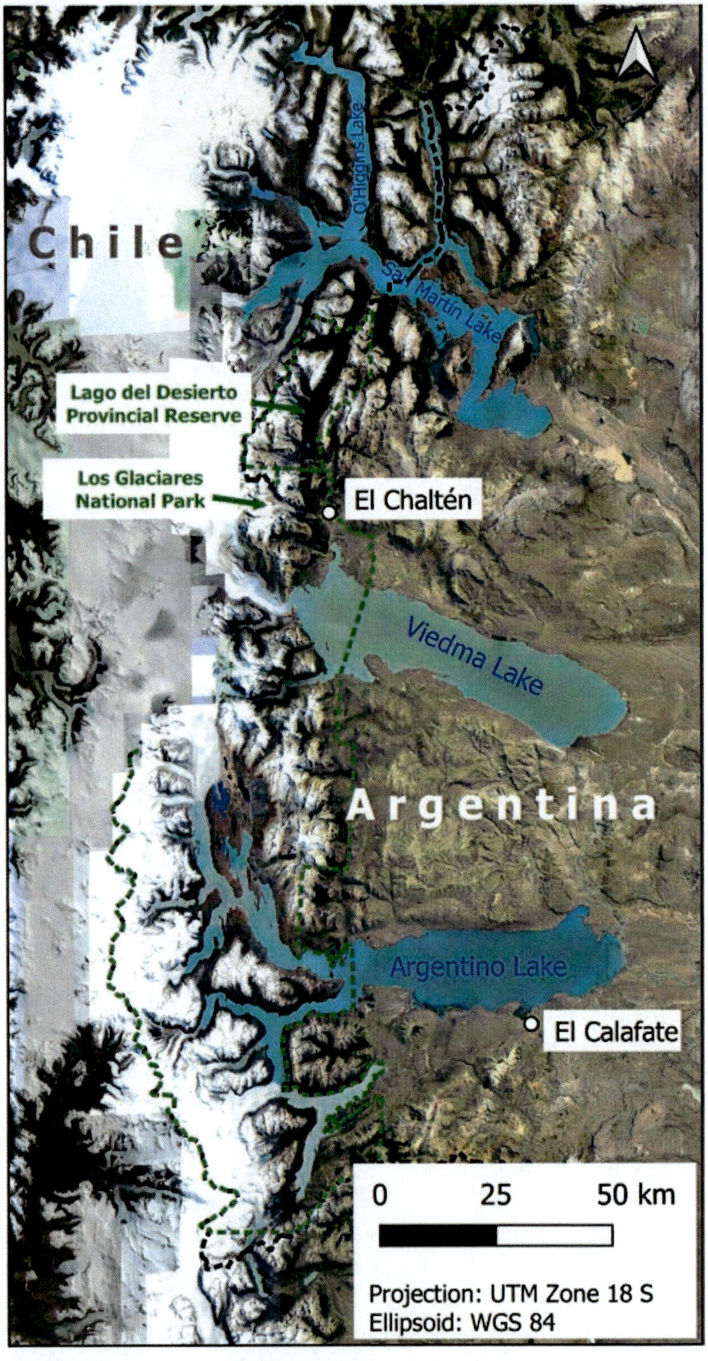

Fig. 9.1 The Los Glaciares National Park and the towns of El Calafate and El Chaltén in Southern Argentine Patagonia. (Vanselow 2021)

Fig. 9.2 Top: The Perito Moreno Glacier, the most famous attraction in the southern area of the Los Glaciares National Park, Argentina. Bottom: The Laguna de los Tres, one of the trekking highlights of El Chaltén in the northern zone of the national park. (Photos by A. Gerique 2019, 2014)

settlers, and halted the possibility of Chilean expansion (Pérez 2014). Colonization did not arrive to the study area until the beginning of the twentieth century, when European pioneers settled down to raise sheep and cows; large cattle companies followed and brought an increasing number of settlers to Lake Viedma (Madsen 1952). These companies installed sawmills and exploited timber. To expand pastures, settlers set enormous forest fires around Lakes Argentino and Viedma. A 1926–1927 fire lasted several months and devastated forest vegetation in the ranches close to the actual study area (Koutché 1947 cited in APN 2019).

National parks and the National Park Directorate (NPD) have played a relevant role in the history of this region. As in the example of Yellowstone and other US national parks, the aim was principally to protect the natural scenic beauty and to provide touristic, recreational amenities for the enjoyment of future generations (Landini 2020; Álvarez 2014). However, in the Argentine case, national parks were also an instrument to strengthen territorial sovereignty, especially after the creation of the NPD in 1934 (Navarro Floria 2008). The NPD became an institutional tool of the state to act directly in borderland areas to not only protect nature but also to promote their effective and permanent occupation, which included the construction of infrastructure (Piantoni 2016). In 1945, the NPD created the LGNP and other national parks in Patagonia and expelled the lumber companies and most of the colonists who grazed sheep (Scarzanella 2002).

Initially, the DPN focused its efforts in the LGNP on organizing activities in its southern zone and providing access to the Perito Moreno Glacier. In the early 1970s, the provincial government began to support tourism. The declaration of the national park as an UNESCO world heritage site in 1981 and the construction of the airport in El Calafate in 2000 substantially increased its visibility and accessibility (Almirón 2007). According to the 2010 Argentinian Census, the department of Lago Argentino was the Argentine department with the highest population growth in the country, registering an increase of more than 150% from 2001–2010.

4.2 *The History Behind the Scenes: The Making of El Chaltén and the Development of Tourism*

Despite its isolation, the northern zone of the LGNP had already received international guests since the beginning of the twentieth century. The German Scientific Society of Buenos Aires' scientific mission explored the area in 1916, and from 1937 on, different international climbing expeditions to the summits of Fitz Roy and Cerro Torre arrived (Halvorsen 1999). Nonetheless, the NPD's first initiative to attract guests to the north took place in 1978 with a concession granted to operate an inn, inaugurated in 1980 in a national park building (Don Guerra oldest local resident, personal communication 2016). On October 12, 1985, ECh was founded by provincial authorities inside the LGNP at the confluence of Fitz Roy and Las Vueltas rivers (Rodríguez 2009). From a conservation point of view, the project seemed senseless; indeed, the NPD opposed the idea vehemently. However, the motivation of the provincial authorities was not tourism; it was the fear that Chile could invade the area. There had already been an armed skirmish in 1965 only kilometers away (Halvorsen 1999), and in 1981, amid a pre-war climate between the two countries, the Argentine Gendarmería (National Guard) had detected a Chilean military unit taking geodesic measurements in Argentine territory (Sanhueza Carvajal 2013). ECh was established in the narrow portion of a valley (Fig. 9.3) to create a living border (*frontera viva*) or, in words of its initiator, as "a real pass of Thermopylae" (Rodríguez 2009:27).

Fig. 9.3 El Chaltén at the confluence of the Fitz Roy and Las Vueltas River inside the Los Glaciares National Park, Argentina, "a real pass of Thermopylae". (Photo Gerique 2019)

In 1989, in compliance with the National Congress, the DPN transferred 135 hectares to the Province of Santa Cruz for the creation of ECh (APN 2019). Shortly afterwards the DPN built the first 12 housing units for personnel of Public Services, Health Center, Local Court, Provincial Police, National Gendarmerie, and Provincial Agrarian Council, among others. Apart from a few ranchers, almost all inhabitants were public servants (Don Guerra, oldest local resident, personal communication 2016). During the 1990s, the Provincial Agrarian Council began to plot land for tourist ventures in ECh. In this way, many of the first inhabitants set up small tourist businesses, becoming the first local entrepreneurs (Raúl Andrade, major of ECh, personal communication, 2016). As early as 1993, ECh was declared Provincial Capital of Trekking due to the great variety of trails; and in 1995, the Argentine Government declared ECh as the annual site for the national trekking festival (STEMEC 2017). At the turn of the century, while much of the country was suffering a deep economic and social crisis, tourism in southern Patagonia offered employment opportunities. The growth of tourists arriving in Santa Cruz Province remained at 7% per year, doubling after the devaluation of Argentina's currency in 2002 (Vacca and Schinelli 2005). While there were only 41 inhabitants in ECh in 1991, population reached 371 in 2001 (INDEC 2010). During the 2000s, provincial and national authorities supported the construction and enlargement of several key infrastructures. By 2010 the population had grown by 440% to 1627 inhabitants (INDEC 2010). As dictated by provincial laws, ECh became a municipality after reaching 1000 inhabitants. While this brought opportunities for self-government, it also increased expenses and the obligations to regulate income (Raúl Andrade, major of ECh, personal communication 2016). By 2015, the local government reported 2325 inhabitants (STEMEC 2017).

4.3 *The Protagonists of Change and Trends Regarding Urban Area*

The residents of ECh are young; most under 60 years old. According to our 2016 survey, the residents who arrived after the starting phase of ECh came mainly (42%) from the Metropolitan area of Buenos Aires (AMBA). Considering that AMBA represents 37% of the country's population (INDEC 2010), 42% does not seem to over-represent the place of origin of the inhabitants. As stated by Mendoza (2017), newcomers were escaping from Buenos Aires and other urban areas seeking tranquility, fresh air, and to be closer to nature. Nevertheless, they also came to work. As in other regions (Rudzitis 1999), nature-based amenities motivated people to come to ECh. One third of the interviewees affirmed that the main reason to move to ECh was nature, followed by another 30% who did it because of job opportunities. Family (15%) was the third most frequent factor. The survey results are summarized on Table 9.1.

The number of inhabitants surges during the busy season (October 12 to Easter week) due to the arrival of *golondrinas* (swallows) seasonal workers (Pérez 2019). To the best of our knowledge, there are no data about these seasonal inhabitants. In our survey, 36% (55 individuals) were part-time residents, while almost two thirds (64%, 96 individuals) lived in EC permanently. In both groups, 94% were Argentines. Except for a retired woman from the U.S., all the other part-time inhabitants were labor migrants or *golondrinas*. The vast majority (94%) were younger than 61 years old, while over two thirds (70%) were between 18 and 35 years old. In short,

Table 9.1 Selected characteristics of the inhabitants of El Chaltén (March 2016; own data)

	Age group (%)			Laborer (%)	Retired (%)	With family in ECh (%)		Region of origin (%)	Main reasons for being in ECh (%)
Residency	16–34	35–60	+60			yes	no		
Permanent	40%	56%	4%	95%	5%	61%	40%	AMBA 42%, St. Cruz 11%	Nature 35%, Work 30%, Family 15%
Non-perm.	70%	24%	6%	98%	2%	13%	87%	AMBA 50%, St. Cruz 11%	Work & nature 31%, Work 29%, Nature 29%

(N = 151, CI = 90%, est. margin error = 7%) AMBA: Metropolitan Area of Buenos Aires, ECh: El Chaltén, St. Cruz: Santa Cruz Province *golondrinas* are mainly young economic migrants that have not started a family yet and who come from the AMBA to work, but for whom nature and outdoor recreation plays an important role. They showed a high degree of mobility; during the low season, over half of them turned into rural-urban migrants and moved back to AMBA and other metropolitan areas, while almost one quarter (23%) spent the time traveling.

Many *golondrinas* aspire after a while to become full-time residents and proprietors in ECh. As ECh is a border town, according to provincial law, anyone who is at least 21 years old, is Argentine, owns no land inside Santa Cruz Province, and has resided at least 5 years in ECh is eligible for a plot to build a home or a business in the municipality (Raúl Andrade, major of ECh, 2016, personal communication).

While some local entrepreneurs offer lodging to their seasonal workers, other *golondrinas* must rent housing (CFEC 2008, unpublished data). Some residents do not own land and are forced to rent housing as well, while they wait to obtain an adjudication of the remaining plots. In addition, the growth of tourist arrivals has stimulated an increase in the accommodation offer (bed places) for tourists, which also requires additional land (Fig. 9.4). It is worth noting the rapid increase of units following the completion of the airport in El Calafate in 2000 and the paving of the road to ECh in 2009. Despite the serious problem of lack of space, in 2016 the municipality created two urban natural reserves reducing the potential buildable urban area to 75.6 hectares (MADS 2019), reinforcing the local commitment to nature conservation. By the end of 2020, there were 26 hostels, 10 hotels, 61 cabins and apartments, 33 guesthouses and lodges, and four campgrounds in ECh (STMEC

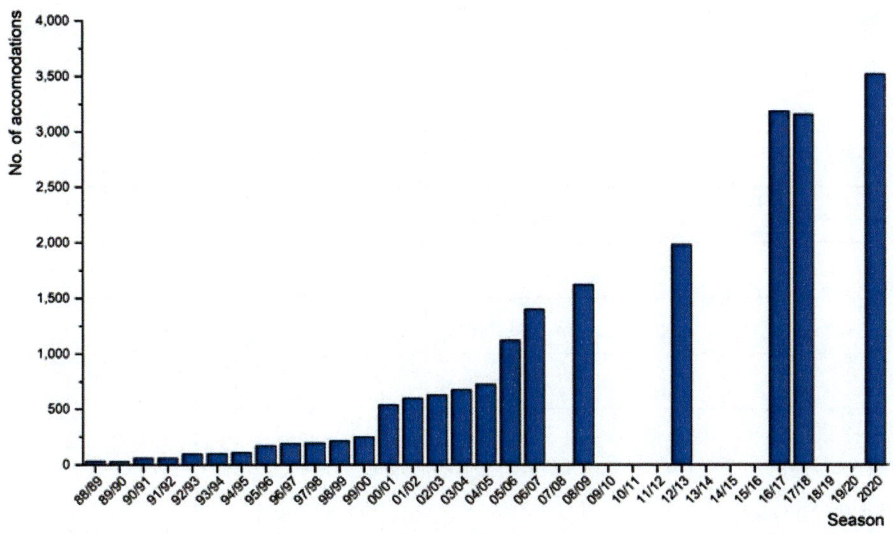

Fig. 9.4 Growth of tourist accommodations (bed places) in El Chaltén between 1988 and 2020 Data: CFEC (2009, unpublished), STMEC, unpublished data cited in La Vanguardia Noticias del Sur 2018, and OET, unpublished data cited in Winfo (2020)

2020). Almost all of these belong to local entrepreneurs with personal ties to the village.

The situation has generated a serious housing and land ownership crisis where local demand has raised prices (Pérez 2019). For several years, many residents have been requesting a response to their claims for land parcels (Fig. 9.5a). In 2016, there were hundreds of uncompleted applications (Raúl Andrade, major of ECh, 2016). The lack of transparency in the process led to accusations of corruption of local officials in the assignment of parcels. For years, rumors circulated about an expansion of the town or the creation of a new urban nucleus to offer land to those on a waiting list (Mendoza 2017). According to our 2016 survey data, 72% of the residents considered that ECh has reached the city limits, while 28% believed that the town could still grow within its borders. The percentages among seasonal inhabitants are similar (74% vs. 26%). Forty percent of the residents and 31% of the seasonal inhabitants who gave an opinion were clearly against a new urban center. However, a majority preferred a new village, to the south along Lake Viedma or in the steppe in front of the national park's entrance (Fig. 9.5b), while others preferred a town to the north, in the valley of the Río Las Vueltas (Fig.9.5c). Only two interviewees mentioned the possibility of expanding the town towards the national park.

Fig. 9.5 Images depicting the conflict between conservation and development in the El Chaltén: (**a**) "SOS land plots now for El Chaltén", a protest sign in the streets of El Chaltén in 2019; (**b**): View of Bahía Túnel (Lake Viedma) in the steppe southeast of El Chaltén; (**c**) The Río Las Vueltas Valley north to El Chaltén. The surroundings could be affected by the creation of a second urban nucleus in the area (Photos: A. Gerique 2019, 2016)

When we asked residents what tourists should remember about ECh, the most repeated words were "nature" (36 times), "landscape" (15), "beauty" (14), "place" (14), and "people" (12). For them, it is important to transmit their community's appreciation for nature and conservation. In addition to the aforementioned creation of two urban protected areas, various measures reflect this assertion, such as an early banning of plastic bags, urban greening with native species, measures against light pollution, or recycling initiatives that go far beyond Argentine standards. While wastewater infrastructure is apparently effective (our analyses in local rivers recorded very low nitrate contamination, max. Peak 9 mg/l), the main challenge seems to be the local carbon footprint. While all goods for local consumption and use must be transported hundreds of kilometers to ECh, waste disposal takes place 133 km away. Moreover, used glass travels 2680 km for recycling in La Matanza, Buenos Aires Province (MEC 2019).

Almost all inhabitants declared that they enjoy the national park several times a year. In addition, most of them consider that the LGNP-staff does a good job in the park and in the town. The national park service not only controls visitors and the park limits (and therefore the urban borders), but it also offers a wide variety of training courses, workshops, and lectures to residents. Only the opposition of part of the population to charging entrance fees for tourists in the northern zone of the LGNP and the presence of dogs belonging to residents inside the park appeared to be sources of conflict between both sides (Alejandro Caparrós, former head of the park rangers in ECh, personal communication 2016, Mendoza 2016). The latter problem can be serious; Banks and Bryant (2007) showed how dog walking along trails in the Australian woodlands led to a 35% reduction in bird diversity and 41% reduction in abundance, both in areas where dog walking is common and where dogs are banned. In the LGNP, dog attacks on Andean deers (*Hippocamelus bisulcus*), a species at high risk of extinction, occur with some frequency (Alejandro Caparrós, former head of the park rangers in ECh, personal communication, 2016).

ECh must also accommodate thousands of tourists. During the 1989/1990 season, 2473 visitors arrived at ECh (CFEC 2009, unpublished). In January 2019, the local Secretary of Tourism estimated 150,000 tourists per season (Ledesma 2019). That is, based on hotel occupancy data from the Secretary of Tourism of El Chaltén (2019, unpublished), the total number of tourists that arrived to ECh in 1.3 days during January 2019 is equivalent to the total number of tourists in the entire 1989/1990 season. Data from the Secretary of Tourism shows an enormous growth of visitors between 1999 and 2018 (Fig. 9.6).

Most tourists who visit ECh are international visitors; however, the trend is changing despite the steady growth. Until 2012, the annual rate of international visitors was close to 70%, more recently the rate decreased to around 50% (APN 2019). Our results show that during March 2016, 57% of the tourists came from overseas (APN 2019). Most of them arrived by bus (82%) and were travelling with their partner (47%), friends (25%), or alone (21%). In addition, 74.5% were younger than 36 years old, and 93% younger than 61. Their educational background was high: 68.5% had a university degree and 10.5% postgraduate studies. Their stay ranged from one to 60 days, with an average of 5 days. Consistent with STEMEC 2017

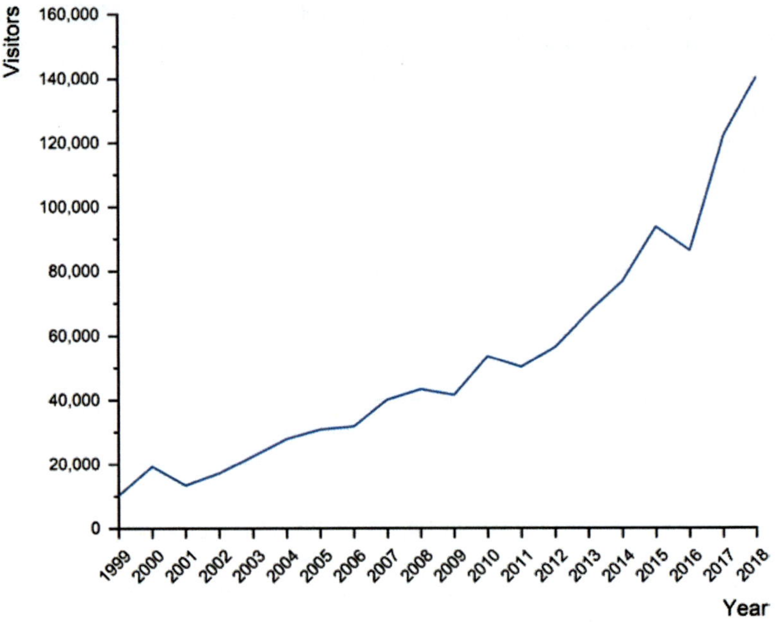

Fig. 9.6 Annual number of visitors in El Chaltén during the period 1999–2018 based on data available from the visitor's register of the national park (STEMEC 2017, and STEMEC 2019 unpublished data). Completing the registry is obligatory for those arriving by public transport during open hours. Following the advice of CFEC (2009), unpublished, STMELC (2017), and APN (2019), the data have been scaled up by 35% to include tourists arriving by other means of transportation or who use the buses that arrive at night

data, the most frequent visit was of 3 days (50.5%). The cost of the trip increases with the spatial distance from the major centers of origin, which limits accessibility for national and international visitors. Therefore, most tourists visit EC only once in their lifetime (APN 2019). However, climbers are the exception and have more frequent visits. Most tourists travel to ECh attracted by the landscape (84%) and/or physical activity (36%). Apart from small infractions such as an unsustainable use of water sources, tourists' behavior inside the national park is, for the most part, perceived as positive (Alejandro Caparrós, former head of the park rangers in ECh, personal communication 2016). Most tourists enjoyed their experience inside the park; the trails obtained the best score under the evaluated park features (Table 9.2). In March 2019, we interviewed 54 additional tourists to discover new trends in their preferences. The results were very similar to those in 2016. We did register an increase in car rental as a means of arrival (15%) to the detriment of the bus option (70%). Tourist satisfaction was again very high except over waste disposal facilities which were perceived as lacking (Table 9.2). About one third (35%) had camped inside the LGNP during their stay.

Almost two thirds stated that their expectations were fulfilled, while more than one third declared that their expectations had been exceeded. The results suggest

Table 9.2 Satisfaction of tourists with their visit to the northern zone of LGNP and ECh, Argentina, during March 2016 and March 2019 (own data)

	Satisfaction with							
Year (sample size)	Costs	Hosting	City design	Hiking trails	Info boards inside the NP	Waste disposal	NP Info material	Cost-Benefit ratio
2016 (n = 150)	2.5	2.0	2.0	1.4	1.6	2.0	1.7	2.2
2019 (n = 54)	2.4	2.0	1.9	1.5	2.0	2.6	2.5	2.0

The tourists were asked to value their satisfaction in a scale between 1 (excellent), 2 (good), 3 (bad), and 4 (very bad)

that despite the increasing tourist numbers, ECh continues to be an attractive destination. The 19/20 season was on track to be the best of all time; the Covid-19 pandemic changed the trend dramatically (OET 2020, unpublished data cited in Winfo 2020).

4.4 The Development of Land Cover and the Impacts of Tourism in the Surroundings of El Chaltén

The size, rugged landscape and elevation conditions, and large number of trails of the LGNP, make it relatively impossible to analyze the impact of ECh and the tourism industry outside the town's urban core. Similarly, site level fieldwork does not allow a landscape level analysis of possible temporal land cover changes. We analyzed these aspects via remote sensing, focusing on the northern part of LGNP from Viedma Lake and Viedma glacier in the South to the Lago del Desierto Provincial Reserve (LDPR) up to the southern shores of Lago del Desierto (Fig. 9.7). The majority of the time series (75%, 442 km^2) showed no significant ($p > 0.05$) NDVI changes over the study period (1985–2020). However, 19% (112 km2) of the time series exhibited a significantly increasing NDVI trend ($p < 0.05$), which suggests an increasing amount of green vegetation, a process usually termed as "greening". In contrast, only 6% (35 km^2) revealed a significantly ($p < 0.05$) decreasing NDVI trend, indicating the loss of vegetation, which is commonly referred to as "browning" (Zhu et al. 2016).

Areas with significant positive and negative change are distributed all over the study area, however positive changes occur more frequently within the borders of LGNP compared to LDPR. Clusters of positive change were detected in the steppe area north of Viedma Lake, in the Túnel river valley, to the south-west of ECh and in the north-eastern corner of LGNP. In contrast, areas that exhibited negative change were found particularly in the southern part of LDPR, but also in some hotspots within the borders of LGNP, especially in the Fitz Roy River valley, to the west of ECh toward Laguna Torre.

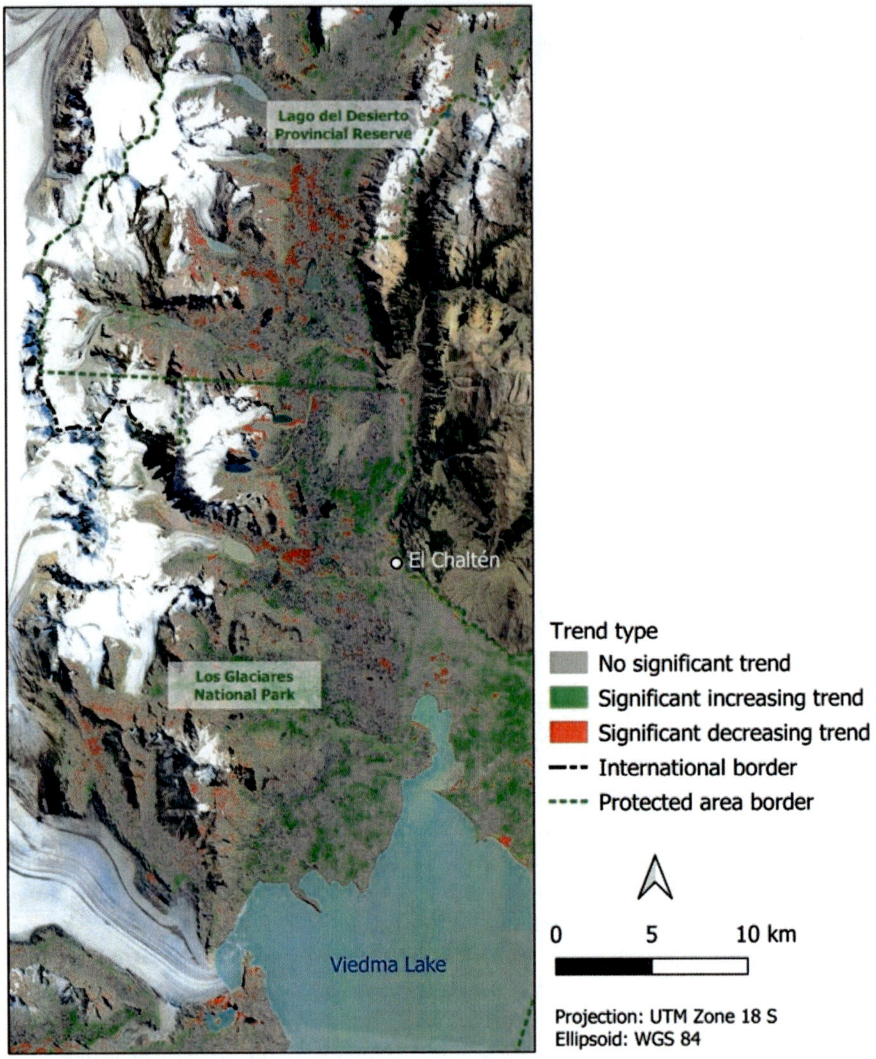

Fig. 9.7 Land cover change as detected by the NDVI time series analysis (Vanselow 2021)

Our results showed significant increasing trends for extensive parts of the study area and 73% of these time series revealed a continuous linear trend. This finding is consistent with global scale observations that show increasing green vegetation cover in many regions of the world in the last decades including the Andes (Zeng et al. 2018; Zhu et al. 2016). Causative drivers of greening trends are of anthropogenic nature including land use and nitrogen depositions, climate change, and improvements in water collection, distribution and diversion infrastructure, and management (López et al. 2020). Climate change is particularly relevant in high latitudes and elevations, both characteristics of the study area (Zhu et al. 2016; Liu et al. 2015). In our study, greening was detected predominantly within the borders of LGNP whereas this development is less distinct in the less strictly protected LDPR, consisting predominantly of private land. This finding suggests that management strategies implemented by the national park positively contribute to vegetation conservation and regeneration.

Browning trends, on the other hand, are by far less extensive compared to greening trends at a global level. This situation is well reflected in our results, since areas exhibiting a greening trend significantly exceeded those with a browning trend. Globally, the main drivers of browning are droughts and fires (Pan et al. 2018). However, direct human impacts such as deforestation and overexploitation of natural resources are important browning factors as well (Lü et al. 2015; Mueller and Dressler 2014). A remarkable outcome of our results is the large share (58%) of non-linear trends within the browning time series. In comparison to greening time series, continuous trends are less common since browning is often driven by discrete disturbances such as fires. Some very dynamic areas such as river floodplains, lakeshores, or dune fields also exhibit browning but were excluded from the interpretation in this study since such areas do not show stable trends.

We illustrate our findings using three examples: Example 1 shows an area 1.5 km to the east of Laguna Torre along a trail. For this location, the smooth function of the variable 'YEAR' exhibits a steep NDVI drop until around 2013, when a minimum is reached and the NDVI starts to increase again. Here, a large fire, which removed large portions of vegetation cover, was caused by a tourist trying to burn toilet paper (Alejandro Caparrós, former head of the park rangers in ECh, personal communication 2016). In recent years, however, our trend analysis indicates that vegetation cover has increased (Fig. 9.8a). Similarly, in 2006 a *gaucho* (i.e., a South American cattle manager) caused an accidental forest fire along the northern border of the national park (APN 2019) that we also detected via remote sensing analysis. This evidence shows that human induced forest fires are still a major concern for national park managers and an important driver of land cover change in the study area. Example 2 represents a campground area (Fig. 9.8b). While the remote sensing analysis results indicated no significant change, our ground truthing evaluation showed that the vegetation below the canopy was substantially affected by tourists, particularly due to trampling. This example demonstrates well the importance of field work and the limitations of non-ground truthed remote sensing analyses. Example 3 is from LDPR. This area is characterized by dispersed settlements affected by the occasional presence of cattle grazing and trampling, and small tourism operations that contribute to vegetation degradation (Fig. 9.8c).

In contrast to the situation in other Patagonian national parks, NPD granted existing *estancias* precarious occupancy and grazing permits inside the park (APN 2019). A family, for example, still uses some park areas as summer and wintering fields (APN 2019), which explains soil compaction, the presence of fodder species such as *Festuca* spp. and *Trifolium repens*, and the presence of cattle inside the national park. However, our results show that the impact caused by these factors is relatively low. The overall built-up area in LGNP is small and concentrated in ECh. This characteristic is not different from other protected areas throughout Argentina

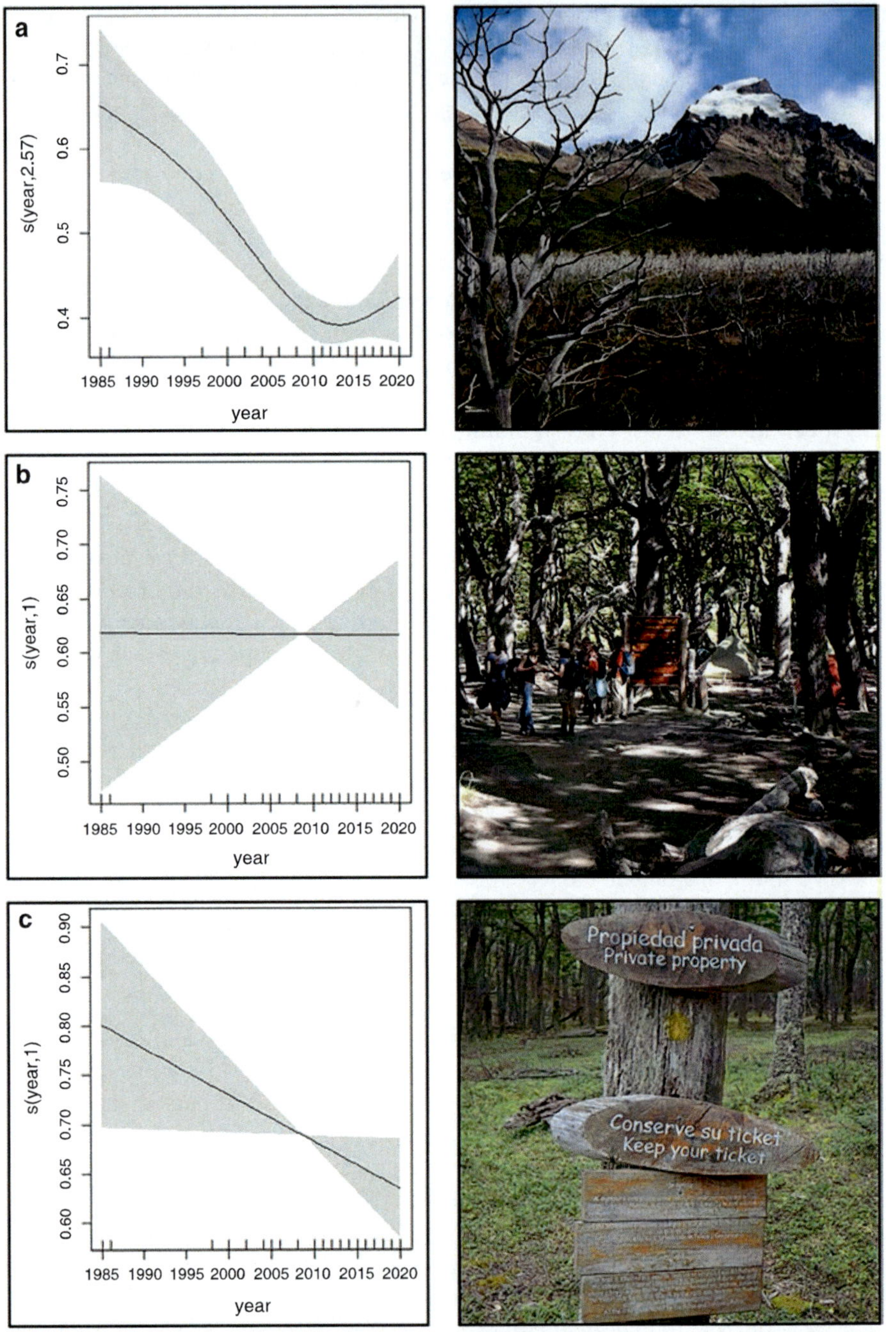

Fig. 9.8 Illustrative examples of land cover change for selected locations (a: Disturbance by a fire started by a tourist near Laguna Torre; b: Impact of trampling on a campground that could not be detected via remote sensing; c: Continuous decreasing trend suggesting successive vegetation degradation on private land in Lago del Desierto Provincial Reserve) (Photos A. Gerique 2014, 2019)

Fig. 9.9 Erosion along a trail in the El Chaltén area. While one path remains open, staff from the national park blocked the path to the right for regeneration (Photo A. Gerique 2019)

(de la Fuente et al. 2020). Nonetheless, the high number of visitors and intensive hiking provoke severe soil erosion along the trails. Every season, national park staff attempt to mitigate this impact by applying a diversity of measures such as closing off trails for regeneration (Fig. 9.9). This problem is outside the scope of our remote sensing analysis, but it can be easily observed through fieldwork.

5 Does 'El Chalten' Represent a Model of Coexistence Between Urbanization, Tourism, and Conservation?

El Chaltén was not the result of careful planning to integrate urban development, tourism, and conservation inside a natural area, but a historical mishap of a conflict between two neighboring nations. However, since 1985, a mix of local, provincial, and state officials, entrepreneurs, amenity migrants, part-time workers, and local tourist capital have created a village of some 2400 inhabitants inside a national park, which, until the arrival of the Covid-19 pandemic in 2020, received more than 150,000 tourists per year. During the last three and a half decades, ECh has not spread outside its boundaries thanks to the effective control of the LGNP-management and the environmental awareness of its inhabitants.

Our results show that, as in other destinations affected by naturbanization processes, most migrants arrived at ECh motivated by nature and landscape beauty, and labor opportunities (cf. Calvache et al. 2016; Prados 2011). A local particularity is the large number of Argentine seasonal workers in ECh who want to move from job-migrants or economic movers to "real" amenity migrants (cf. Mitchell and Shannon 2017; Moss 2009). The lack of space and the difficult access to land parcels for foreigners seems to have prevented the arrival of international amenity migrants. Also, to date, the village has apparently avoided an influx of major foreign investors and land speculators; most businesses are in local hands.

The number of tourist arrivals increased steadily until the Covid-19 pandemic. Despite the overcrowding, the satisfaction of tourists with their stay in ECh and the LGNP attractiveness do not seem to have diminished. However, the rate of international tourists has decreased since 2012. This change reduces the potential impact of possible external shocks like the pandemic but increases the exposure to the ups and downs of Argentina's economy. As the results of the remote sensing time series analysis show, the main anthropogenic changes in the northern zone of the LGNP are related to two unintentional fires, with only one of these fires connected to tourism. The human footprint is greater in the areas outside the park, despite being within a sparsely inhabited provincial reserve (LDPR). Other problems, such as soil compaction, the existence of non-native species, and overgrazing, are the most conspicuous consequence of the presence of *estancias* established prior to the national park. Fieldwork reveals that the main impacts provoked by ECh and its industry inside the LGPN are erosion along the trails and the destruction of the understory in campgrounds. Future research should thus focus on the carrying capacity of the trails and campgrounds inside the LGNP as well as the impact of tourists on flora and fauna.

Our mix-methods analysis shows that limiting the concentration of housing to a small buildable zone has reduced infrastructure costs and facilitated water, energy, and waste management. There exist other less obvious advantages of a small buildable area, as a comparison with La Angostura shows. With a population of 15,710 inhabitants (INDEC 2010) and an area of 8000 ha (IGC 2007), La Angostura is a famous tourist destination located within the Nahuel Napi National Park, Northern Patagonia. Like ECh, La Angostura's population grew very fast (by 114% between 2001 and 2010, INDEC 2010). González et al. (2009) showed how the increased number of visitors and external investments led to an oversupply of bed places with negative impacts on profit. In addition, by having a larger area for development, La Angostura's urban morphology provoked a diffuse urban growth that affected the landscape and local attractiveness. Moreover, as in other mountain regions (cf. Perlik 2006), land shortage brought about economic displacement. The original inhabitants retreated to less suitable areas to live in, as prices increased dramatically in the most attractive areas. The combination of a small urban core with the absence of foreign tourist investors in ECh represents an advantage that may have helped avoid the results of La Angostura's urban expansion.

Fieldwork has shown that the main problem that affects ECh and that could endanger both biodiversity and natural landscape features in the northern area of the LGNP is the social demand for land. Ideally, the issue should be resolved at the local level, by allowing a higher ratio of dwellings per plot and by applying creative solutions (such as owners' cooperatives and a moratorium on the creation of new lodging places). A new settlement could have immediate effects on the environment, whether it is built in the valley north of ECh, along Lake Viedma, or in the steppe. By building on private land, the danger of urban sprawl into adjacent areas would be accentuated, increasing environmental and landscape damage, and hampering monitoring. Urban growth would also increase traffic to and from ECh, disturbing the tranquility of the valley, especially if the settlement is founded north of town. As in La Angostura, in the medium term, the commodification of the aesthetic landscape would provoke economic displacement. ECh would probably become a (high-class) tourist resort with the consequent replacement of most of its population with those who are better off economically. Alternatively, if the new village were to reduce the number of future visitors to ECh, local communities could potentially increase political pressure to see more benefits from the natural resources, but in ways less consistent with current conservation goals (Stewart et al. 2016). Any of these changes could result in UNESCO placing the LGNP on its list of endangered heritage sites or even in removing the status as World Natural Heritage.

The complexity of naturbanization processes makes each case unique. Nevertheless, the example of ECh shows that it is possible to combine urban development, tourism, and conservation within a protected area. The key seems to be the presence of clear limits, a good urban infrastructure, adequate park management, and a population that identifies with the conservation of the surrounding natural environment. For the model to be sustainable over time, institutions and inhabitants must limit the numbers of tourist arrivals if they continue to grow. Moreover, they should accept that the urban area resembles a privileged island in the Patagonian landscape that cannot grow beyond its limits, even if this means that there is no longer space for some of those who wish to reside in ECh.

Ultimately, this research demonstrates the utility of combining qualitative and quantitative survey data with quantitative analysis of remote sensing indices (NDVI) to inform socio-environmental planning of naturbanization in a tourist village embedded in a montane national park. The importance of adding field data and local knowledge of both the physical and human geography of this region is necessary to contextualize sub-canopy dynamics, fire instances, and grazing patterns that could threaten the sustainability of tourism.

Acknowledgements We thank the Franconian Geographical Society FGG for the financial support. Our Gratitude goes to the inhabitants of El Chaltén for their hospitality and generous participation in the interviews. We especially thank Romina Bernard, Don Guerra, Daniel Rodriguez, Raúl Andrade, Alejandro Caparrós, Dante Ardenghi, and all those anonymous tourists who shared their precious time with us. We are grateful to Marina llg, Nina Knode, Jana Peters, Franziska Wirth, and Lilian Reiss for their enthusiastic assistance during fieldwork.

References

Almirón A (2007) Valorización turística de un área protegida. El caso del destino Los Glaciares en el extremo sur patagónico Revista Párrafos Geográficos 6(1), http://igeopatorg/parrafosgeograficos/ Accessed 19 Feb 2021

Álvarez G (2014) El lugar de los parques nacionales en la representación de una Patagonia turística, discusión y habilitación del paisaje patagónico durante el siglo XX. Magallania 42(1):53–76. https://www.redalyc.org/articulo.oa?id=506/50631449004. Accessed 21 Feb 2021

Ampuero C, Norambuena M, Cáceres AP (2015) Etapas de crecimiento de El Calafate, Provincia de Santa Cruz. Patagonia Austral Andina Argentina Boletín de Estudios Geográficos 115

APN, Administración de Parques Nacionales (2019) Plan de Gestión del Parque Nacional Los Glaciares

Bailey KM, McCleery RA, Binford MW et al (2015) Land-cover change within and around protected areas in a biodiversity hotspot. J Land Use Sci 11(2):154–176. https://doi.org/10.1080/1747423X.2015.1086905

Banks PB, Bryant JV (2007) Four-legged friend or foe? Dog walking displaces native birds from natural areas. Biol Lett 3:611–613. https://doi.org/10.1098/rsbl.2007.0374

Belardi JB, Espinosa SL, Carballo Marina F et al (2020) Al pie del cerro Fitz Roy: cazadores recolectores en el Noroeste del lago Viedma, provincia de Santa Cruz. Revista del Museo de Antropología 13(3):295–306. https://doi.org/10.31048/1852.4826.v13.n3.29017

Bender O, Kanitscheider S (2013) Amenity migration in the southern Andes and the southern European Alps – a key factor for sustainable regional development. Mitteilungen der Österreichischen Geographischen Gesellschaft 155:105–124

Bender O, Roth CE, Job H (2017) Protected areas and sustainable development in the Alps. Econ Mont J Prot mount areas res manage (9):5–16. https://doi.org/10.1553/eco.mont-9-sis5

Benson M, O'Reilly K (2016) From lifestyle migration to lifestyle in migration: categories, concepts and ways of thinking. Migr Stud 4(1):20–37. https://doi.org/10.1093/migration/mnv015

Berry BJL (1980) Urbanization and Counterurbanization in the United States. AAPSS 451:14–20

Borrero LA, McEwan C (1997) The peopling of Patagonia. In: McEwan C, Borrero LA, Prieto (eds) Patagonia – natural history, prehistory, and ethnography at the uttermost end of the earth. Princeton University Press, Princeton, pp 32–45

Borsdorf A, Hidalgo (2009) Searching for fresh air, tranquility and rural culture in the mountains. A new lifestyle for Chileans? Erde 140(3):275–292

Calvache F, Prados MJ, Lourenço JM (2016) Assessment of National Parks affected by naturbanization processes in southern Europe. J Environ Plan Manag 59(9):1629–1655. https://doi.org/10.1080/09640568.2015.1083416

CBD Convention on Biological Diversity (2022) Nations adopt four goals, 23 targets for 2030 in landmark UN biodiversity agreement. Official CBD Press Release 19 December 2022 – Montreal

Champion AG (ed) (1989) Counterurbanization – the changing pace and nature of population deconcentration. Edward Arnold, London

Chape S, Spalding MD, Jenkins MD (eds) (2008) The World's protected areas: status, values, and prospects in the twenty-first century. University of California Press, Berkeley

de la Fuente B, Bertzky B, Delli G et al (2020) Built-up areas within and around protected areas: global patterns and 40-year trends. Glob Ecol Conserv 24. https://doi.org/10.1016/j.gecco.2020.e01291

DeFries R, Karanth KK, Pareeth S (2010) Interactions between protected areas and their surroundings in human-dominated tropical landscapes. Biol Conserv 143:2870–2880. https://doi.org/10.1016/j.biocon.2010.02.010

del Valle RC, Prados Velasco MJ (2019) People and settlements in Andalusian national parks. Residential development and quality of life as a trigger of naturbanization processes. Investigaciones Geográficas 71:9–25. https://doi.org/10.14198/INGEO2019.71.01

DSDG United Nations Division for Sustainable Development Goals (2015) Sustainable Development Goals https://sustainabledevelopment.un.org/sdgs. Accessed 15 Feb 2021

Elmqvist T, Fragkias M, Goodness J et al (eds) (2013) Urbanization, biodiversity and ecosystem services: challenges and opportunities: a global assessment. Springer. https://doi.org/10.1007/978-94-007-7088-1
Feilhauer H, Dahlke C, Doktor C et al (2014) Mapping the local variability of Natura 2000 habitats with remote sensing. Appl Veg Sci 17(4):765–779. https://doi.org/10.1111/avsc.12115
Forkel M, Carvalhais N, Verbesselt J et al (2013) Trend change detection in NDVI time series: effects of inter-annual variability and methodology. Remote Sens 5:2113–2144. https://doi.org/10.3390/rs5052113
Franco NV, Borrero LA, Brook GA et al (2018) Changes in technological organization and human use of the space in the south of Patagonia (Argentina) during the late Holocene. In: Robinson E, Sellet F (eds) Lithic technological organization and paleoenvironmental change: global and diachronic perspectives. Springer, Nueva York, pp 301–320
Fraser RH, Olthof I, Pouliot D (2009) Monitoring land cover change and ecological integrity in Canada's National Parks. Remote Sens Environ 113:1397–1409. https://doi.org/10.1016/j.rse.2008.06.019
Geldmann J, Barnes M, Coad L et al (2013) Effectiveness of terrestrial protected areas in reducing habitat loss and population declines. Biol Conserv 161:230–238. https://doi.org/10.1016/j.biocon.2013.02.018
Gemelli A (2007) Plan de Turismo Sustentable para la Provincia de Santa Cruz 2008–2016. Consejo Federal de Inversiones Provincia de Santa Cruz. Informe Final, Buenos Aires
Giusti M, Prados Velasco MJ (2013) Naturbanización en la Pampa Argentina: urbanización vs. racionalidad ecológica: el caso de Chascomús (Provincia de Buenos Aires, Argentina). Seminario Internacional de Investigación en Urbanismo, Barcelona-Buenos Aires. Barcelona, pp 619–635
Godoy C (2000) El gran libro de la provincia de Santa Cruz. Tomo 1 y 2. Milenio ediciones y Alfa Centro Literario, Neuquén
González R, Otero A, Nakayama L et al (2009) Las movilidades del turismo y las migraciones de amenidad: problemáticas y contradicciones en el desarrollo de centros turísticos de montaña. Rev Geogr Norte Gd 44:75–92. https://doi.org/10.4067/S0718-34022009000300004
Halvorsen P (1999) Entre el río de las vueltas y los hielos continentales. Editorial Vinciguerra, Buenos Aires
IGC Instituto de Gestión de Ciudades (2007) Plan VLA07. Plan de ordenamiento territorial y amibental de Villa La Angostura. Documento Base de Trabajo. Municipalidad de Villa La Angostura, Argentina
INDEC Instituto Nacional de Estadística y Censos (2010) Censo Nacional de Población, Hogares y Viviendas 2010. https://www.indec.gob.ar/indec/web/Nivel4-Tema-2-41-135. Accessed 18 Feb 2021
Janoschka M (2009) The contested spaces of lifestyle Mobilities: regime analysis as a tool to study political claims in Latin American retirement destinations. Erde 140(3):1–20
Jones N, McGinlay J, Dimitrakopoulos PG (2017) Improving social impact assessment of protected areas: a review of the literature and directions for future research. Environ Impact Assess Rev 64(1):1–7. https://doi.org/10.1016/j.eiar.2016.12.007
Kondo M (2012) Protecting the idyll but not the environment: second homes, amenity migration and rural exclusion in Washington state. Landsc Urban Plan 106:174–182. https://doi.org/10.1016/j.landurban.2012.03.003
Kordel S, Pohle P (2018) International lifestyle migration in the Andes of Ecuador: how migrants from the USA perform privilege, import rurality and evaluate their impact on local community. Sociol Rural 58(1):126–146. https://doi.org/10.1111/soru.12133
Koutché V (1947) Parque Nacional Los Glaciares. Informe inédito
La Vanguardia Noticias del Sur (2018) Crecen las estadísticas de visitas en El Chaltén, 3rd of January https://wwwlavanguardianoticiascomar/nota/18401-crecen-las-estadisticas-de-visitas-en-el-chalten/ Accessed 15 Feb 2021
Landini G (2020) Parques Nacionales y la configuración de un destino turístico: transformaciones socio territoriales en San Martín de los Andes, Argentina (1937–1955). Anuario de Historia Regional y de las. Fronteras 26(1):71–110. https://doi.org/10.18273/revanu.v26n1-2021003

Ledesma J (2019) El Chaltén, elegido como uno de los destinos más hospitalarios del mundo El Cronista Clase 10 January 2019. https://www.cronista.com/clase/break/El-Chalten-elegido-como-uno-de-los-destinos-mas-hospitalarios-del-mundo-20190110-0010.html. Accessed 21 Feb 2021

Liu Y, Li Y, Li S (2015) Motesharrei, S. spatial and temporal patterns of global NDVI trends: correlations with climate and human factors. Remote Sens 7:13233–13250. https://doi.org/10.3390/rs71013233

Loeffler R, Steinicke E (2007) Amenity Migration in the U.S. Sierra Nevada. Geogr Rev 97(1):67–88. https://doi.org/10.1111/j.1931-0846.2007.tb00280.x

López S, López-Sandoval MJ, Jung J-K (2020) New insights on land use, land cover, and climate change in human–environment dynamics of the equatorial Andes. Ann Am Assoc Geogr 111(4):1110–1136

Lü Y, Zhang L, Feng X, Zeng Y et al (2015) Recent ecological transitions in China: greening, browning, and influential factors. Sci Rep. https://doi.org/10.1038/srep08732

MADS Ministerio del Ambiente y Desarrollo Sustentable, Consejo Agrario Provincial de Santa Cruz y Municipalidad de El Chaltén (2019) Plan de Forestación con Especies Nativas

Madsen A (1952) Bocetos de la Patagonia Vieja. Ciordia y Rodríguez Editores, Buenos Aires

Mathelvet R, Thomson JD, Folke C et al (2016) Protected areas and their surrounding territory: socioecological systems in the context of ecological solidarity. Ecol Appl 26(1):5–16. https://doi.org/10.1890/14-0421

McIntyre M (2009) Rethinking amenity migration: integrating mobility, lifestyle and social-ecological systems. Erde 140(3):229–250

MEC (2019) Jornadas de acopio de vidrio para reciclar. Municipalidad de El Chaltén https://wwwelchaltengobar/jornadas-de-acopio-de-vidrio-para-reciclar/ Accessed 18 Feb 2021

Medina V (2017) Las movilidades poblacionales y su impacto territorial en la estructura espacial de las ciudades turísticas. El caso de San Carlos de Bariloche. EURE 43(129):71–92. https://doi.org/10.4067/S0250-71612017000200004

Mendoza M (2016) Educational policing: park rangers and the politics of the green (E)state in Patagonia. J Lat Am Caribb Anthropol 21(1):173–192. https://doi.org/10.1111/jlca.12195

Mendoza M (2017) Post-neoliberal labor in Patagonia: informality and citizenship in the green economy. Dialect Anthropol 41:55–76. https://doi.org/10.1007/s10624-017-9446-9

Mitchell CJA, Shannon (2017) Establishing the routes to rural in-migrant proprietorship in a Canadian tourism region: a mobilities perspective. Popul Space Place 24(3). https://doi.org/10.1002/psp.2095

Moss LAG (2009) The mountain amenity migration phenomenon, why it is happening and our response. In: Moss LAG, Glorioso RS, Krause A (eds) Understanding and managing amenity -led migration in mountain regions. The Banff Centre, Banf

Mueller T, Dressler G, Tucker et al. (2014) Human land-use practices Lead to global long-term increases in photosynthetic capacity. Remote Sens 6:5717–5731. doi:https://doi.org/10.3390/rs6065717

Nagendra H, Lucas R, Honrado JP et al (2013) Remote sensing for conservation monitoring: assessing protected areas, habitat extent, habitat condition, species diversity, and threats. Ecol Indic 33:45–59. https://doi.org/10.1016/j.ecolind.2012.09.014

Navarro Floria P (2008) El proceso de construcción social de la región del Nahuel Huapi en la práctica simbólica y material de Exequiel Bustillo (1934–1944). Revista Pilquen, Sección Ciencias Sociales. Año IX(9):1–14

Neumann RJ (1997) Protected area buffer zones and the politics of land in Africa. Dev Change 28(39):559–582

Oliva M, García-Frapolli E, Porter-Bolland L et al (2019) Early detection of conflicts for the Management of Protected Areas: the case of charcoal production in the Los Petenes biosphere reserve, Mexico. Environ Manag 64:52–63. https://doi.org/10.1007/s00267-019-01170-x

Pan N, Feng X, Fu B et al (2018) Increasing global vegetation Browning hidden in overall vegetation greening: insights from time-varying trends. Remote Sens Environ 214:59–72. https://doi.org/10.1016/j.rse.2018.05.018

Pérez AE (2014) Sociedades originarias de Patagonia: pasado y presente. Fundación de Historia Natural Félix de Azara, Buenos Aires

Pérez S (2019) Geopolíticas del cotidiano en la frontera patagónica: Las dinámicas del habitar en torno a los Campos de Hielo Patagónicos. Revista Austral de Ciencias Sociales 37:187–207. https://doi.org/10.4206/rev.austral.cienc.soc.2019.n37-11

Perlik M (2006) Chapter 15: The specifics of amenity migration in the European Alps. In: Moss LAG (ed) The amenity migrants: seeking and Sustaining Mountains and their cultures. CABI Publishing, Wallingford

Perlik M (2011) Alpine gentrification: the mountain village as a metropolitan neighbourhood. Rev Geogr Alp:99–1. https://doi.org/10.4000/rga.1370

Petrova S (2014) Communities in transition: protected nature and local people in eastern and Central Europe. Routledge studies in environmental policy and practice. Taylor and Francis, Kindle-Version

Piantoni G (2016) La fundación de Parques Nacionales y sus agencias de control en Argentina y Brasil: una propuesta comparativa en la primera mitad del siglo XX. Cuadernos del Sur – Historia 45:105–129

Prados MJ (2009) Conceptual and methodological framework of naturbanization. In: Prados MJ (ed) Naturbanization: new identities and processes for rural-natural areas. Taylor and Francis, London, pp 11–28

Prados MJ (2011) Naturbanización. Algunos ejemplos en áreas de montaña y periurbana Treballs de la Societat Catalana de Geografia 71–72:179–200

Rochelle S, Morrison C, Castley JG (2015) Birdwatching and avitourism: a global review of research into its participant markets, distribution and impacts, highlighting future research priorities to inform sustainable avitourism management. J Sustain Tour 23(8–9):1257–1276. https://doi.org/10.1080/09669582.2014.924955

Rodríguez DR (2009) La fundación de El Chaltén y su evolución. Tecnología electrónica del futuro, Río Gallegos

Rudzitis G (1999) Amenities increasingly draw people to the rural west. Rural Dev Perspect 14(2):9–13

Sanhueza Carvajal R (2013) Desbrozando mitos: América Latina y Chile ante la guerra del Atlántico Sur. Revista Política y Estrategia 11:59–78

Scarzanella E (2002) Las bellezas naturales y la nación: Los parques nacionales en Argentina en la primera mitad del siglo XX. Rev Eur Estud Latinoam Caribe 73:5–21

Simpson GL (2018) Modelling palaeoecological time series using generalized additive models. Front Ecol Evol:6. https://doi.org/10.3389/fevo.2018.00149

Soliku O, Schraml U (2018) From conflict to collaboration: the contribution of co-management in mitigating conflicts in mole National Park. Ghana Oryx 54:1–11. https://doi.org/10.1017/S0030605318000285

STEMEC Secretaría de Turismo Municipalidad de El Chaltén (2017) El Chaltén de pueblo de frontera a destino turístico. Santa Cruz. https://www.yvera.tur.ar/publicaciones/documentos/01082314-d5d6-4527-804f-3c5855d505f1.pdf. Accessed 20 Feb 2021

STEMEC Secretaría de Turismo Municipalidad de El Chaltén (2020) El Chaltén, Guía de Servicio y Senderos. https://elchalten.tur.ar/wp-content/uploads/2020/12/EL-CHALTEN-TURISMO-MAPA.pdf. Accessed 23 Feb 2021

Stewart EJ, Wilson J, Espiner S et al (2016) Implications of climate change for glacier tourism. Tour Geogr 18(4):377–398. https://doi.org/10.1080/14616688.2016.1198416

Stolton S, Dudley N, Belokurov A et al (2019) Lessons learned from 18 years of implementing the management effectiveness tracking tool (METT): a perspective from the METT developers and implementers. Parks 25(2):79–92. https://doi.org/10.2305/IUCN.CH.2019.PARKS-25-2SS.en

Terborgh J, Peres CA (2002) The problem of people in parks. In: Terborgh J, van Schaik C, Davenport et al (eds) Making parks work: strategies for preserving tropical nature. Island Press, Washington, DC

Tulla A-F, Stoica IV, Pallarès-Blanch M et al (2017) Can Naturbanization promote environmentally friendly built-up areas? A comparison between Cadí-Moixeró (Catalonia, Spain) and Comana

(Romania) natural parks. European Countryside 9(4):679–708. https://doi.org/10.1515/euco-2017-0039
UNEP-CBD (2010) The Strategic Plan for Biodiversity 2011–2020 and the Aichi Biodiversity Targets. Conference of the Parties to the Convention on Biological Diversity. Tenth meeting, Nagoya, Japan, 18–29 October 2010. UNEP/CBD/COP/DEC/X/2
UNEP-WCMC (2021). Protected Area Profile for Latin America and Caribbean from the World Database of Protected Areas. March 2021. https://www.protectedplanet.net/region/SA. Accessed 1 Mar 2021
UNEP-WCMC and IUCN (2021) Discover the world's protected areas. www.protectedplanet.net/en. Accessed 1 Mar 2021
UNESCO United Nations Educational and Scientific Organization (2021) World heritage list. Los Glaciares National Park. https://whc.unesco.org/en/list/145/. Accessed 17 Feb 2021
Urry J (2000) Mobile Sociology. Br J Sociol 51(1):185–203
Vacca C, Schinelli D (2005) Caracterización económica de la Provincia de Santa Cruz y la perspectiva de un nuevo perfil productivo. 7° Congreso Nacional de Estudios del Trabajo. ASET Asociación Argentina de Estudios del Trabajo
van Noorloos F, Steel G (2015) Lifestyle migration and socio-spatial segregation in the urban(izing) landscapes of Cuenca (Ecuador) and Guanacaste (Costa Rica). Habitat Int 54. https://doi.org/10.1016/j.habitatint.2015.08.014
Verbesselt J, Hyndman R, Newnham G et al (2010) Detecting trend and seasonal changes in satellite image time series. Remote Sens Environ 114:106–115. https://doi.org/10.1016/j.rse.2009.08.014
Wang Y, Mitchell BR, Nugranad-Marzilli J et al (2009) Remote sensing of land-cover change and landscape context of the National Parks: a case study of the northeast temperate network. Remote Sens Environ 113:1453–1461. https://doi.org/10.1016/j.rse.2008.09.017
Watson JEM, Venter O, Lee J et al (2018) Protect the last of the wild. Nature 563(7729):27–30. https://doi.org/10.1038/d41586-018-07183-6
Welling JT, Árnason P, Ólafsdottír R (2015) Glacier tourism: a scoping review. Tour Geogr 17(5):635–662. https://doi.org/10.1080/14616688.2015.1084529
Winfo (2020) Turismo Calafate Winfo, 30th of January. Estadísticas provinciales indican «el mejor enero de los últimos años» con mayoría de extranjeros. https://winfo.com.ar/viajes/2020/01/estadisticas-provinciales-indican-el-mejor-enero-de-los-ultimos-anos-con-mayoria-de-extranjeros/. Accessed 16 Feb 2021
Wood SN (2017) Generalized additive models: an introduction with R (second edition). Chapman and Hall/CRC, Boca Raton
Worboys GL (2015) Concept, purpose and challenges. In: Worboys GL, Lockwood M, Kothari A et al (eds) Protected area governance and management. ANU Press, Canberra, pp 9–42
Zeng Z, Peng L, Piao S (2018) Response of terrestrial evapotranspiration to Earth's greening. Curr Opin Environ Sustain 33:9–25. https://doi.org/10.1016/j.cosust.2018.03.001
Zhu Z, Piao S, Myneni RB et al (2016) Greening of the earth and its drivers. Nat Clim Chang 6:791–795. https://doi.org/10.1038/nclimate3004
Zunino HM, Matossian B, Hidalgo R (2012) Poblamiento y desarrollo de enclaves turísticos en la Norpatagonia chileno-argentina. Migración y frontera en un espacio binacional. Rev Geogr Norte Gd 53:137–158. https://doi.org/10.4067/S0718-34022012000300009

Chapter 10
GIS Approaches to Environmental Justice in Mexico's Oil and Gas Production Zones with Implications for Latin America

Matthew Fry and Andrew Hilburn

Abstract This chapter demonstrates how to integrate geographic information systems (GIS) data into environmental justice (EJ) research in Latin America's oil and natural gas extraction regions. It discusses two forms of distributional justice analysis: traditional and benefit sharing. Two recent GIS-based EJ studies conducted in a major oil and gas production area in Mexico demonstrate the challenges, workarounds, and types of geospatial data that can be used to conduct distributional environmental justice studies. We also evaluate the availability of open-access oil and natural gas, and demographic data among six hydrocarbon-producing countries in the region and provide guidelines and suggestions for conducting GIS-based EJ studies in Latin America. Finally, this chapter contends that integrating geospatial and empirical EJ analysis with qualitative EJ approaches not only strengthens EJ analysis in oil and natural gas production zones in Latin America, but also can be used to legitimize political claims and increase community engagement, as well as facilitate interdisciplinary dialog.

Keywords Environmental justice · Extraction · Natural gas · Oil

1 Introduction

Policy shifts throughout Latin America during the 1990s initiated a massive boom in mining and fossil fuel extraction activities (Bridge 2008; Bebbington and Bury 2013). Since then, a growing body of research examines local environmental,

M. Fry (✉)
Department of Geography, University of Utah, Salt Lake City, United States, Utah
e-mail: m.fry@utah.edu

A. Hilburn
Department of Social Sciences, Texas A&M International University, Laredo, TX, USA
e-mail: ahilburn@tamiu.edu

S. López (ed.), *Socio-Environmental Research in Latin America*, The Latin American Studies Book Series, https://doi.org/10.1007/978-3-031-22680-9_10

community, and livelihood impacts in oil and natural gas extraction areas (e.g., Billo 2015; Lu 2012; Orta-Martínez and Finer 2010; Widener 2007; Kimerling 1994), with human rights violations, wealth plundering, and land dispossession important topics of environmental justice (EJ) scholarship in the region (Carruthers 2014; Urkidi and Walter 2011). Recent research based on geographic information systems (GIS) in the U.S. analyzes marginalized populations' exposure to environmental and health risks associated with oil and natural gas production activities (e.g., Clough and Bell 2016; Johnston et al. 2016; Fry et al. 2015; Ogneva-Himmelberger and Huang 2015); however, the application of GIS-based EJ approaches in Latin America has been limited, often due to a lack of publicly-available geospatial data on fossil fuel extraction activities in the region. While the data availability issue still exists, many countries now offer a wider array of spatial and demographic data free for download. In this chapter, we identify some methods and data that can be used to integrate GIS into EJ research to enhance analyses of inequalities in fossil fuel extraction areas in Latin America.

From a methodological perspective, EJ studies can employ qualitative or quantitative approaches. Qualitative approaches in the U.S. context commonly rely on ethnography and interviews (e.g., Malin and Demaster 2016; Filteau 2016; Schafft and Biddle 2014; Schafft et al. 2013; Perry 2013), Q-methodology (e.g., McLaughlin and Cutts 2018), and archival analysis (e.g., Carre 2012). Globally, networks like 'Oilwatch' and 'EJAtlas' assemble and categorize EJ case study reports of conflicts and territorial rights in oil regions (e.g., Temper et al. 2018; Anguelovski and Alier 2014). The majority of hydrocarbon EJ studies in international contexts tend to be qualitative, featuring broad, regional overviews (e.g., Temper et al. 2018; O'Rourke and Connolly 2003), narratives on the violent dispossession of rural and Indigenous peoples' livelihoods by the arrival of exploration and production activities (e.g., Lu 2012; Orta-Martínez and Finer 2010; Akpabio and Akpan 2010; Obi 2010; Widener 2007), and grassroots resistance to infrastructure siting (e.g., Carruthers 2014; Widener 2007). Thus, the EJ qualitative approach provides insight into how people experience environmental injustices, especially in place-based contexts. Unfortunately, the approach often lacks quantitative validation (Schroeder et al. 2008), which can be used to legitimize political claims (Hale 2004), increase community engagement (Reyes-García et al. 2012), and facilitate participatory capacity building (Torso et al. 2020). Therefore, empirical analysis of EJ issues using geospatial technologies integrated with traditional qualitative EJ approaches such as interviews and community participation can strengthen EJ analysis in oil and natural gas production zones, as well as facilitate interdisciplinary dialog among scholars.

To demonstrate current limitations and opportunities for quantitative EJ research in Latin America, this chapter reviews the EJ perspective and discusses two GIS-based EJ studies conducted in the Tampico-Misantla Basin, a major oil and gas production area in Mexico. The two studies demonstrate the challenges, work-arounds, and types of geospatial data that can be used to conduct EJ studies. We then broaden our scope and examine oil and natural gas regions in Latin America, evaluating the availability of open-access data in those regions and provide suggestions for workarounds in places where the data are lacking. Lastly, we suggest some best practices when designing and implementing GIS-based EJ studies.

2 Empirical Environmental Justice Studies

Environmental injustices occur whenever social groups, and in particular socially marginalized groups, bear disproportionate environmental risks, have unequal access to environmental goods, or are not given the opportunity to participate in environmental decision-making (Shrader-Frechette 2002). Determining distributional environmental injustices, which are premised on the spatial allocation of risks and benefits, is obviously well suited for GIS analysis. For the placement of oil and natural gas wells, analyzing the demographics of populations near extractive activities, revenue allocation, and the dynamics of deciding where, how, and whether to drill are key considerations (Fry et al. 2015; Kroepsch et al. 2019). Due to higher incidences of exposure, people living near oil and gas activities are more likely to experience environmental and health risks.

Exposure to environmental and public health risks are common in oil and natural gas production zones. Environmental degradation occurs during land clearing and drilling (Matthew et al. 2017). Toxic spills, waste dumping, and water contamination pose long-term public health risks to people living in oil extraction areas (Aguilera et al. 2010). Hazardous emissions from drilling and production activities include particulate matter, ozone, and volatile organic compounds, which also pose health risks to nearby populations (McKenzie et al. 2017, 2018; Brown et al. 2014; Helmig et al. 2014). Studies show higher incidences of miscarriages, birth defects, and preterm births among those same populations (Smith et al. 2013; McKenzie et al. 2014; Whitworth et al. 2018). Communities in oil extraction areas also experience nuisances like noise and light pollution, and increased truck traffic and road damage (Blair et al. 2018; Hays et al. 2017; Rahm et al. 2015; Adgate et al. 2014). For these reasons and others, living in close spatial proximity to active oil and natural gas wells, pipelines, flaring, and compressor stations can be harmful for physical and mental health, and overall wellbeing (Mayer 2017). Nevertheless, there are potential benefits for people who live in close proximity to oil and natural gas production activities, including access to jobs, royalties, access fees, and increased tax revenues (Brown et al. 2016; Hardy and Kelsey 2015; Weber et al. 2014; King 2012). Yet, it often remains unclear if these benefits accrue among local populations, because census or other socio-economic datasets do not generally conform to the risk areas for oil and gas activities. It is also difficult to discern whether those employed and receiving wages are local residents or outsiders (Kroepsch et al. 2019). Other potential remunerations include oil industry community development infrastructure, monetary donations, and other industry investments in extraction areas.

Geospatially, distributive environmental injustices occur in two ways: (1) when community proximity to oil and gas production activities, and thus risk, correlates with demographic measures of inequality including age, education level, race, or income level; and (2) when those who are affected by the greatest risks do not receive any measurable benefits (Kroepsch et al. 2019). The first type is a 'traditional distributive injustice' and the second is a 'benefit-sharing distributive injustice' (Fig. 10.1). Traditional distributive studies typically apply geospatial analysis to explore associations among socio-demographic spatial enumerative units (e.g., census tracts) and active oil and gas wells and infrastructure (see e.g., Hilburn and Fry 2019; Clough and Bell 2016; Johnston et al. 2016; McKenzie et al. 2016; Ogneva-Himmelberger and Huang 2015). Although results from these studies are mixed, Johnston et al. (2016) and Ogneva-Himmelberger and Huang (2015) find evidence that minority and lower-income rural residents in some U.S. hydrocarbon production areas, such as the Marcellus and Eagle Ford Shales, experience traditional distributional injustices. In benefit-sharing studies, identifying and tracking the benefits associated with oil and gas production activities is a central consideration (Fry and Hilburn 2020; Fry et al. 2015).

To date, most distributional justice studies use GIS vector-based buffers informed by existing environmental and public health research, ranging from 100 m to 5 km around active wells, to demarcate areas most affected by production activities (Fig. 10.1; see e.g., Hilburn and Fry 2019; Czolowski et al. 2017; Clough and Bell 2016; McKenzie et al. 2016; Ogneva-Himmelberger and Huang 2015). After spatially stratifying populations into "affected" and "unaffected" categories, these studies employ spatial-statistical testing like differences of means tests to determine if racially, socially, and economically marginalized populations live in greater proximity to active wells. Proximity to other production infrastructure like pipelines or compressor stations can also be used to test for injustices (Clough 2018). Ogneva-Himmelberger and Huang (2015) define two spatial models to interpret populations near oil and gas wells. The first model implies that the siting of hydrocarbon

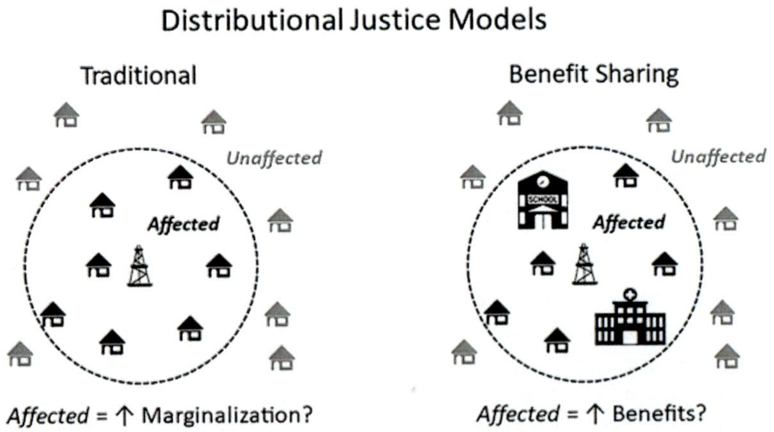

Fig. 10.1 Distributional justice models showing a GIS buffer around an active oil well, with homes/communities affected by potential negative health and environmental effects located inside the buffer, and unaffected homes/communities outside the buffer

infrastructure near historically underrepresented populations exposes them to environmental and public health risks. The second model includes the same race and social status criteria, as well as proximity buffers to evaluate degree of exposure. The latter model is the more prevalent in all manners of EJ studies today (Maantay and Maroko 2018; Mennis 2002) including those related to oil and gas (e.g., McKenzie et al. 2017; Czolowski et al. 2017; Clough and Bell 2016; Johnston et al. 2016). Another EJ approach uses spatial interpolation methods, such as kriging or inverse distance weighting, to model and predict risks instead of predetermining them based on distance buffers (e.g., Maantay and Maroko 2018; Meng 2015).

There are limitations and challenges to GIS-based EJ studies. First, assigning buffer distances can be problematic (Czolowski et al. 2017). Studies base buffer distances on public health research findings, regulatory distances, whole unit measures (e.g., 1 km), or draw from earlier EJ studies (Clough and Bell 2016; Ogneva-Himmelberger and Huang 2015). A second issue arises with population data that requires enumeration unit scaling where polygon size variations force researchers to asymetrically map data (Clough and Bell 2016). A third and related issue deals with acquiring comparable oil and natural gas well data sets across different extraction regions (Clough and Bell 2016, Ogneva-Himmelberger and Huang 2015).

Many quantitative EJ studies also base their determination of environmental injustices solely on statistical analyses and often do not fully consider historical and socioeconomic contexts, complimentary data sets, or site visits, which are strengths of qualitative EJ studies. Fortunately, with the increasing availability of geospatial data for oil and gas infrastructure, population demographics, and other variables (Taylor 2017), there are currently more opportunities to conduct distributional justice analyses in regions across the Global South than in the past, providing new prospects for quantitative validation of EJ issues (Schroeder et al. 2008), facilitating community-researcher participatory capacity building (Torso et al. 2020), and creating greater opportunities for dialogue among qualitative and quantitative analytical cultures (see e.g., Jasanoff 1993).

3 Two Environmental Justice Studies in the Tampico-Misantla Basin, Mexico

To demonstrate how to conduct GIS-based EJ studies in Latin America, we review two projects that use publicly available geospatial data to examine rural communities in the Tampico-Misantla Basin of eastern Mexico (Fig. 10.2). Historically, the Tampico-Misantla Basin has been an important terrestrial oil production zone and remains the third most productive in Mexico overall. Running in an arc from north-southeast and extending across 57,492 km^2, the basin spans the territorial boundaries of five Mexican states: Veracruz, Tamaulipas, San Luis Potosí, Hidalgo, and Puebla. The largest population centers within the basin are the port city of Tampico in Tamaulipas State at its northern end and Poza Rica, Tuxpan, and Papantla, in

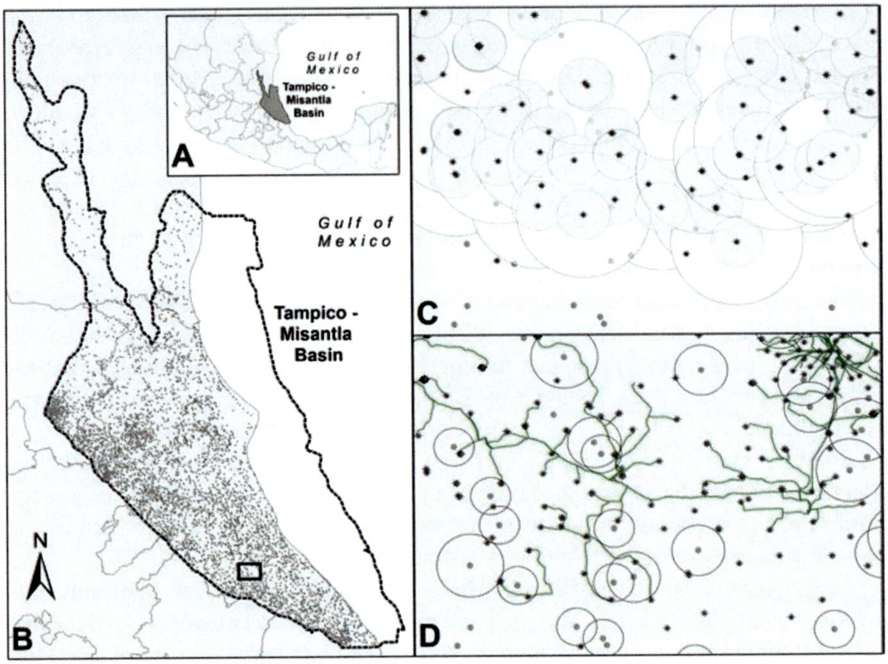

Fig. 10.2 The location of the Tampico-Misantla Basin in Mexico (**a**) and *localidades* (communities) in the Basin (**b**). Buffers measuring 1 and 3 km around active oil and gas wells demarcate affected and unaffected areas used in the Hilburn and Fry (2019) study, communities are gray dots (**c**). Varying buffer sizes around communities reflect different community areas and affected households (**d**) used in Fry and Hilburn (2020), pipelines are green, active oil and gas wells are black stars

Veracruz State at its southern terminus. Over four-thousand rural communities, featuring nucleated urban town landscapes surrounded by fields and pasture, dot the region.

The Tampico-Misantla Basin has been central to Mexico's terrestrial oil and natural gas production since the early 1900s. Geologically, it consists of an assemblage of largely conventional hydrocarbon plays from the Lower Jurassic to the early Tertiary. In Mexico, the "field" is the fundamental territorial-geologic unit of hydrocarbon production and the Tampico-Misantla Basin is partitioned into 111 fields. One particular field, the Poza Rica Field, was the main producer for Mexican terrestrial oil from the 1930s to the 1970s. During this time, the City of Poza Rica and the surrounding area experienced massive infrastructural, industrial, and urban development, setting off rapid social and environmental changes (Salas Landa 2016). These changes are largely encompassed by the conversion of the tropical rainforest and Indigenous Totonaco agroecosystems into a hydrocarbon, cattle, and cash crop maize/citrus-producing landscape with a hierarchy of urban settlements with Poza Rica serving as a de facto regional capital (King 2007; Santiago 2006). Although oil production in the region began to decline in the late 1960s while Mexico's production focus shifted to offshore deposits, Poza Rica and its environs have remained an administrative center for Mexican energy firms. Additionally, declining production in Mexico's offshore sector since the early 2000s has recently ignited interest in a sub-basin called the Chicontepec Basin and its Aceite Terciario del Golfo (ATG) deposits. Although the most recent unconventional projects have stalled and have not delivered expected returns (Fry 2018), the Tampico-Misantla Basin remains a focus for Mexico's terrestrial energy sector.

3.1 A Traditional Distributional Justice Study

Hilburn and Fry (2019) employed similar spatial and statistical models as those used in U.S.-based EJ studies, including distance buffers around active wells and other infrastructure to model exposure to negative externalities from oil and gas production in the southern part of the Tampico-Misantla Basin. Based on distances used in other EJ studies, buffers measuring 1 and 3 km demarcate affected and unaffected areas around the wells (Fig. 10.2). In Mexico, oil and gas well geospatial data are available from the Comisión Nacional de Hidrocarburos (CNH; National Hydrocarbons Commission) and population data from the Instituto Nacional de Estadística, Geografía é Informática (INEGI; National Institute of Statistics, Geography, and Informatics). INEGI provides population geospatial unit data. The smallest possible enumeration unit available is the *localidad* and is represented as the centroid point of a small settlement. The study eliminated large urban areas to prevent skewing the sample (Johnston et al. 2016); it also omitted settlements with <30 residents due to incomplete data.

The study used the Consejo Nacional de Población (CONAPO; Population National Council) marginality index derived from INEGI data to determine if a disproportionate number of marginalized people live near active oil and gas wells (Hilburn and Fry 2019). CONAPO (2010) defines marginalization as "the suite of social problems (disadvantages) of a community [or] locality and makes reference to a group of people or families" (p. 11). The CONAPO dataset uses variables from the 2000 and the 2010 INEGI census data, including quality of life and material wealth-poverty indicators such as percent illiterate, percent not finishing primary school (low education attainment), percent households with no integrated bathroom, percent households without electricity, percent households without piped water, average household size, percent households with dirt floors, and percent households with a refrigerator. The weighted marginalization index serves as a proxy measure of economic and social marginalization: the higher the value the more marginalization. These data can be linked to the point-based INEGI locality data using a unique identifier code.

Hilburn and Fry's (2019) spatial model involved a sample of settlements stratified by an affected/unaffected variable (i.e., settlements affected by potential negative health and environmental effects and settlements unaffected by those effects, see Fig. 10.1) used in a non-parametric difference of means evaluation (Wilcoxon test). A statistically significant difference of means – indicating greater marginality index values of the affected population (see Fig. 10.1) – became the criterion for determining an environmental injustice. The analysis, however, found that communities close to oil and gas wells have lower marginality indices than those farther from wells, a seemingly contradictory picture to what a traditional distributional justice perspective would predict for those communities. To further confound initial presumptions, the results show marginalization values among affected communities decreasing between 2000 and 2010, which suggests that communities near active wells were better off in 2010 than they were in 2000. Notably, despite lower and decreasing rates of marginalization among affected communities, Hilburn and Fry (2019) did not measure environmental or health impacts among those communities, which the literature suggests might be substantial (Kroepsch et al. 2019; McKenzie et al. 2018; Mayer 2017). Likewise, it is important to recognize that communities on both sides of the buffers in the study, i.e., affected *and* unaffected, fall below Mexico's national marginalization median.

With these caveats in mind while interpreting the geospatial results, Hilburn and Fry (2019) looked for evidence of benefits among communities in close proximity to oil and gas wells to explain the decreasing marginalization rates. In fact, the state oil company, *Petróleos Mexicanos* (Pemex), does invest in affected communities through social development projects that focus on town infrastructure, education, and public health which seem to address several variables measured by CONAPO including literacy, education attainment, and sanitation. Field observations in June 2018 among affected communities in the Tampico-Misantla Basin also indicated the presence of social development projects such as schools, health clinics, community kitchens, rural sanitation, recreational facilities, roads, bridges, and parks. To determine if communities most affected by oil and natural gas activities did in fact receive more social development benefits, the researchers conducted a benefits-sharing distributional analysis among the same communities, which is the subject of the following section.

3.2 A Benefit-Sharing Distributional Justice Study

Fry and Hilburn (2020) tested the normative ideal of distributional justice, that local communities "who bear the greatest environmental and health risks [receive] the most financial benefits" (Fry et al. 2015, p. 100) among communities affected by hydrocarbon production activities in the Tampico-Misantla Basin. According to this perspective, areas in the Basin with the highest levels of oil and natural gas activities, manifest in extraction infrastructure and production volumes, should also receive the most social development donations and infrastructure from Pemex. Like other oil and natural gas companies, Pemex implemented community development programs beginning in the 1990s to better engage local communities and to reduce conflicts resulting from production activities (Watts 2005; Frynas 2005). Today, Pemex uses community development projects to improve "the quality of life of families in oil regions" and enable "communities to enjoy the benefits that the oil industry entails" (Pemex Sustainability Report 2016, p. 69). In their study, Fry and Hilburn (2020) used oil and gas infrastructure as proxies for *both* benefits and risks distribution, with social development infrastructure (e.g., clinics, schools, bridges) representing benefits and wells, pipelines, compressors, separator stations, and oil production rates representing risks. And because social development infrastructure is spatially fixed, it overcomes some of the EJ challenges associated with tracking wage and employment benefits.

Mexico's National Hydrocarbons Commission (CNH) website provides geospatial data on petroleum and natural gas production, as well as the locations for wells and pipelines. Mexico's INEGI provides census data. For social development donations and infrastructure, Fry and Hilburn (2020) used rapid appraisal techniques (Takasaki et al. 2000) to identify Pemex and oil services contractors' social development infrastructure and public works projects. They used Pemex Annual Sustainability Reports for information on social development infrastructure and donations.

Instead of buffers around oil and natural gas infrastructure, the study placed buffers around the INEGI generated centroid for each rural community (Fig. 10.2). To reflect population differences in those communities, varying buffer sizes were used to reflect different community areas and affected households. Field observations and satellite imagery confirmed that buffer areas with radiuses of 750 m, 1000 m, and 1250 m contained most community houses and buildings for different scaled populations. Within 114 rural community buffer areas, the researchers calculated population density, the number of wells drilled, total well count, pipeline length (meters), the number of hydrocarbon installations (compressor stations, separator stations, etc.), the number of social development infrastructure, the oil and gas production, and the number of non-Pemex, state-funded public works.

Fry and Hilburn (2020) used Pearson product-moment correlation to determine if there were significant relationships between the number of social development projects, the explanatory variable, and eight response variables: population; marginalization; drilling activity (i.e., number of wells drilled between 2006 and 2017); total well count; pipeline length; number of compressors, separator stations, and other infrastructure; oil and natural gas production; and non-Pemex public works. Because of the high number of communities with zero social development infrastructure, they also ran Pearson correlation tests without those communities in order to make a comparison among communities where social development infrastructure is present in relation to counts of wells and pipeline lengths as well as production.

The results of the analysis showed a weak positive correlation between production and social development infrastructure and a weak negative correlation between the latter and population. In other words, there are slightly more social development infrastructure where production is comparatively higher, and there are slightly less where populations are relatively low. However, the high number of locations with no social development infrastructure likely influenced these relationships. Otherwise, there was no statistical difference in the level of social marginality, drilling activity, well count, pipelines, compressors, state public works, or oil production among sampled communities. However, a descriptive statistical analysis showed slightly more unevenness in the distribution of costs and benefits, with more costs in some communities and less benefits in others, and vice versa. Although identifying the causes for the uneven distribution of benefits is complicated (Pirker et al. 2007), Fry and Hilburn (2020) contend that this is a result of (1) the targeting of social developments to areas where large investments were being made into oil production that ultimately never materialized; (2) partisan politics, political turnover, and misappropriation of funding; (3) variations in community capacity to organize; and (4) access to urban areas and state-financed public works projects.

4 Feasibility of GIS-Based Environmental Justice Studies in Latin America's Oil Production Zones

This section outlines some general rules-of-thumb to guide GIS-based EJ studies in Latin America's hydrocarbon production areas. It then provides information on publicly available geospatial data that can be used to examine EJ in oil and natural gas production areas. We discuss opportunities to improve GIS-based EJ research and provide workarounds when GIS data are not available.

While there is no single optimal rubric to assess EJ using geospatial data and statistical testing, there are some best practices for framing quantitative EJ inquiries. The first step is to assess data availability followed by the determination of the distributional justice framing: traditional vs benefit sharing (Fig. 10.1). Because data sources vary by country and region in quality, date produced, scale, spatial resolution, and accessibility, the researcher needs to be flexible in determining which data to include. They should also remain open to repeatedly searching for updated data. Once available data options are identified, the researcher should ideate on how those data can be used or modified. The type of distributional analysis will determine which data to use and how. Researchers can then identify pathways to answer their research questions using those data. Flexibility and creativity with an eye on the extant literature are real virtues in determining which spatial models, analyses, and tests to use with the available data. It is important to keep in mind that methodological insights may be found outside of EJ studies.

After determining the initial approach toward assessing EJ, researchers should take special care to define their spatial and statistical model. They also need to assess whether spatial overlap between the source of potential risks and the affected communities signifies an environmental injustice or not. In other words, it is important to carefully delineate the spatial model and statistical test results, and then to clearly explain any limitations. Limitations should not be regarded as weaknesses, rather they help generate research questions to round out analysis. Validation by both literature review, government document analysis, and on-site field work are the final step in this process. In this final step, researchers can reflect on the utility of the EJ approach, while also determining how or if the results make sense.

Overcoming the challenges of data acquisition is a key criterion to conducting GIS-based EJ research in Latin America and regions across the development spectrum. In Mexico, census data and hydrocarbon data have been available on-line since about the year 2000. To identify socio-spatial data available among other Latin American countries at large-scale and fine resolution, on inland oil and natural gas production, we conducted an on-line search in December 2020. Table 10.1 provides information on government agencies and types of data currently available for download.

Coupling GIS analysis with field observations, public records, and key actor interviews can improve quantitative EJ research. Such integration requires dialog with qualitative EJ approaches. Field observations can help identify physical infrastructure. These can also be used to conduct rapid appraisal of home improvements and other signs of infrastructural benefits. Accessing local government records can reveal the magnitude of social development monetary investments and expenditures in affected communities, as well as other material donations like library books, computers, or medicine. Interviews with key actors, including community leaders, can reveal the number of local oil workers, possible environmental liabilities, and local perceptions of the impacts of development projects, as well as negative health, environmental, and livelihood effects. These participatory, observational, archival, and interview methods also work well if GIS data are limited, since these data can

Table 10.1 GIS data available in Latin America. Only countries with data are shown

Country	Data hosting agency	Formats	Type
Mexico	Instituto Nacional de Estadística y Geografía (INEGI)	.csv, .shp, .pdf	Population and household (sex, age, place of birth, health coverage, social security, educational characteristics, marital status, employment characteristics, tenure, housing conditions)
	Consejo Nacional de Población	.csv	Marginality index
	Comisión Nacional de Hidrocarburos (CNH)	.shp	Production (well, field, operator, state, concession area); petroleum activity (active and terminated wells); economic indicators; contract areas; registered operators; reserves
Argentina	Instituto Geografía Nacional; Base de Asentamientos Humanos	.csv, .kml, .shp, GeoJSON	Habitat and social infrastructure; political and territorial units
	Instituto Nacional de Estadística y Censos	.csv	Population and household (sex, age, place of birth, health coverage, social security, educational characteristics, marital status, employment characteristics, tenure, housing conditions)
	Secretaría de Gobierno de Energía	.csv, .shp, .pdf	Production (well, field, concession areas); permits; environmental studies; geologic formations; upstream investments; petroleum distribution; registered operators; reserves

(continued)

Table 10.1 (continued)

Country	Data hosting agency	Formats	Type
Bolivia	Infraestructura de Datos Espaciales del Estado Plurinacional de Bolivia (GeoBolivia)	OpenLayers, .kml, .gml	Production (well, pipelines)
	Instituto Nacional de Estadística (INE)	Openlayers	Enumeration units by departamento, provincia, and municipio – arrays of population and household statistics, poverty, education, social security, politics, and health
Brazil	Portal de mapas interatios	.shp; .csv	State and municipal level agricultural, industrial, demographic, etc.
	Agência Nacional do Petróleo	.shp	Exploration blocks, production fields, concessions, blocks, wells
	Instituto Brasileiro de Geografia e Estatística	.xlsx; .ods	Housing unit characteristics, emigration; household demographics; resident characteristics; mortality
Colombia	Departamento Administrativo Nacional de Estadistica (DANE)	.shp, GeoPackage, Spatialite, GeoJSON	Geospatial population data at numerous spatial scales already integrated with national census data; rural agropecuary data
	Agencia Nacional de Hidrocarburos (ANH)	.shp	Wells and pipelines, oil and gas fields and production zones, sedimentary basins, seismographic data
Ecuador	Sistema Nacional de Información	Various forms	A clearinghouse with links to many different agencies that handle geospatial data

be used to build geospatial database attributes. Observations lend an often-missing interpretive dynamic to strictly quantitative research and can also serve to ground truth infrastructure for identification on Google Earth imagery or other remotely sensed data products.

5 Conclusion

This chapter reviewed GIS-based EJ research, provided an overview of the methods and GIS data used in two EJ studies in an oil and natural gas extraction region in Mexico, and synthesized available geospatial and population data for conducting EJ studies in Latin America's oil production zones. With the rise of fully functional freeware packages such as QGIS and online training resources, GIScience is becoming less expert-driven and increasingly participatory. Additionally, the greater availability of geospatial data on oil and gas infrastructure, population demographics, and other variables, permits this type of analysis in many regions across the Global South. Although EJ analyses are not new to Latin America, empirical studies on risk and benefit distributions can add a quantitative dimension to those studies (Schroeder et al. 2008), which can be used to legitimize political claims (Hale 2004), increase community engagement (Reyes-García et al. 2012), and facilitate participatory capacity building (Torso et al. 2020). As well, instead of census data on income and race, studies in Latin America could rely on social development indicators and combined marginalization indexes.

GIS platforms are well suited for two types of distributional EJ studies: traditional and benefit-sharing. For the first, using well and population enumeration data, researchers can compare discreetly-defined exposure to negative externalities among marginalized and lower-income communities. For the second, determining if people who live in oil and gas extraction areas receive benefits from the hydrocarbons that lie beneath their communities, can be achieved through an analysis of oil and natural gas company social development infrastructure. Another advantage of the benefit-sharing distributional justice analysis is that it provides a way for academics and other stakeholders to hold oil and extractive industries accountable for their social development obligations. As our case studies show, different datasets, theoretical frameworks, and methodological approaches can be used to assess and analyze different forms of EJ.

Because of expansive neoliberal privatization reforms of formerly state-owned natural resources in Latin America over the past three decades and inclusion of social development projects as supposed forms of compensation (Bebbington and Bury 2013), there is a clear need for socio-spatial assessments of extractive industry social and environmental impacts. GIS and other geotechnologies can play a critical role in EJ analysis of these industries and their community impacts. Combined with case studies and place-based analysis, GIS-based EJ studies can bring greater attention and continuity to the plights of the people and communities most impacted and marginalized by oil and natural gas extractive activities.

References

Adgate J, Goldstein B, McKenzie L (2014) Potential public health hazards, exposures and health effects from unconventional natural gas development. Environ Sci Technol 48:8307–8320

Aguilera F, Laffon B, Méndez J, Pásaro E (2010) Review on the effects of exposure to spilled oils on human health. J Appl Toxicol 30(4):291–301

Akpabio E, Akpan N (2010) Governance and oil politics in Nigeria's Niger Delta: the question of distributive equity. Hum Ecol 30(2):111–121

Anguelovski I, Alier JM (2014) The 'environmentalism of the poor' revisited: territory and place in disconnected glocal struggles. Ecol Econ 102:167–176

Bebbington A, Bury J (2013) Subterranean struggles: new dynamics of mining, oil, and gas in Latin America, vol 8. University of Texas Press, Austin

Billo E (2015) Sovereignty and subterranean resources: an institutional ethnography of Repsol's corporate social responsibility programs in Ecuador. Geoforum 59:268–277

Blair BD, Adgate JL, Brindley S et al (2018) Residential noise from nearby oil and gas well construction and drilling. J Expo Sci Environ Epidemol 28(6):538–547

Bridge G (2008) Global production networks and the extractive sector: governing resource-based development. J Econ Geogr 8(3):389–419
Brown D, Bonaparte H, Lewis C, Weinberger B (2014) Understanding exposure from natural gas drilling puts current air standards to the test. Rev Environ Health 29(4):277–292
Brown J, Fitzgerald T, Weber J (2016) Asset ownership, windfalls, and income: evidence from oil and gas royalties. Research working paper RWP 16-12, Federal Reserve Bank of Kansas City
Carre N (2012) Environmental justice and hydraulic fracturing: the ascendancy of grassroots populism in policy determination. J Soc Change 4(1):1–13
Carruthers DV (2014) Introduction: popular environmentalism and social justice in Latin America. In: Carruthers D (ed) Environmental justice in Latin America: problems, promise, and practice. MIT Press, Cambridge, MA
Clough E (2018) Environmental justice and fracking: a review. Curr Opin Environ Sci Health 3:14–18
Clough E, Bell D (2016) Just fracking: a distributive environmental justice analysis of unconventional gas development in Pennsylvania, USA. Environ Res Lett 11(2):025001
CONAPO (2010) Indice de Margniación por Localidad 2010. Comisión Nacional de Población, Mexico City. http://www.conapo.gob.mx/work/models/CONAPO/indices_margina/2010/documentoprincipal/Capitulo01.pdf
Czolowski E, Santoro R, Srebotnjak T, Shonkoff S (2017) Toward consistent methodology to quantify populations in proximity to oil and gas development: a National Spatial Analysis and review. Environ Health Perspect 125(8):086004
Filteau M (2016) "If you talk badly about drilling, You're a pariah": challenging a capitalist patriarchy in Pennsylvania's Marcellus Shale region. Rural Sociol 81(4):519–544
Fry M (2018) Shifting volumetric imaginaries of oil potential in Mexico's Chicontepec Basin: investible asset, reserve replacement, and oilfield services zone. J Lat Am Geogr 17(3):73–101
Fry M, Hilburn A (2020) The distributional justice of oil industry social development projects and oil field production activities. Extr Ind Soc 7:647–659
Fry M, Briggle A, Kincaid J (2015) Fracking and environmental (in)justice in a Texas city. Ecol Econ 117:97–107
Frynas JG (2005) The false developmental promise of corporate social responsibility: evidence from multinational oil companies. Int Aff 81(3):581–598
Hale CR (2004) Rethinking Indigenous politics in the era of the indio permitido. North Am Congr Lat Am 38:16–24
Hardy K, Kelsey TW (2015) Local income related to Marcellus Shale activity in Pennsylvania. Community Dev 46(4):329–340
Hays J, McCawley M, Shonkoff SB (2017) Public health implications of environmental noise associated with unconventional oil and gas development. Sci Total Environ 580:448–456
Helmig D, Evans J, Park JH, Thompson C (2014) Highly elevated atmospheric levels of volatile organic compounds in the Uintah Basin, Utah. Environ Sci Technol 48:4707–4715
Hilburn A, Fry M (2019) Empirical environmental justice research in hydrocarbon extraction areas: examining applicability outside the US using a Mexican case study. Appl Geogr 107:63–71
Jasanoff S (1993) Bridging the two cultures of risk analysis. Risk Anal 13:123–123
Johnston J, Werder E, Sebastian D (2016) Wastewater disposal wells, fracking, and environmental injustice in southern Texas. Am J Public Health 106(3):550–556
Kimerling J (1994) The environmental audit of Texaco's Amazon oil fields: environmental justice or business as usual? Harv Hum Rights J 17:199–224
King A (2007) Trade and totomoxtle: livelihood strategies in the Totonacan region of Veracruz, Mexico. Agr Hum Values 24(1):29–40
King GE (2012) Hydraulic fracturing 101: what every representative, environmentalist, regulator, reporter, investor, university researcher, neighbor and engineer should know about estimating frac risk and improving frac performance in unconventional gas and oil wells. In: SPE hydraulic fracturing technology conference. Society of Petroleum Engineers, The Woodlands

Kroepsch AC, Maniloff PT, Adgate JL et al (2019) Environmental justice in unconventional oil and natural gas drilling and production: a critical review and research agenda. Environ Sci Technol 53(12):6601–6615

Lu F (2012) Petroleum extraction, Indigenous people and environmental injustice in the Ecuadorian Amazon. In: Gordon F, Freeland G (eds) International environmental justice: competing claims and perspectives. ILM Publishers, London

Maantay J, Maroko A (2018) Brownfields to Greenfields: environmental justice versus environmental gentrification. Int J Environ Res Public Health 10:2233–2250

Malin S, DeMaster K (2016) A devil's bargain: rural environmental injustices and hydraulic fracturing on Pennsylvania's farms. J Rural Stud 47:278–290

Matthew D, Moran Nathan T, Taylor Tabitha F, Mullins Sehrish S, Sardar Maureen R, McClung (2017) Land-use and ecosystem services costs of unconventional US oil and gas development. Front Ecol Environ 15(5):237–242. https://doi.org/10.1002/fee.1492

Mayer A (2017) Quality of life and unconventional oil and gas development: towards a comprehensive impact model for host communities. Extr Ind Soc 4:923–930

McKenzie LM, Guo R, Witter RZ, Satvitz DA, Newman LS, Adgate JL (2014) Birth outcomes and maternal residential proximity to natural gas development in rural Colorado. Environ Health Perspect 122:412–417

McKenzie LM, Allshouse W, Burke T, Blair B, Adgate J (2016) Population size, growth, and environmental justice near oil and gas wells in Colorado. Environ Sci Technol 50:11471–11480

McKenzie LM, Allshouse WB, Byers TE, Bedrick EJ, Serdar B, Adgate JL (2017) Childhood hematologic cancer and residential proximity to oil and gas development. PlosOne 12(2):e0170423

McKenzie LM, Blair B, Hughes J, Allshouse WB, Blake NJ, Helmig D, Milmoe P, Halliday H, Blake DR, Adgate JL (2018) Ambient nonmethane hydrocarbon levels along Colorado's northern front range: acute and chronic health risks. Environ Sci Technol 52(8):4514–4525

McLaughlin D, Cutts B (2018) Neither knowledge deficit nor NIMBY: understanding opposition to hydraulic fracturing as a nuanced coalition in Westmoreland County, Pennsylvania (USA). Enviro Manage 62(2):305–322

Meng Q (2015) Spatial analysis of environment and population at risk of natural gas fracking in the state of Pennsylvania, USA. Sci Tot Environ 515-16:198–206

Mennis J (2002) Using geographic information systems to create and analyze statistical surfaces of population and risk for environmental justice analysis. Soc Sci Q 3(1):281–297

O'Rourke D, Connolly S (2003) Just oil? The distribution of environmental and social impacts of oil production and consumption. Annu Rev Environ Resour 28:587–617

Obi C (2010) Oil extraction, dispossession, resistance, and conflict in Nigeria's oil-rich Niger Delta. Can J Dev Stud 30(1–2):219–236

Ogneva-Himmelberger Y, Huang L (2015) Spatial distribution of unconventional gas wells and human populations in the Marcellus Shale in the United States: vulnerability analysis. Appl Geogr 60:165–174

Orta-Martínez M, Finer M (2010) Oil frontiers and Indigenous resistance in the Peruvian Amazon. Ecol Econ 70:207–218

Pemex Sustainability Report (2016) http://www.pemex.com/en/responsibility/sustainable/Documents/20171123_IS_2016_ENG_FINAL_assurance%20letter.pdf. Accessed 25 May 2021

Perry S (2013) Using ethnography to monitor the community health implications of onshore unconventional oil and gas developments: examples from Pennsylvania's Marcellus Shale. New Solut 23(1):33–53

Pirker K, Rodríguez A, Manuel J, Guzmán HI (2007) El acceso a la información para la contraloría social: El caso de las donaciones y donativos de Pemex a Tabasco. Fundar, Centro de Análisis e Investigación, México

Rahm D, Fields B, Farmer JL (2015) Transportation impacts of fracking in the eagle ford shale development in rural south Texas: perceptions of local government officials. J Rural Community Dev 10(2):78–99

Reyes-García V, Orta-Martínez M, Gueze M, Luz AC, Paneque-Gálvez J, Macía MJ, Pino J, TAPS Bolivian Study Team (2012) Does participatory mapping increase conflicts? A randomized evaluation in the Bolivian Amazon. Appl Geogr 34:650–658

Salas Landa MS (2016) Crude residues: the workings of failing oil infrastructure in Poza Rica, Veracruz, Mexico. Environ Plan A 48(4):718–735

Santiago MI (2006) The ecology of oil: environment, labor, and the Mexican revolution, 1900–1938. Cambridge University Press, Cambridge

Schafft K, Biddle C (2014) School and community impacts of hydraulic fracturing within Pennsylvania's Marcellus Shale region, and the dilemmas of educational leadership in Gasfield boomtowns. Peabody J Educ 89(5):670–682

Schafft K, Borlu Y, Glenna L (2013) The relationship between Marcellus Shale gas development in Pennsylvania and local perceptions of risk and opportunity. Rural Sociol 78(2):143–166

Schroeder R, Martin K, Wilson B, Sen D (2008) Third world environmental justice. Soc Nat Resour 21(7):547–555

Shrader-Frechette K (2002) Environmental justice: creating equality, reclaiming democracy. Oxford University Press, Oxford

Smith KR, Frumkin H, Balakrishnan K, Butler C, Chafe ZA, Fairlie I, Kinney P, Kjellstrom T, Mauzerall DL, McKone TE, McMichael AJ (2013) Energy and human health. Annu Rev Public Health 34:159–188

Takasaki Y, Barham BL, Coomes OT (2000) Rapid rural appraisal in humid tropical forests: an asset possession-based approach and validation methods for wealth assessment among forest peasant households. World Dev 28(11):1961–1977

Taylor L (2017) What is data justice? The case for connecting digital rights and freedoms globally. Big Data Soc 4(2):1–14

Temper L, Demaria F, Scheidel A, Del Bene D, Martinez-Alier J (2018) The Global Environmental Justice Atlas (EJAtlas): ecological distribution conflicts as forces for sustainability. Sustain Sci 13(3):573–584

Torso K, Cooper CM, Helkey A, Meyer C, Kern AL, Wardropper CB (2020) Participatory research approaches in mining-impacted hydrosocial systems. Hydrol Sci J 65(14):2337–2349

Urkidi L, Walter M (2011) Dimensions of environmental justice in anti-gold mining movements in Latin America. Geoforum 42(6):683–695

Watts MJ (2005) Righteous oil? Human rights, the oil complex, and corporate social responsibility. Annu Rev Environ Resour 30:373–407

Weber JG, Burnett JW, Xiarchos IM (2014) Shale gas development and housing values over a decade: evidence from the Barnett Shale. IAEE energy and the economy conference, New York, NY, USA, June 15–18. Available at: https://web.archive.org/web/20200320221217id_/http://www.usaee.org/usaee2014/submissions/OnlineProceedings/Weber%20et%20al%20Barnett%20Housing%20Values.pdf

Whitworth KW, Marshall AK, Symanski E (2018) Drilling and production activity related to unconventional gas development and severity of preterm birth. Environ Health Perspect 037006(1):8

Widener P (2007) Benefits and burdens of transnational campaigns: a comparison of four oil struggles in Ecuador. Mobilization 12(1):21–36

Chapter 11
Contributions to Socio-environmental Research through Participatory GIS in Archaeology

Alina Álvarez Larrain and Jason Nesbitt

Abstract This chapter reviews the contributions of an engaged archaeological framework to past and present research about human-environmental relationships in Latin America. We argue that archaeology can play an expanded role in advancing our understanding of long-term socio-environmental systems by promoting greater integration between scientific research and broader societal needs and local spatial knowledge within the context of sustainability. We further suggest that participatory approaches can bridge some of the conceptual divides that separate archaeologists and anthropologists from Indigenous and local communities and contribute to the decolonization of the discipline. Specifically, the aim is to show how participatory mapping and participatory geographic information systems can be suitable tools to engage archaeological studies of human-environment interactions from an integrative research perspective.

Keywords Landscapes · Human–environment interactions · Remote sensing and GIS · Participatory mapping · Local spatial knowledge · Traditional ecological knowledge

---

A. Á. Larrain (✉)
Consejo Nacional de Investigaciones Científicas y Técnicas (CONICET), Instituto de Datación y Arqueometría, Centro de Desarrollo Tecnológico "Gral. Manuel Savio", Palpalá, Jujuy, Argentina
e-mail: alarrain@indya.unju.edu.ar

J. Nesbitt
Department of Anthropology, Tulane University, New Orleans, LA, USA

S. López (ed.), *Socio-Environmental Research in Latin America*, The Latin American Studies Book Series, https://doi.org/10.1007/978-3-031-22680-9_11

1 Introduction

A major topic of contemporary archaeological inquiry is the examination of the relationship between people and the environment (Kintigh et al. 2014). A crucial component of this research centers on a variety of spatial analytical methods (Gillings et al. 2020) that have been developed by archaeologists over the last century. Among the analytical techniques that archaeologists utilize, remote sensing (RS) has become increasingly important. Yet, as stated in the introduction of this volume, a challenge that RS studies face is its integration with the social sciences and humanities. In this sense, archaeology has contributed to narrowing this gap due to its long relationship with RS and spatial analysis (Clarke 1977; Hodder and Orton 1976) in addition to common research questions that focus on long-term perspectives on human-environment relationships. This characteristic of archaeology has helped connect social, historical, and ecological perspectives, providing new insights into socio-environmental system dynamics (Van der Lleeuw and Redman 2002).

Latin American archaeology first incorporated RS into its practice between the 1930s and1960s. Aerial photography was used in some of the first studies of pre-Hispanic settlement patterns, a type of spatial analysis that examines how archaeological sites are distributed on the landscape (Willey 1953; González 1957; Kosok 1965; Denevan 1993). In the 1990s, archaeological studies adopted satellite imagery and geographic information systems (GIS) as fundamental tools for mapping and sophisticated spatial analysis (Grau Mira 2006; Llobera 2007; Kosiba and Bauer 2013). More recently, the introduction of unmanned aerial vehicles (UAVs), light detection and ranging (LiDAR) technologies, and photogrammetry for surveying large archaeological areas and the production of high-resolution digital elevation models led to a revolution in the way spatial and socio-environmental studies were conducted (Chase et al. 2017; Canuto et al. 2018; Greco 2018; Iriarte et al. 2020; VanValkenburgh et al. 2020).

Contemporary archaeology goes well beyond the reconstruction of past societies for the sole purpose of increasing historical and anthropological knowledge. The data generated by archaeology coupled with climate proxies, local spatial knowledge (LSK), and traditional ecological knowledge (TEK), provide a long-term multi-scalar understanding of human-environment interactions (c.f. Sandweiss and Kelley 2012, Nesbitt 2016, 2020, Caramanica et al. 2020, Haldon et al. 2020, Kohler and Rockman 2020, Pauketat 2020). For example, satellite imagery and GIS are being used to estimate the extent of anthropogenic soils known as Amazonian Dark Earths (ADEs) or *terras pretas* (Heckenberger et al. 2007, 2008; Palace et al. 2017). Similarly, LiDAR has demonstrated the scale of settlement in Mesoamerica and parts of Amazonia, suggesting exceptionally large, dense populations prior to European contact (Canuto et al. 2018; Iriarte et al. 2020). These technologies coupled with LSK and TEK can determine the impact that current Indigenous activities, including cultivation practices, have on these soils and the surrounding environment (Schmidt 2013).

This research is not undertaken in a vacuum and has real implications for Indigenous descendent communities. With respect to past and present human-environment dynamics, there is an increased awareness of the role archaeological research plays in relation to the needs and demands of local stakeholders around the world (Colwell-Chanthaphonh and Ferguson 2008; Atalay 2012; Douglass et al. 2019; Gupta et al. 2020), including most countries in Latin America (Duin et al. 2014; McAnany and Rowe 2015; Álvarez Larrain et al. 2019; Herrera Wassilowsky and Curet 2019). This awareness of local demands has reoriented archaeological investigation to create projects in which knowledge is exchanged and shared. In this sense, engaged archaeology is contributing to an increased understanding of the adaptability and resilience of communities, using the past for better planning of the future, improvement of local ways of life, and promoting environmental justice (McIntosh et al. 1999; Heckenberger 2014; Caramanica et al. 2020; Douglass and Cooper 2020; Fisher 2020; Heckenberger 2020; Kohler and Rockman 2020; Rick and Sandweiss 2020; Rockman and Hritz 2020).

Local and Indigenous people's engagement with archaeological studies is not new; the record of LSK and TEK as it relates to the study of past environments', especially in the tropical lowlands of Central America, Andes, and Amazonia, is very robust (Erickson 1998; Kendall 2005; Gnecco and Ayala Rocabado 2010; Schmidt 2013; McAnany and Rowe 2015). Local communities are claiming archaeological landscapes and objects as part of their ancestral cultural heritage and vital resources not only for territorial protection, but for achieving a good way of living. We argue that archaeology can play an expanded role in advancing our understanding of long-term human-environmental systems by promoting greater integration between scientific research and broader societal needs within the context of sustainability (Fisher 2020). The aim of this chapter is to show how participatory mapping (PM) and participatory geographic information systems (PGIS) can be suitable tools to engage archaeological studies of human–environment interactions from an integrative research perspective.

In the following section, we discuss the trajectory of PGIS practice in the mapping of Latin American landscapes and territories that established a precedent for the use of PGIS in archaeological research. We review current PGIS applications in archaeology and some relevant methodological approaches. Finally, we discuss the opportunities and limitations of PGIS to address issues in human-environment dynamics in Latin America, offering some reflections about new epistemological considerations on the application of geospatial technologies.

2 PGIS and Archaeology in Latin America

2.1 Trajectory of the Practice

PGIS emerged 50 years ago in land claim cases of Indigenous peoples in North America. The primary purpose of these mapping efforts was the protection of ancestral lands and recognition of local communities' decision making in resource management in the face of commercial development projects (Freeman 1976; Brody 1988; Tobias 2000; Flavelle 2002; Chapin et al. 2005; Candler et al. 2006). Today, PGIS incorporates practices of PM with simple procedures such as sketch or mental maps, to the use of modern spatial information technologies including RS, GIS, UAVs, global positioning systems (GPS), 3D Projection-Augmented Landscape Models (3DPALM), and open access web platforms and software (Abbot et al. 1998; Corbett and Keller 2006; Sieber 2006). In many cases, PGIS is used as a spatial database and a political tool to represent LSK and TEK within conventional spatial information formats, providing legitimacy to local claims (Abbot et al. 1998; Dunn 2007; Corbett 2009; McCall and Dunn 2012; McCall 2014).

In Latin America, PGIS was more recently introduced and is closely connected with Participatory-Action (PA) research approaches (Fals Borda 1999), in which Indigenous and rural communities affected by socio-environmental problems actively participate in all steps of the research process, including planning and decision-making, which in turn generate opportunities for social improvement. These projects have been undertaken mainly by geographers and anthropologists in cooperation with conservation activists and non-governmental organizations (NGOs).

Although diverse applications of PGIS in Latin America exist (e.g., in urban planning, public health, hazard management, monitoring infringements, among others) (Risler and Ares 2013; McCall 2014), we focus on emerging mapping efforts led by Indigenous, descendant, and traditional communities directed toward territorial protection, including the preservation of their environmental and cultural heritage. These PGIS applications have a variety of goals, which include people's desire to: (1) map ancestral lands; (2) inform territory/land claims; (3) record traditional uses of territories; (3) manage natural resources and contribute to biodiversity conservation; (4) preserve Indigenous cultural resources and heritage; and (5) support land-use planning (Knapp and Herlihy 2002; Herlihy and Knapp 2003; Chapin et al. 2005; Stocks 2005; McCall 2014; Kelly et al. 2017; Álvarez Larrain and McCall 2020; Sletto et al. 2020). An extensive review of the literature (Álvarez Larrain and McCall 2019a) and recent compilations resulting from various PM conferences in Latin America (Sletto 2011; Salamanca and Espina 2012; Sletto et al. 2020), show that the practice of PGIS related to environmental and territorial protection has become widespread in the region since the 1990s. A detailed review of this extensive work is beyond the scope of this chapter, but we highlight here it's contribution to socio-environmental research and management.

Latin America has become the next predatory frontier of the globalized neoliberal world. As McCall (2014) states, the people who inhabit these lands face increasing external pressures on their territories derived from farmland and pastureland expansion; accelerated exploitation of minerals, privatization of drinking water, and petroleum production for external markets; illegal deforestation; road development, real estate business growth, and hydroelectric dam development. In the past, PGIS projects in the region were intimately related to the need to secure land tenure and natural resources by communities inhabiting the tropical lowlands in countries such as Brazil, Colombia, Ecuador, Honduras, Nicaragua, and Panama, and protecting their territories from extractivism and outsider encroachment (Offen 2009). As Stock (2003: 345) notes, "For us, land is a factor of production, a commodity on the market. It can be bought, sold and accumulated"; but as McCall (2014) claims, local and Indigenous communities are heavily dependent on the land and its resources for cultural survival. Losing their lands is tantamount to losing their health and socio-environmental sustainability, along with their livelihoods, cultures, and identities. Currently, the remaining stands of tropical rain forest are disappearing at alarming rates due to the pressure of loggers, cattle ranchers, and entrepreneurs involved in the extraction of natural resources as part of the ongoing integration of remote areas into the global economy (e.g., Plotkin 2020).

In the late 1980s, environmentalists were challenged to shift their focus on preserving flora and fauna in human-excluded protected areas, and align with the "sustainable development" paradigm (Ruggerio 2021). As it became more apparent that the most biodiverse and healthiest ecosystems often coincide with the home of Indigenous peoples, the international conservation community began to view participatory management strategies as key elements to the conservation of the forested lowlands (Chapin et al. 2005; Stocks 2005; Kelly et al. 2010; Ramirez-Gomez et al. 2013; Sletto et al. 2013). A precursor of the idea that land use mapping by Indigenous people could bolster land tenure claims and biodiversity protection was a 1992 map of Central America and Southern Mexico produced by the anthropologist Mac Chapin showing the connection between Indigenous lands and the last remaining forest in the region (Chapin and Threlkeld 2001, 2008). Since then, a new wave of PM projects of Indigenous lands emerged to better understand Indigenous land use and integrate the knowledge, practices, and beliefs of these communities into socio-environmental research and conservation planning (TMCC 1997; IBC and Chase Smith 2012; Cerra 2014; Nenquimo 2018; ACT 2019). Climate change and the global ecological crisis have taken this partnership even further. For example, scientific studies have established that Indigenous reserves in the Brazilian Amazon have been most effective at avoiding deforestation, and that forest cover and carbon stocks are best maintained by collective long-term use-rights management systems (Duin 2018). In Central America, over 200 protected areas were established by 1990 and, virtually all the largest ones are home to substantial Indigenous populations (Smith 2008: 86). In some cases, PGIS projects have contributed to Indigenous self-determination by assisting communities obtain legal land tenure and encouraging concrete involvement in planning and resource management. Securing legal protection for their homelands and resources is perhaps, the most fundamental challenge Indigenous peoples face in preserving their way of life (Denniston 1994), but achieving land ownership has not necessarily guaranteed the conservation of the ecosystems that they rely on. For example, the socio-environmental mapping carried out in Peru by the *Instituto del Bien Común* with the Yánesha peoples of the Amazon showed that mining and forestry concessions were granted, and protected areas created by the state in previously recognized Indigenous territories (Smith et al. 2003; IBC and Smith 2012).

PGIS projects undertaken in other South American countries have focused on recovering ancestral lands lost by the configuration of national territories, as is the case with the Diaguita and Mapuche communities of Chile and Argentina (Cerra 2011; Arias 2012; Hirt 2012; Arenas 2013; Cerra 2014; Melin Pehuen et al. 2019). However, as analyzed by Cerra (2014), although the demarcation of territorial borders can be recognized by the states, the complexity of territorial representations is not necessarily captured. The narratives and memories rooted in the landscapes, which provide primary support for Indigenous claims and environmental management practices, are usually not mapped in a Cartesian space.

Many Indigenous peoples and local communities have a close connection to place. One of the main purposes of PGIS has been the recording of information that documents the occupation history of their ancestral territory before the State and the non-Indigenous (and often powerful) segments of society. These landscapes are commonly in intimate relationship with natural resources and have helped depict the way of life of these populations. Indigenous communities want to emphasize the vitality of their contemporary culture, as well as their connection to the past. Teague (2011) observes that people alter the land through cultural practices and, therefore, the material characteristics of cultural landscapes are evidence of the long-standing relationship of a community with land. Such evidence is important in establishing traditional use rights that can be legally recognized. Yet, it is still necessary to reflect on what happens with the intangible, symbolic, mythological, and memory spaces that connect people to land (Basso 1996), which has an intimate relationship with environmental practices. Archaeology, as a discipline focused on the study of the past, can contribute to the creation of other kinds of historical narratives through the examination of past and present human-environmental dynamics.

2.2 Current PGIS in Archaeology and Methodological Approaches

In recent years, archaeology has reoriented its objectives and research focus. While maintaining its strengths in reconstructing the deep history of Indigenous cultural groups, it is also trying to shed itself of much of its colonial baggage. As a result, archaeologists have allocated significant efforts to create collaborative partnerships that address the interests and demands of local communities to recover their ancestral relationships with the landscape and participate in decision-making regarding the local environment and archaeological heritage.

With respect to mapping Indigenous cultural heritage, archaeological documentation has been based upon the fundamentals of Western cartography. Scholars have created decolonizing research methodologies by forming partnerships and recognizing the authority and rights of local stakeholders, including Indigenous, descendant, and local communities to manage their natural and cultural environments (e.g., Davis et al. 2021 for a case study example). Within Latin American archaeology, there is an emerging literature that discusses the application of PGIS on community-based research (Álvarez Larrain and McCall 2019a, 2019b). Broadly speaking, we can identify two main trends within PGIS-based research: (1) the recording of archaeological evidence and TEK to account for ancestry as a strategy against extractivism and its encroachment into Indigenous lands (Heckenberger et al. 2003; Heckenberger 2004; Manasse 2008; Quesada 2009; Heckenberger 2014; Manasse and Vaqué 2014); and (2) research that incorporates LSK and multivocality in archaeological interpretations and heritage protection (Endere and Curtoni 2006; Martínez Celis 2013; Duin et al. 2015; Álvarez Larrain et al. 2019; Martin Silva et al. 2019; Álvarez Larrain et al. 2020; McAnany 2020).

One of the pioneering PGIS archaeology projects in Latin America, "The Upper Xingu Indigenous History Project" in the Xingu National Park in Brazil's Mato Grosso region, was initiated in 2001. For the Xinguano/Kuikuru people, archaeology has been a critical tool that has allowed them to defend their cultural and land resources by learning about the ancient places where they dwelled, linking oral histories to their deeper past (Heckenberger 2004). Archaeological studies based on RS and excavations have shown large, permanent, and densely distributed settlements and significant landscape modification, including *terras pretas*, widespread across large areas of the Amazon region and hidden under the forest canopy (Heckenberger 2020). Archaeologists have asserted that Pre-Columbian societies developed sophisticated systems of land-use and management and engineered landscapes creating mosaics of bio-historical diversity (Heckenberger et al. 2003; Palace et al. 2017), supporting the idea that the current Amazon tropical forest is the product of millennia of anthropogenic activities (e.g., Balée 2014; Walker 2018; Balée et al. 2020). The PGIS project developed research strategies focused on partnership and training with descendent communities whose livelihoods depend on the integrity of the Xingu socio-ecological landscapes. It can be quickly assessed from open-access satellite images that the Xingu National Park contains devastated forests along its boundaries caused by cattle ranching, soy farming, urbanization, and hydro-electric projects. The project involved satellite image analysis, GPS ground-based mapping, and GIS that incorporated ecological data and TEK, and was designed to work in conjunction with broader global environmental agendas, such as biodiversity conservation and tropical forest degradation reduction programs. More recently, this collaborative project initiated the development of a GIS dashboard that enables local communities use mobile applications to monitor who is traveling within the park, track real-time health patients, and visualize the impacts of the Covid-19 pandemic (Fig. 11.1). In countries like Argentina, with a long history of neglect and state violence against native communities, PGIS has led to a new Indigenous cartography of ancestral territories, with the goal of developing a useful legal tool for land claims cases. From this perspective, archaeological evidence constitutes an important construct of cultural identity for these communities. PGIS appeared at a time when different social actors claimed cultural and archaeological heritage, which was negatively affected by modern economic activities such as mining, real estate transactions, tourism, and commercial development embedded in neoliberal logic (Manasse 2008; Montini 2008; Quesada 2009; Manasse and Vaqué 2014).

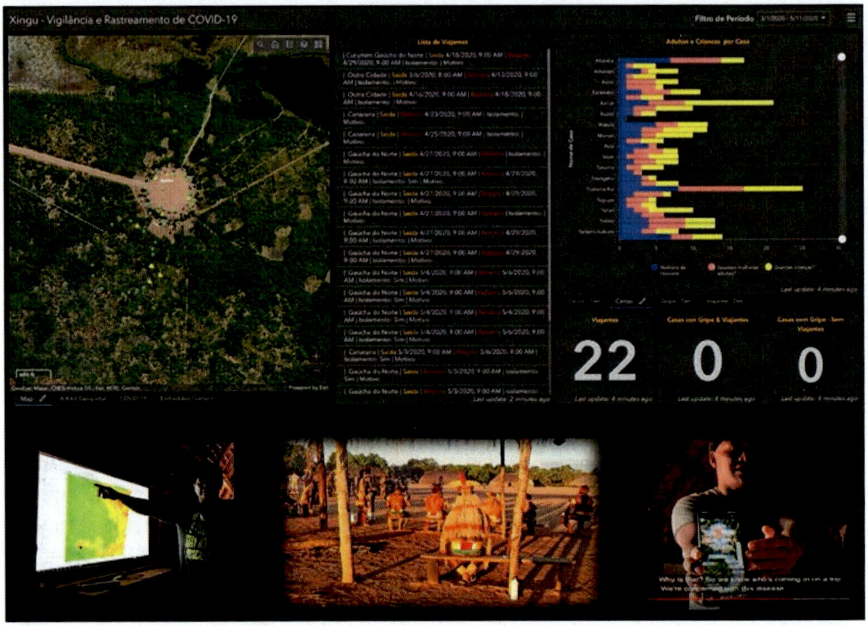

Fig. 11.1 COVID-19 dashboard of the Kuikuru people living in the Xingu National Park in Brazil's Mato Grosso region (Source: https://storymaps.arcgis.com/stories/d13c50b64ada4e53856b3d4d64a08bcb)

Working with local families in Antofalla, NW Argentina, Quesada (2009) used satellite images to identify the places where families practiced traditional economic activities, indicating the location of toponyms and places of ritual practices. The author shows the continuous use of the territory by the Indigenous community in a landscape that was depicted as empty on standard (Western) maps, which had clear geopolitical implications. During the mapping process, Quesada observed that satellite images were oriented in relation to the position of the "morning sun" or "afternoon sun" (2009: 163). In the mapping of a Diaguita community territory in Cafayate, NW Argentina, Cerra (2014) indicates that the maps were drawn according to the upslope direction of hills. Barrera Lobatón (2009) expresses this dichotomy between virtual Euclidean space vs. lived space, and the importance of the narratives in the ways of representing that experienced space. In the former, estrangement is required, in the latter that is impossible.

As Montini (2008) points out, the most visible impact of the privatization of space is the breakdown of the production system of the local populations, as is the curtailment of access to natural resources that is crucial to cultural and material life. As it can be seen in the mapping processes undertaken by the Mapuche people, the way of understanding the territory assumes an occupation dynamic that does not necessarily translate into a continuous human presence throughout the territorial space related to subsistence. There are forbidden spaces for people that have significant ancestral and ritual meaning, but are equally necessary to sustain the balance of life in a territory (Rodriguez de Anca et al. 2013). As Smith et al. (2012: 121) state: "There are many dimensions of local Indigenous knowledge and landscapes that are difficult or impossible to include in a computer database, or that lose their meaning when taken out of context. Stories about legendary battles between mountains, the calls of birds associated with specific places, the awe that ancient archaeological sites inspire, or the tensions arising from boundary disputes are things that cannot be reduced to zeros and ones". Though the authors are referring to the Mexican context, this statement could apply to a wide range of Indigenous populations in the Americas (see for example Basso 1996; Bray 2015). In addition, territorial overlaps can be a central feature of traditional land tenure that contradicts conventional notions of private property.

Recent archaeological research within the Maya region in Guatemala also demonstrates some of the trends towards PGIS in socio-environmental studies with descendant communities. Maya archaeology has pioneered RS and satellite technology to record archaeological evidence hidden by forest cover. As noted earlier, this technology- especially LiDAR- has created a wealth of new settlement data that was not readily available through conventional surveys (Canuto et al. 2018). On the other hand, challenges about partnership building around community mapping are critically reviewed in relation to the politics of heritage, the nation state (Chase et al. 2020) and identity among Maya communities. The Guatemalan Community Mapping Project was originally conceived to represent shrines and archaeological sites. Later, community members expanded the scope of the project to include natural and cultural resources (McAnany et al. 2015; McAnany and Rowe 2015). Technical workshops on computer and field-based mapping and GPS to identify sites, and on how to display points on Google Earth, were part of the approach.

In French Guiana and Suriname, a growing number of Wayana people have begun to appreciate archaeology as a tool to describe their past and as a powerful asset to rewrite their history (Duin et al. 2014, 2015; Duin 2017, 2018). In Wayana villages, archaeologists and community members employed ecological indicators and pottery fragments together with oral history to construct a GIS model for predicting the location of archaeological sites. Wayana elders and their oral accounts provide insight into settlement patterns that are not on any historical maps. Wayana social memory indicates that many people used to live in areas along tributary rivers, which are largely depicted as vacant on official maps. In contrast, community-based research demonstrated that this region was far more densely populated. In a similar context and in relation with oral history, the Amazon Conservation Team (ACT) presents "Terrastories", a Mapbox platform application built to enable local communities to locate and map their oral storytelling traditions (Fig. 11.2). Terrastories is designed to be entirely offline-compatible, so that remote communities can access the application without needing internet connectivity (ACT 2019: 37)[7].

A recent experience with high school students in Yocavil, a Diaguita-Calchaqui territory in NW Argentina, combined aerial photography mapping, field trips, and UAV flights to register meaningful cultural sites (Álvarez Larrain et al. 2020). The aim was to approach the archaeological landscapes of Yocavil from a local spatial

Fig. 11.2 "Terrastories", a Mapbox platform application built to enable local communities to locate and map their own oral storytelling traditions about places (Source: https://terrastories.io/demo/)

knowledge viewpoint to assess the significance of archaeological sites in the history and culture of the valley, as well as in relation to the knowledge generated by archaeologists. The main activity consisted of a field trip to a location using a satellite image enlargements of the valley's southern portion. Students travelled to a pre-Hispanic town with some defense features located on a low hill near the school. Once at the top of the hill Álvarez Larrain and colleagues employed the technique of scale mapping and asked students to mark the places they knew including modern towns, archaeological sites, or other types of meaningful places on the images. A few days after the workshop, a field trip was organized with one of the students and his teacher to record a meaningful site. The researchers were surprised when they recognized that this settlement presented a clear Inka architectural pattern, which had important implications for the reconstruction of the expansion of the Inka Empire from its capital, Cuzco, Peru. Subsequently, a drone survey was carried out to make the preliminary map of the core part of the settlement by photogrammetry and obtain video footage (Fig. 11.3).

The history of the Indigenous peoples of Latin America is subject to complex internal and external socio-political processes (e.g., genocide, disease, territorial and population displacements, and confinement) that began in the late fifteenth century and endures until today. Archaeology and anthropology have operated for well over a century within these contexts (e.g., Trigger 1984; Habu et al. 2008). In recent years, archaeologists have made important strides toward decolonizing archaeology through a variety of more inclusive practices that aim at benefiting the communities with which they work. In our view, PGIS is one such approach that bridges some of the divides that separate the views of academics and those of Indigenous actors.

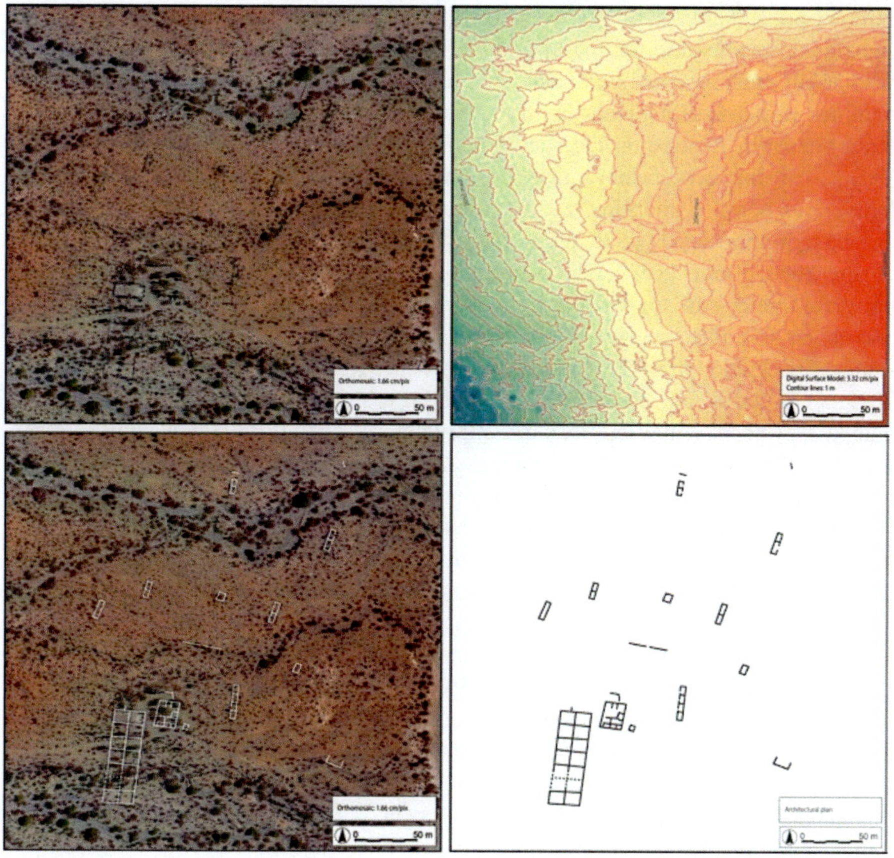

Fig. 11.3 Orthomosaic, digital surface model, and preliminary architectural and topographical plan of Inka site in Yocavil, NW Argentina, generated by UAV and photogrammetry (Author: Catriel Greco)

3 Discussion

Modern Western cartography supported by RS and GIS has been widely criticized for promoting predominantly scientific knowledge generation while neglecting other epistemological perspectives. These critiques are now widely recognized in several disciplinary fields (Rundstrom 1990; Harley 1992; Rundstrom 1995; Harley 2005; Wood 2010; Hacıgüzeller 2012). By the mid-1990s, increased criticism led to the introduction of PGIS, but not without new dilemmas and challenges (Poole 1995; Abbot et al. 1998; Stocks 2003; Elwood 2006; Wainwright and Bryan 2009; McCall and Dunn 2012).

One of the primary concerns is whether the use of these approaches implies a subordination of native epistemologies to Western cartographic norms with the consequent potential loss of Indigenous ways of knowing. McAnany et al. (2015: 7) warn, “Does the act of mapping homogenize communities for the purpose of external presentation, particularly when legal land claims are at stake? [...] These questions move us into the realm of situational ethics and, as such, yield no ready answer but must be contextualized". While we cannot ignore that Indigenous communities live in a globalized world and under nation state policies, the answer to these questions depend on the specific contexts wherein they exist. But precisely because of the power that maps have in establishing the hegemonic society, Indigenous peoples, marginalized communities, and other subordinate groups can use this rhetorical power to represent alternative visions of worlds and futures (Offen 2009; Sletto 2010). This "counter-mapping" (Peluso 1995) can become a powerful strategy to support the fight for Indigenous rights and sustainable livelihoods across Latin America (Denniston 1994; Gavazzi and Spyer Resende 1998; Finley-Brook and Offen 2009; Sletto 2009, 2010; Risler and Ares 2013; Kelly et al. 2017; ACT 2019). Nevertheless, it is essential to consider the purpose of different mapping projects. This step does not imply compromising Indigenous epistemologies, but rather co-constructing a process and a (temporary) product that serves as tools for dialogue with the larger society.

With the rapid technological advancement of geographic information science, new issues in PM approaches arise including developing local expertise in the handling of these technologies, the cost of software and computer equipment, hardware and software maintenance, among others. PGIS could involve technical processes and deter local participation. In Latin America, these conditions usually lead to a dependence on external technical expertise, as well as on financial resources managed by universities, government agencies, or international organizations including NGOs, that have agendas that may interfere with those of a given Indigenous community. For example, the archaeological mapping project with Maya communities in Guatemala discussed earlier (McAnany et al. 2015: 11), contracted a British-based company to provide secure networked storage for data points, and to programmable GPS units with community customized icons to make data collection user-friendly. This original proposal had to be shelved because the partnership with the software company proved too costly to renew even for an U.S. archaeological project.

Another relevant issue is the need of Western scientific standards to achieve broader "credibility", in conjunction with the issues that arise through the public exposure of TEK and LSK, which can lead to misinterpretations and misuse. As Sletto (2011) points out, hand-drawn maps that are cheap to reproduce and easy to read are usually adopted by local people for general use, meanwhile their computer or GIS counterparts are handled by external agents, which restricts accessibility, especially of the people who are supposed to be represented and served in those maps. The actions of academic researchers and practitioners in helping communities to develop, use, and make their maps public has been extensively criticized and debated (Rundstrom 1995; Johnson et al. 2006; Wainwright and Bryan 2009). To be effective, maps that show Indigenous land or traditional uses, for example, must position those claims within the state's territory and modern Western logic. Stocks (2003) notes that the first regional mapping of Indigenous communities in eastern Honduras and Nicaragua in the 1990s, under the sponsorship of Cultural Survival and the World Bank, was done without establishing an agreement on boundaries between territories. They were both relatively short-term projects that produced maps with territorial overlaps that highlight areas of shared use or close kinship connections between communities. However, the maps were used by politicians to support the argument that if the Indigenous people could not agree on their boundaries -understood as fixed and concrete territorial limits-, the state had no reason to promote Indigenous land legalization. The pressure to demarcate those limits exacerbated by the capitalist logic of private property led to conflicts between communities.

A recent review of mapping projects in Latin America refers to this type of participatory approach as *radical cartographies*. Defined as the employment of modern technology with a radical use by communities (Sletto et al. 2020), PGIS practitioners have chosen "to bend spatial information technology to suit social needs and cultural expression" (McAnany et al. 2015: 7). These approaches involve an understanding that there are other epistemological conceptualizations of space that do not respond to Cartesian logic, and the search for creative ways to represent those epistemologies that are facilitated by digital devices and interfaces (Kelly et al. 2017). Mapping tools are now cheaper and more available, and a growing number of NGOs, academics, and lay citizens are engaged in innovative PGIS procedures.

To that end, counter-mapping should question preconceived knowledge and ideas of space, and thereby, lead to an epistemological and ontological re-definition of cartography itself. Rather than fully rejecting Western cartography and GIS, cartographers should use these as tools for cultural re-appropriation by employing them critically and aligning them with Indigenous cartographic traditions (Hirt 2012). Resisting the homogenization of spatial representations requires innovation in tool engineering, mapping methods, and forms of representation. Technological developments not only enable outreach to communities, but also facilitates intergenerational dialogues with younger generations who have fully embraced this technology for new ways of learning, working, and socializing. For example, among the Pemon people of southern Guyana and western Brazil, who currently reside within the Canaima National Park, the younger members are more engaged in tourism, tend to avoid traditional activities such as hunting, gardening, and fishing, and are also reluctant to embrace TEK because it is considered 'old-fashioned' or even environmentally destructive (Sletto 2009). Possible avenues for future research include the creation of new narratives and graphic techniques for incorporating spiritual and cultural experiences of place, recording intangible cultural heritage, and capturing the wealth of environmental knowledge and practical skills that are transmitted from one generation to the next.

There has been, in fact, a technological revolution since the first PM projects, with the growing incorporation of GIS and other recent innovations that seek to incorporate dynamic information such as stories, memories, pictures, and sounds. GIS maps can show either fluid or fuzzy boundaries as patterns of dynamic land use and toponyms, in addition to fixed boundaries required by cadastral land tenure systems (Kelly et al. 2017). Another technological frontier is the internet, which has led to rapidly developing web-based mapping technologies that allow Indigenous people and their allies to use apps such as Google Earth, Google Maps, and other user-driven mapping interfaces. These apps facilitate participation in mapping projects, to publish online maps, and other data such as photographs and video. New internet technologies can make possible representations that are truer to the complex articulations of space, time, and culture that characterize Indigenous landscapes. For example, in the Cauca region of Colombia, the Fundación Minga developed innovative mapping projects resulting in "speaking maps", which are living maps that facilitate critical conversations and provide endogenous documentation of local knowledge among residents in Afro-Colombian communities. These talking maps demonstrated how memories and social relationships shape different conceptions of the territory (Velasco Alvarez 2020).

As Hacıgüzeller (2017) observes, digital cartographic interaction involves actions (switching layers on or off, changing symbology, or scaling maps up or down) carried out in playful ways. This makes it easy to consider digital maps as fleeting phenomena enacted through contextually situated practices, a consideration embraced by this performative turn in cartography. Approaches are currently being employed in 3D Projection-Augmented Landscape Models (3DPALM), that use physical 3D landscapes enhanced with geo-simulation models in GIS to support participatory planning, education, and cross-cultural knowledge exchange (Fig. 11.4). In Mexico, researchers used 3D-printed plastic tiles and the projection

Fig. 11.4 A 3D Projection-Augmented Landscape Model using geo-simulation in GIS and 3D-printed plastic tiles

of multiple terrain layers to assists engagement with local farmers to explore sustainable agricultural practices. Models can also be displayed on a sand box for a more performative interaction (Fisher et al. 2019). In this regard, a major challenge is to venture out from conventional maps and replace them with cartographic representations that more richly depict how Indigenous people perceived the landscapes that they inhabit.

4 Conclusions

This chapter reviewed current applications of PGIS in archaeological research in collaboration with Indigenous, descendant and traditional communities of Latin America. Over the last several decades, Latin American and foreign archaeologists developed an array of techniques for examining the relationships between pre-Hispanic Indigenous peoples and the landscape wherein they exist. Yet, in recent years, archaeology has been called upon to do much more with this information, including engaging in research problems that are of interest to the Indigenous communities with whom they work.

The integration of LSK and TEK and geospatial science, mediated through the construction of common research agendas, enriches both types of epistemologies. For local and Indigenous people, archaeology can contribute to the identification of historical settlement locations unknown to them. This can be a critical tool for learning about the places of ancestors lost from the collective memory, a type of knowledge that is relevant to defending their cultural resources and exploring and preserving their heritage for future generations. Identity and culture are intrinsically linked to territory and local resources. For archaeological research, people's environmental knowledge gathered over many generations can significantly contribute to the way one interprets a site (Green et al. 2003). Further, oral tradition can greatly enrich our understanding of the different meanings of places. As a result, future endeavors in this field will require researchers to be aware of the role that the archaeologist plays within communities that have been disproportionately affected by social and environmental injustices that impact land rights, relationships with the landscape, and cultural heritage. In understanding these relationships, archaeology will continue to play a fundamental role as one of the few disciplines equipped to reconstruct the past and provide material evidence of human-environmental interactions to understand the present.

PGIS ceased to be just a tool for the generation of documents that can be read by external stakeholders, and over time, it became a process for the organization and internal decision-making of the communities according to their own understanding of human-environment relations. Today, it is used to document environmental degradation and resource extraction, as well as create comprehensive development plans and resource management strategies. In other words, PGIS has been adopted by communities as a tool for research and planning, contributing to their self-determination. The main challenge is how to effectively use these maps in their internal affairs and in discussions with state administrators and decision makers. The "post- mapping" phase should be one of the most important and challenging parts of PGIS-related work. Maps should not be the end goal of the collaborative process but the beginning. Perhaps the most relevant questions for PGIS are: why do we want to create a map? What is its purpose? How will it be used, and by whom? PGIS should be understood as a social and political process that goes beyond the creation of maps and that involves more than purely technological processes. It should serve to strengthen community identity, allowing Indigenous people to take control of their lands not only legally, in terms of territorial rights, but through the development of social roots and sustainable management practices.

References

Abbot J, Chambers R, Dunn C et al (1998) Participatory GIS: opportunity or oxymoron? Particip Learn Action Notes:27–34

ACT/Amazon Conservation Team (2019) Mapping and recording place-based oral histories: a methodology. First Digital Edition. https://www.amazonteam.org/wp-content/uploads/2019/06/ACT_OralHistories_Guide_2019_ENGLISH-draft.pdf

Álvarez Larrain A, McCall MK (2019a) Participatory mapping and participatory GIS for historical and archaeological landscape studies: a critical review. J Archaeol Method Theory 26:643–678. https://doi.org/10.1007/s10816-018-9385-z

Álvarez Larrain A, McCall MK (2019b) La cartografía participativa como propuesta teórico-metodológica para una arqueología del paisaje latinoamericana. Un ejemplo desde los Valles Calchaquíes (Argentina). Antípoda 36:85–112. https://doi.org/10.7440/antipoda36.2019.05

Álvarez Larrain A, McCall MK (2020) Paisajes culturales, arqueología y mapeo participativo. Aportes y enseñanzas desde el continente americano. In: Urquijo PS, Boni FA (eds) Huellas en el paisaje. Geografía historia y ambiente en las Américas, CIGA-UNAM, Morelia, pp 341–375

Álvarez Larrain A, Greco C, McCall MK (2019) Saberes y percepciones locales sobre los paisajes arqueológicos: experiencias de cartografía participativa en Yocavil (Noroeste argentino). Boletín Antropológico 37:2542–3304

Álvarez Larrain A, Greco C, Tarragó M (2020) Participatory mapping and UAV photogrammetry as complementary techniques for landscape archaeology studies: an example from North-Western Argentina. Archaeol Prospect:1–15. https://doi.org/10.1002/arp.1794

Arenas P (2013) La participación de Tucumán en el relevamiento territorial de la Ley 26160: una mirada desde las prácticas. Población y Soc 20:125–136

Arias PD (2012) Nosotros vamos a dibujar nuestro propio espacio territorial. Reapropiación del territorio y apropiación de la Cartografía en la Zonal Pewence Corpus. Arch virtuales la alteridad Am 2(1)

Atalay S (2012) Community-based archaeology: research with, by, and for Indigenous and local communities. University of California Press, Berkeley

Balée W (2014) Historical ecology and the explanation of diversity: Amazonian case studies. In: Verdade LM, Lyra-Jorge MC, Piña CI (eds) Applied ecology and human dimensions in biol Conserv. Springer, Berlin, pp 19–33

Balée W, de Oliveira VH, dos Santos R et al (2020) Ancient transformation, current conservation: traditional Forest management on the Iriri River. Brazilian Amazonia Hum Ecol 48:1–15. https://doi.org/10.1007/s10745-020-00139-3

Barrera Lobatón S (2009) Reflexiones sobre Sistemas de Información Geográfica Participativos (sigp) y cartografía social. Cuad Geogr Rev Colomb Geogr:9–23. https://doi.org/10.15446/rcdg.n18.12798

Basso KH (1996) Wisdom sits in places. Landscape and language among the Western apache. University of New Mexico Press, Albuquerque

Bray T (2015) The archaeology of Wak'as: explorations of the sacred in the pre-Columbian Andes. University Press of Colorado, Boulder

Brody H (1988) Maps and dreams: Indians and the British Columbia frontier. Douglas & McIntyre, Vancouver, BC, Canada

Candler C, Olson R, Deroy S, Broderick K (2006) PGIS as a sustained (and sustainable?) practice: first nation experiences in treaty 8 BC. Canada Particip Learn Action:51–57

Canuto MA, Estrada-Belli F, Garrison TG et al (2018) Ancient lowland Maya complexity as revealed by airborne laser scanning of northern Guatemala. Science 80:361. https://doi.org/10.1126/science.aau0137

Caramanica A, Huaman Mesia L, Morales CR et al (2020) El Niño resilience farming on the north coast of Peru. Proc Natl Acad Sci 117:24127–24137. https://doi.org/10.1073/pnas.2006519117

Cerra MC (2011) Mapeando representaciones: cerros, centros y fronteras. Comunidad Diaguita-Calchaquí "El Divisadero" (Cafayate, Salta). In: Rodríguez L (ed) Resistencias, conflictos y negociaciones. El valle Calchaquí desde el período prehispánico hasta la actualidad. Prohistoria Ediciones, Rosario, pp 211–229

Cerra MC (2014) En el devenir de las políticas públicas: Programa de Relevamiento Territorial de comunidades indígenas. Comunidad diaguita-calchaquí 'El Divisadero'. Intersecc en Antropol 15:103–114

Chapin M, Threlkeld B (2001) Indigenous landscapes: a study in Ethnocartography. Center for the Support of Native Lands, Arlington

Chapin M, Threlkeld B (2008) Mapping Indigenous Lands. A practical Guidebook. Center for the Support of Native Lands, Arlington

Chapin M, Lamb Z, Threlkeld B (2005) Mapping Indigenous lands. Annu Rev Anthropol 34:619–638. https://doi.org/10.1146/annurev.anthro.34.081804.120429

Chase A, Chase D, Chase A (2017) LiDAR for archaeological research and the study of historical landscapes:89–100. https://doi.org/10.1007/978-3-319-50518-3_4

Chase A, Chase D, Chase A (2020) Ethics, new colonialism, and Lidar data: a decade of Lidar in Maya archaeology. J Comp Appl Archaeol 3(1):51–62

Clarke DL (1977) Spatial archaeology. Academic, Boston

Colwell-Chanthaphonh C, Ferguson TJ (2008) Collaboration in archaeological practice: engaging descendant communities. Altamira Press, Lanham

Corbett J (2009) Good practices in participatory mapping. International Fund for Agricultural Development (IFAD), Rome

Corbett J, Keller P (2006) Using community information systems to communicate traditional knowledge embedded in the landscape. Particip Learn Action 54:21–27

Davis DS, Buffa D, Rasolondrainy T, Creswell E, Anyanwu C, Ibirogba A, Randolph C, Ouarghidi A, Phelps LN, Lahiniriko F, Maharesy C, Manahira G, Douglass K (2021) The aerial panopticon and the ethics of archaeological remote sensing in sacred cultural spaces. Arch Prosp 28(3):305–320. https://doi.org/10.1002/Arp.1819

Denevan WM (1993) The 1931 Shippee-Johnson aerial photography expedition to Peru. Geogr Rev 83:238–251

Denniston D (1994) Defending the land with maps. World Watch:27–32

Douglass K, Cooper J (2020) Archaeology, environmental justice, and climate change on islands of the Caribbean and southwestern Indian Ocean. Proc Natl Acad Sci U S A 117:8254–8262. https://doi.org/10.1073/pnas.1914211117

Douglass K, Morales EQ, Manahira G et al (2019) Toward a just and inclusive environmental archaeology of Southwest Madagascar. J Soc Archaeol 19:307–332. https://doi.org/10.1177/1469605319862072

Duin RS (2017) Indigenous archaeologies and histories in Suriname: from pre-Columbian archaeology and ethno-archaeology to a community-based archaeology of engagement. In:

Duin R, White C (eds) Archaeology in Suriname. Anton de Kom Universiteit van Suriname, Paramaribo, pp 73–108
Duin RS (2018) Archaeological agenda in the Guianas. In: Encyclopedia of Global Archaeology, pp 1–28
Duin RS, Toinaike K, Alupki T, Opoya A (2014) Engaged Archaeolog participatory mapping with Indigenous people of t upper Maroni river bas. Amaz Rev Antropol 6:332–357
Duin RS, Toinaike K, Alupki T, Opoya A (2015) Archaeology of engagement: Indigenous people, social memory, and making history in the upper Maroni Basin (northern Amazonia). Curr Anthropol 56:753–761
Dunn CE (2007) Participatory GIS - a people's GIS? Prog Hum Geogr 31:616–637. https://doi.org/10.1177/0309132507081493
Elwood S (2006) Critical issues in participatory GIS: deconstructions. Trans GIS 10:693–708
Endere ML, Curtoni RP (2006) Entre Lonkos y "ólogos". La participación de la comunidad indígena Rankülche de Argentina en la investigación arqueológica. Arqueol Suramericana 2:72–92
Erickson C (1998) Applied archaeology and rural development: Archaeology's potential contribution to the future. In: Whiteford MB, Whiteford S (eds) Crossing currents. Continuity and Change in Latin America. Prentice Hall, Upper Saddle River, pp 34–45
Fals Borda O (1999) Orígenes universales y retos actuales de la IAP (Investigación Acción Participativa). Análisis Político 38:71–88
Finley-Brook M, Offen K (2009) Bounding the commons: land demarcation in northeastern Nicaragua. Bull Lat Am Res 28:343–363. https://doi.org/10.1111/j.1470-9856.2009.00302.x
Fisher C (2020) Archaeology for sustainable agriculture. J Arch Res 28:393–441. https://doi.org/10.1007/s10814-019-09138-5
Fisher R, Heckbert S, León Villalobos JM, Sutton S (2019) Augmenting physical 3D models with projected information to support environmental knowledge exchange. Appl Geogr 112:1–10. https://doi.org/10.1016/j.apgeog.2019.102095
Flavelle A (2002) Mapping our land. Lone Pine Publishing, Alberta
Freeman M (1976) Report, Inuit land use and occupancy project (3 vols.). Department of Indian Affairs and Northern Development, Ottawa, Canada
Gavazzi RA, Spyer Resende M (1998) Atlas geográfico indígena do Acre. Comissao Pró-Índio do Acre, Rio Branco
Gillings M, Hacigüzeller P, Lock G (eds) (2020) Archaeological spatial analysis: a methodological guide. Routledge, London
Gnecco C, Ayala Rocabado P (2010) Pueblos indígenas y arqueología en América Latina. Ediciones Uniandes, Bogotá
González AR (1957) La fotografía y el reconocimiento aéreo en las investigaciones arqueológicas del NOA. An Arqueol y Etnol:41–62
Grau Mira I (ed) (2006) La Aplicación de los SIG en la Arqueología del Paisaje. Universidad de Alicante, Alicante
Greco C (2018) Photogrammetric survey with UAV of strategic villages of Yocavil (Catamarca, Argentina). In: Álvarez Larrain A, Greco C (eds) Political landscapes of the late intermediate period in the Southern Andes. The Pukaras and their Hinterlands. Springer, Cham, pp 35–60. https://doi.org/10.1007/978-3-319-76729-1. In: Whiteford MB, Whiteford S (eds) Crossing Currents. Continuity and Change in Latin America, Prentice Hall, Upper Saddle River, NJ, USA, pp. 34–45
Green LF, Green DR, Neves E (2003) Indigenous knowledge and archaeological science. J Soc Archaeol 3:366–398. https://doi.org/10.1177/14696053030033005
Gupta N, Blair S, Nicholas R (2020) What we see, what we Don't see: data governance, archaeological spatial databases and the rights of Indigenous peoples in an age of big data. J F Archaeol 45:S39–S50. https://doi.org/10.1080/00934690.2020.1713969
Habu J, Fawcett C, Matsunaga JM (eds) (2008) Evaluating multiple narratives: beyond nationalist, colonialist, imperialist archaeologies. Springer, New York
Hacıgüzeller P (2012) GIS, critique, representation and beyond. J Soc Archaeol 12:245–263. https://doi.org/10.1177/1469605312439139

Hacıgüzeller P (2017) Archaeological (digital) maps as performances: towards alternative mappings. Nor Archaeol Rev 50:149–171. https://doi.org/10.1080/00293652.2017.1393456

Haldon J, Chase AF, Eastwood W et al (2020) Demystifying collapse: climate, environment, and social agency in pre-modern societies. Millennium 17:1–33. https://doi.org/10.1515/mill-2020-0002

Harley JB (1992) Rereading the Maps of the Columbian Encounter Author(s). Ann Assoc Am Geogr 82:522–536

Harley JB (2005) La Nueva Naturaleza de los Mapas. Ensayos sobre la Historia de la Cartografía. Fondo de Cultura Económica, Buenos Aires

Heckenberger M (2004) Archaeology as Indigenous advocacy in Amazonia. Pract Anthropol 26:35–39. https://doi.org/10.17730/praa.26.3.j3x06m427k558232

Heckenberger M (2014) Bio-historical diversity, sustainability and collaboration in the Xingu. Anuário Antropológico:69–96. https://doi.org/10.4000/aa.1250

Heckenberger M (2020) Xingu garden cities: Amazonian urban landscapes, or what? In: Farhat G (ed) Landscapes of preindustrial urbanism. Dumbarton Oaks, Washington, DC, pp 225–261

Heckenberger M, Kuikuro UT, Russell JC et al (2003) Amazonia 1492: pristine Forest or cultural parkland? Science 301(80):1710–1714

Heckenberger M, Russell JC, Toney JR, Schmidt MJ (2007) The legacy of cultural landscapes in the Brazilian Amazon: implications for biodiversity. Philos Trans R Soc B Biol Sci 362:197–208. https://doi.org/10.1098/rstb.2006.1979

Heckenberger M, Russell JC, Fausto C et al (2008) Pre-Columbian urbanism, anthropogenic landscapes, and the future of the Amazon. Science 321(80):1214–1217. https://doi.org/10.1126/science.1159769

Herlihy PH, Knapp G (2003) How much participation equals participatory? Hum Organ 62:303–314

Herrera Wassilowsky A, Curet LA (2019) Arqueología para América Latina en el siglo XXI. Antípoda Rev Antropol Arqueol 1:3–10. https://doi.org/10.7440/antipoda36.2019.01

Hirt I (2012) Mapping dreams/dreaming maps: bridging Indigenous and Western geographical knowledge. Cartogr Int J Geogr Inf Geovisualization 47:105–120. https://doi.org/10.3138/carto.47.2.105

Hodder I, Orton C (1976) Spatial analysis in archaeology. Cambridge University Press, Cambridge, UK

IBC/Instituto del Bien Común, Smith RC (2012) Cartografía Socio Ambiental en la gestión de la Amazonía peruana. Fundación Avina and Fundación Gordon y Betty Moore, Lima

Iriarte J, Robinson M, de Souza J et al (2020) Geometry by design: contribution of Lidar to the understanding of settlement patterns of the mound villages in SW Amazonia. J Comput Appl Archaeol 3:151–169. https://doi.org/10.5334/jcaa.45

Johnson JT, Louis RP, Pramono AH (2006) Facing the future: encouraging critical cartographic literacies in Indigenous communities. Acme 4:80–98

Kelly J, Herlihy P, Derek S et al (2010) Indigenous territoriality at the end of the social property era in Mexico. J Lat Am Geogr 9:161–181. https://doi.org/10.1353/lag.2010.0026

Kelly J, Herlihy P, Tappan T et al (2017) From cognitive maps to transparent static web maps: tools for Indigenous territorial control in La Muskitia, Honduras. Cartogr Int J Geogr Inf Geovisualization 52:1–19

Kendall A (2005) Applied archaeology: revitalizing Indigenous agricultural technology within an Andean community. Public Archaeol 4:205–221. https://doi.org/10.1179/pua.2005.4.2-3.205

Kintigh KW, Altschul JH, Beaudry MC, Drennan RD, Kinzig AP, Kohler TA, Limp WF, Maschner HDG, Michener WK, Pauketat TR (2014) Grand challenges for archaeology. Am Ant 79:5–24. https://doi.org/10.7183/0002-7316.79.1.5

Knapp G, Herlihy P (2002) Mapping the landscape of identity. Conf Lat Am Geogr Yearb 27:251–268

Kohler TA, Rockman M (2020) The IPCC: a primer for archaeologists. Am Antiq 85:627–651. https://doi.org/10.1017/aaq.2020.68

Kosiba S, Bauer AM (2013) Mapping the political landscape: toward a GIS analysis of environmental and social difference. J Archaeol Method Theory 20:61–101. https://doi.org/10.1007/s10816-011-9126-z

Kosok P (1965) Life, land and water in ancient Peru. Long Island University Press, New York
Llobera M (2007) Reconstructing visual landscapes. World Archaeol 39:51–69. https://doi.org/10.1080/00438240601136496
Manasse B (2008) Articulación de saberes: mapeando territorio indígena desde las evidencias del pasado. In: IX Congreso Argentino de Antropología Social. MS, Posadas, Misiones, pp 1–14
Manasse B, Vaqué ML (2014) Relevamiento arqueológico en territorio de la comunidad indígena diaguita de El Mollar, Tucumán, Argentina. Arqueologia 20:183–205
Martin Silva VB, Zabala ME, Fabra M (2019) Cartografía Social como recurso metodológico para el análisis patrimonial. Experiencias de mapeo en Miramar (Córdoba, Argentina). Perspect Geográfica 24:125–148. https://doi.org/10.19053/01233769.8631
Martínez Celis D (2013) Inventario participativo de sitios con arte rupestre en Sutatausa (Cundinamarca). Una experiencia de apropiación social del patrimonio cultural. www.%0Arupestreweb.info/inventariorupestresutatausa.html
McAnany PA (2020) Imagining a Maya archaeology that is anthropological and attuned to Indigenous cultural heritage. Heritage 3:318–330. https://doi.org/10.3390/heritage3020019
McAnany PA, Rowe SM (2015) Re-visiting the field: collaborative archaeology as paradigm shift. J F Archaeol 40:499–507. https://doi.org/10.1179/2042458215Y.0000000007
McAnany PA, Rowe SM, Cholotio IQ et al (2015) Mapping Indigenous self-determination in highland Guatemala. Int J Appl Geospatial Res 6:1–23. https://doi.org/10.4018/ijagr.2015010101
McCall MK (2014) Mapping territories, land resources and rights: communities deploying participatory mapping/pgis in Latin America. Geogr Dep Univ Sao Paulo 94. https://doi.org/10.11606/rdg.v0i0.534
McCall MK, Dunn CE (2012) Geo-information tools for participatory spatial planning: fulfilling the criteria for "good" governance? Geoforum 43:81–94. https://doi.org/10.1016/j.geoforum.2011.07.007
McIntosh RJ, Tainter JA, McIntosh SK (eds) (1999) The way the wind blows: climate, history, and human action. Columbia University Press, New York
Melin Pehuen M, Mansilla Quiñones P, Royo Letelier M (2019) Cartografía cultural del wallmapu: Elementos para descolonizar el mapa en territorio mapuche. LOM Ediciones, Santiago
Montini G (2008) "Revalorizando nuestras raíces": una Arqueología del Oeste del Valle de Las Carreras, Tafí del Valle, Tucumán. Universidad Nacional de Tucumán, Argentina
Nenquimo O (2018) Mapeo territorial Waorani. Revista de la Universidad de México 838/839, Dossier Mapas:67
Nesbitt J (2016) El Niño and second-millennium BC monument building at Huaca Cortada (Moche Valley, Peru). Antiquity 90:638–653. https://doi.org/10.15184/aqy.2016.70
Nesbitt J (2020) Ancient agriculture and climate change on the north coast of Peru. Proc Natl Acad Sci 202017725:24617–24619. https://doi.org/10.1073/pnas.2017725117
Offen K (2009) O mapeas o te mapean: Mapeo indígena y negro en América Latina. Tabula Rasa:163–189
Palace MW, McMichael CNH, Braswell BH et al (2017) Ancient Amazonian populations left lasting impacts on forest structure. Ecosphere 8:1–19. https://doi.org/10.1002/ecs2.2035
Pauketat T (2020) When the rains stopped: evapotranspiration and ontology at ancient Cahokia. J Anthropol Res:410–438. 0091-7710/2020/7604-0002$10.00
Plotkin MJ (2020) The Amazon: what everyone needs to know. Oxford University Press, Oxford
Poole P (1995) Geomatics: who needs it? Cult Surv Q 18
Quesada MN (2009) Discursos cartográficos y territorios indígenas en Antofalla. Intersecc en Antropol 10:155–166
Ramirez-Gomez S, Brown G, Tjon Sie Fat A (2013) Participatory mapping with Indigenous communities for conservation: challenges and lessons from Suriname. Electron J Inf Syst Dev Ctries 58:1–22. https://doi.org/10.1002/j.1681-4835.2013.tb00409.x
Rick TC, Sandweiss DH (2020) Archaeology, climate, and global change in the age of humans. Proc Natl Acad Sci 117:8250–8253. https://doi.org/10.1073/pnas.2003612117

Risler J, Ares P (2013) Manual de mapeo colectivo: recursos cartográficos críticos para procesos territoriales de creación colaborativa. Tinta Limón, Buenos Aires

Rockman M, Hritz C (2020) Expanding use of archaeology in climate change response by changing its social environment. Proc Natl Acad Sci 117(15):8295–8302. https://doi.org/10.1073/pnas.1914213117

Rodriguez de Anca A, Villareal J, Cristina Valdez (2013) Otros mapas, otras territorialidades. Reflexiones a partir de experiencias de mapeo cultural participativo en Territorio Mapuce. In: I Jornadas Norpatagónicas de Experiencias Educativas en Ciencias Sociales para la Escuela Secundaria. II Jornadas Provinciales de Geografía, Ciencias Sociales y Educación, Río Negro

Ruggerio CA (2021) Sustainability and sustainable development: a review of principles and definitions. Sci Total Environ 786:147481. https://doi.org/10.1016/j.scitotenv.2021.147481

Rundstrom RA (1990) A cultural interpretation of Inuit map accuracy. Geogr Rev 80:155–168. https://doi.org/10.2307/215479

Rundstrom RA (1995) GIS, Indigenous peoples, and epistemological diversity. Cartogr Geogr Inf Syst 22:45–57. https://doi.org/10.1559/152304095782540564

Salamanca C, Espina R (2012) Mapas y derechos: Experiencias y aprendizajes en América Latina. Editorial de la Universidad Nacional de Rosario, Rosario

Sandweiss DH, Kelley AR (2012) Archaeological contributions to climate change research: the archaeological record as a Paleoclimatic and Paleoenvironmental archive. Annu Rev Anthropol 41:371–391. https://doi.org/10.1146/annurev-anthro-092611-145941

Schmidt M (2013) Amazonian dark earths: pathways to sustainable development in tropical rainforests? Bol do Mus Para Emilio Goeldi. Ciencias Humanas 8:11–38. https://doi.org/10.1590/S1981-81222013000100002

Sieber R (2006) Public participation geographic information systems: a literature review and framework. Ann Assoc Am Geogr 96:491–507. https://doi.org/10.1111/j.1467-8306.2006.00702.x

Sletto B (2009) "Indigenous people don't have boundaries": Reborderings, fire management, and productions of authenticities in indigenous landscapes. Cult Geogr 16:253–277. https://doi.org/10.1177/1474474008101519

Sletto B (2010) Self-Management in Indigenous Spatial Representations and the importance of capacity-building and consciousness-raising: the case study of Ethnocartographic project Inna Kowantok, sector 5 Pemon (Kavanayen- Mapauri). La Gran Sabana Antropológica TOMO LIII:43–75

Sletto B (2011) Cartografía participativa y derechos al territorio y los recursos. In: Memorias del foro internacional Bogotá. Instituto de Estudios Latinoamericanos, Universidad de Texas en Austin, Washington DC

Sletto B, Bryan J, Torrado M, et al. (2013) Territorialidad, mapeo participativo y política sobre los recursos naturales: la experiencia de América Latina. Cuad Geogr Rev Colomb Geogr 22:193–209. https://doi.org/10.15446/rcdg.v22n2.37014

Sletto B, Bryan J, Wagner A, Hale C (2020) Radical cartographies. Participatory mapmaking from Latin America. University of Texas Press, Austin

Smith DA (2008) Ethnogeography of the Mayangna of Nicaragua. Ethno- and historical geographic studies in Latin America: essays honoring William V. Davidson, edited by P H Herlihy, K Mathewson, and C S revels, 85-104. Geoscience publications, Department of Geography and Anthropology, LSU, Baton Rouge, LA

Smith RC, Benavides M, Pariona M, Tuesta E (2003) Mapping the past and the future: Geomatics and Indigenous territories in the Peruvian Amazon. Hum Organ 62:357–368. https://doi.org/10.17730/humo.62.4.3d82x31cywm8wl25

Smith DA, Herlihy PH, Ramos Viera A et al (2012) Using participatory research mapping and GIS to explore local geographic knowledge of Indigenpus Lansdcapes in Mexico. Focus Geogr 55:119–125

Stocks A (2003) Mapping dreams in Nicaragua's Bosawas reserve the advent of geographic information systems (GIS) and global positioning system (GPS) technology. Hum Organ 62:344–356

Stocks A (2005) Too much for too few: problems of Indigenous land rights in Latin America. Annu Rev Anthropol 34:85–104. https://doi.org/10.1146/annurev.anthro.33.070203.143844

Teague H (2011) Mapeo participativo y las políticas ambientales. Cartografía Participativa y derechos al territorio y los recursos. Memorias del Foro Internacional Bogotá, Sletto B (ed.) 81-117. University of Texas, Austin

TMCC/Toledo Maya Cultural Council (1997) Maya atlas: the struggle to preserve Maya land in southern Belize. North Atlantic Books, Berkeley

Tobias TN (2000) Chief Kerry's moose. A guidebook to land use and occupancy mapping, research design and data collection. The Union of British Columbia Indian chiefs y Ecotrust Canada, Vancouver

Trigger BG (1984) Alternative archaeologies: nationalist, colonialist, imperialist. Man 19:355–370

Van der Lleeuw S, Redman C (2002) Placing archaeology at the center of socio-natural studies. Am Ant 67(4):597–605. https://doi.org/10.2307/1593793

VanValkenburgh P, Cushman KC, Butters LJC et al (2020) Lasers without lost cities: using drone Lidar to capture architectural complexity at Kuelap, Amazonas, Peru. J F Archaeol 45:S75–S88. https://doi.org/10.1080/00934690.2020.1713287

Velasco Alvarez, AC (2020) social Polygraghy: territory as a living memorial to culture and nature. Sletto B, Bryan J, Wagner a, Hale C (ed) radical cartographies. Participatory mapmaking from Latin America. 35-44. University of Texas Press, Austin, TX

Wainwright J, Bryan J (2009) Cartography, territory, property: postcolonial reflections on Indigenous counter-mapping in Nicaragua and Belize. Cult Geogr 16:153–178. https://doi.org/10.1177/1474474008101515

Walker J (2018) River, island and field: a historical ecology of the Bolivian Amazon. University of new Mexico Press, Albuquerque

Willey G (1953) Prehistoric settlement patterns in the Virú Valley, Peru. Bureau of American Ethnology Bulletin 155, Washington, DC

Wood D (2010) Rethinking the power of maps. Guilford, New York

Chapter 12
Comparing Volunteered Data Acquisition Methods on Informal Settlements in Mexico City and São Paulo: A Citizen Participation Ladder for VGI

Alexandre Pereira Santos, Vitor Pessoa Colombo, Katharina Heider, and Juan Miguel Rodriguez-Lopez

Abstract In the early 2000s, Web 2.0 technologies prompted an explosion in geographic data that include Volunteered Geographic Information (VGI), a set of methods that brings user contribution to the center of data acquisition. These methods increase the capacity of community-driven and local initiatives to create geographic information and close existing data gaps in authoritative sources. Informal settlements constitute an example of where a major vacuum exists, as maps are often incomplete, outdated, or imprecise. However, quality issues regarding VGI frequently arise, as do questions on citizen participation and empowerment. This study explores how different VGI approaches support citizen participation and user empowerment, in tandem with the opportunities and limitations of VGI to map informal settlements in Latin America. We propose a VGI comparison framework to evaluate citizen participation in two informal settlement mapping projects in São Paulo and Mexico City. Such a framework includes four categories: (1) required material resources; (2) required geographic information system (GIS) literacy; (3) user agency; and (4) involvement of research subjects. The results demonstrate that higher citizen involvement in São Paulo stems from the inclusion of residents through participatory mapping methods. Conversely, the Mexico City's case demonstrates how crowdsourcing may happen irrespective of and contrary to the goals from those represented in the data. We suggest that VGI is a powerful tool for generating timely and precise data on informal settlements, but research subjects should have agency over geographic information collected about them.

A. P. Santos (✉) · K. Heider · J. M. Rodriguez-Lopez
University of Hamburg, Center for Earth System Research and Sustainability (CEN), Climate Change and Security Research Group (CLISEC), Hamburg, Germany

V. Pessoa Colombo
École Polytechnique Fédérale de Lausanne, Communauté d'Études pour l'Aménagement du Territoire (CEAT), Lausanne, Switzerland

S. López (ed.), *Socio-Environmental Research in Latin America*, The Latin American Studies Book Series, https://doi.org/10.1007/978-3-031-22680-9_12

Keywords Citizen participation · Participatory mapping · Volunteered geographic information · Informal settlements · Latin America

1 Introduction

From the early 2000s, there was an explosion of available geographical information made possible by Web 2.0 technologies, including Volunteered Geographic Information (VGI). VGI is a set of methods based on users' contributions to the acquisition of geographic information (Goodchild 2007). With the introduction of VGI, consumers of geographic information (formerly passive) can become active data producers. These methods marked Geographic Information (GI) production, which transitioned from being highly technical and opaque to the average citizen to become a synonym of inclusion in an increasingly digital society. This transition took place due to the advent of geotagged big data, characterized by the ubiquitous use of global navigation satellite systems (e.g., GPS), the surge in geo-marketing, and the massive adoption of personal location sensors (Sui et al. 2013b; Yan et al. 2020). Despite the undeniable advantages of the availability of GI, this explosion of data generation brought about problems such as unwanted surveillance and breaches of privacy (Bertone and Burghardt 2017), including commercial use and political misuse of volunteered information (e.g., the Cambridge Analytica scandal) (Sharma 2019), and unwarranted governmental or private surveillance (Ricker et al. 2015).

One response to the privacy and data-ownership concerns is to take control of the means of production, editing, and dissemination of information. Open and free data activism, along with collaborative stances at intellectual production (e.g., collective intelligence, peer-production, co-creative labor), constitute efforts in this direction (Yan et al. 2020). VGI falls within this scope, most notably because of its emphasis on blurring the boundaries between users and consumers of information that create, enlarge, review, and otherwise contribute to the information. Examples of this phenomenon encompass open GI platforms such as OpenStreetMap (Harvey 2013). VGI presents a hybridization of roles between those who record and collect GI, those who use it, and those represented by it. This relationship is not inherently fairer, but the distributed ownership and agency provide active roles to citizens that otherwise would be passive subjects in the different mapping efforts.

This chapter adopts a broad definition of VGI, which includes participatory, collaborative, and open-sourced GI methods. By doing this, we deliberately opt not to break VGI away from techniques such as Public Participatory Geographic Information Systems (PPGIS), as proposed by some authors (Verplanke et al. 2016). Instead, we explore the differences between techniques within a mapping methods spectrum, in which participation is in the center.

This chapter presents a study on the application of VGI to map informal settlements in Latin America. The questions that structure this research are: *(1) How do different VGI approaches support citizen participation and user empowerment? (2) What are the opportunities and limitations of VGI in mapping informal settlements in Latin America beyond current authoritative data acquisition procedures?* These questions stem from the realization that authoritative sources such as registries, census, or urban planning documents do not adequately portray informal, illegal, peripheral, or otherwise deprived settlements. A recent stream of community or volunteer-driven mapping experiences made possible by Web 2.0 interaction creates novel GI sources, closing some existing gaps in authoritative data sources. These applications also present issues of empowerment, privacy, and citizenship, on which this investigation focuses. Methods and tools employed within the VGI spectrum directly impact citizen participation and empowerment (Corbett and Keller 2005; Reynard 2018), which are two of its main premises and require clarification. To address these issues and clarify differences in terms of methods and expected outcomes of VGI, we propose a framework for assessing citizen participation in VGI and applying it to two case studies: peripheral urbanization in Mexico City, and participatory mapping in inner-city slums in São Paulo. The novelty in this research resides in our focus on user empowerment as the driver for a user participation ladder in VGI.

This chapter addresses citizen participation and empowerment questions from a comparative perspective in the VGI experiences and research spectrum. The following section provides an overview of the theoretical questions regarding user participation in VGI and the lack of data on informal settlements. The methods section presents a comparison framework for VGI applications based on citizen participation that collects environmental and socioeconomic data in these settlements at varying resolutions. In the results section, we present the analysis of two contrasting case studies using the citizen participation framework. The discussion section reflects on the breadth of the VGI spectrum, notably the empowerment of users, volunteers, and citizens through the VGI applications. It also discusses the potential of VGI to provide high-quality GI about informal settlements in developing countries. We conclude with remarks on the necessity of interdisciplinarity and participatory processes in research and policy development, most notably when socioeconomic inequality is a relevant factor.

1.1 Lack of Information about Informal Settlements

VGI presents advantages to GI's democratization, most notably through a more equitable data acquisition's distribution and by reducing the distance between producers and users of GI. As with other Web 2.0 technologies, it dramatically expands the role of information in everyday life for millions of people, which increases the pace at which data are produced and used (Sui et al. 2013b; Yan et al. 2020), as seen in location-based devices. However, despite the increased integration predominant in the developed world, differences persist across regions and demographics. Overall, men have more facilitated access than women, and developed economies present much better access than less-developed countries (LDCs, as defined by UN-DESA 2021). Men in developed countries would be the upper end of the technology accessibility spectrum, as close to 90% of them have access to the web. At the opposing end, only 14% of women in LDCs have access to it (International Telecommunications Union 2019). This stark contrast exemplifies the differences in place, gender, income, and other socioeconomic factors that determine the ability to access, produce, and disseminate GI (Corbett and Keller 2005; Sui et al. 2013b).

At the urban scale, the most vulnerable areas are frequently under-represented or absent from official sources (Souza 2012; Camboim et al. 2015; Kuffer et al. 2018; Mahabir et al. 2018). Deprived areas, such as slums, squatters, or informal settlements, often miss key geographic features in commonly available data sources (Hachmann et al. 2018). The missing elements may be settlement size, incomplete boundaries, total population, number, and location of buildings and enterprises (Patel and Baptist 2012; Hachmann et al. 2018). Initiatives such as Missing Maps (Scholz et al. 2018) and the Muungano wa Wanavijiji Non-Governmental Organization (NGO) (Lines and Makau 2018) seek to counter these problems and demonstrate the breadth of existing challenges. The lack of cartographic representation of socially vulnerable settlements furthers their symbolic and physical exclusion. The lack of cartographic information may present severe challenges for research and policy, may hinder development and access to fundamental civil rights (Patel et al. 2012), negatively influence self- and outside perception of communities (e.g., in political instances) (Corbett and Keller 2005), and lead to biases against communities (Watson 2009).

In this context, our research helps level off the playing field by increasing data transparency. It provides NGOs, public institutions, international organizations, and researchers with a straightforward way of visualizing irregular settlements' structure and spatial dynamics (e.g., urban expansion) over time. We assume that transparency promotes good governance and fair transactions. When information is not openly available, local elites' incentives for exploitation and opportunistic behavior increase (e.g., due to control of information such as land market dynamics, regulations, and political clout). Currently, available information about the conditions and dynamics of informal settlements is not sufficient or robust, which politicians and local officials routinely mismanage or exploit (Rodriguez Lopez et al. 2017a).

At the community level, the lack of data commonly means underrepresenting a population, its business, culture, and assets (Corbett and Keller 2005), increasing the difficulty to access credit, for example. Land tenure is a critical issue, as the lack of tenure rights often stems from outdated or incomplete registries. These issues may stoke conflicts (Hachmann et al. 2018), sapping long-term agency from communities and endanger small-scale businesses and services (Patel et al. 2012). When population or household data are missing, public infrastructure planning often underestimates the demand for services and investment. Public and private interventions also face increased uncertainty. Planning is less precise, procurement and contracting often occur based on broad assumptions, and projects need longer development cycles, as they compensate for inexistent essential information (Pedro et al. 2017; Pedro and Queiroz 2019). In these cases, the costs for implementing public goods or services increase and public officials often divert resources from the desired results to the initial phases of planning.

From the city management perspective, it is notorious that the lack of information severely hinders urban planning (Zhang 2019). Along with political and economic factors, lack of information and the limited cognition caused by it, fuel a *tabula rasa* approach to design. In this approach, urban master plans and spatial projects often circumvent, exclude, or seek to replace informal settlements entirely (Watson 2009). Strategic policies are frequently ineffective when essential information is missing (Patel et al. 2012), especially when considering the undocumented and dynamic nature of land-use in informal settlements that challenge conventional land-use tools like zoning and cadastral plans (Hachmann et al. 2018). The lack of information may lead to misconceptions, creating myths or partial truths that disrupt public policy effects or make them poorly adapted to the intended population groups (Patel et al. 2012).

The lack of demographic data and GI on informal settlements also has negative public health implications. For instance, coarse spatio-temporal resolutions of health and demographic data challenges the implementation of targeted interventions to prevent or mitigate outbreaks (WHO 2010; Elsey et al. 2016). In addition, the informal settlements' socioeconomic and spatial characteristics exacerbate the risks of communicable and non-communicable diseases (Ezeh et al. 2017; Corburn et al. 2020). Physical and social factors are key health determinants (Barton and Grant 2006). In this sense, combining the georeferenced settlement and health data becomes crucial to plan effective interventions (Friesen et al. 2020). Poor health data (e.g., coarse, lacking precision or outdated) significantly challenge planning and implementing interventions that are critical to tackling urban health inequities.

In this regard, monitoring systems that provide longitudinal data on slums (e.g., NUHDSS in Nairobi, Kenya) play a critical role in health decision-making at the intra-urban scale by providing health data with the appropriate spatio-temporal resolution. Monitoring systems like these can benefit from VGI by integrating local communities' contributions, which may provide critical insights to combat health emergencies like the COVID-19 pandemic, for example. Furthermore, much of the literature focuses on the spatial-time scan (e.g., the nature of the non-linear dynamics), early warning systems (Hohl et al. 2020), or resilience (Scheffer et al. 2001, 2012). At the same time, there is a lack of research addressing changes that affect the structure of social and environmental systems (i.e., irreversible regime shifts). The COVID-19 pandemic, for example, presents regime shifts in many critical areas (e.g., health, social interaction, policy, political debate, among others). VGI could complement existing information and work in tandem with other sources of information to represent system states and processes with increased spatial and temporal resolution. These improvements can play a significant role in the coming decades, notably when considering populations often missing from official sources (e.g., the squatters, slum dwellers, and others).

1.2 Citizen Participation in VGI

This work proposes a description of the broad spectrum of VGI techniques and methods from the perspective of citizen participation, focusing on user agency. This stance emphasizes the "volunteered" in VGI, which is essential in differentiating this group of techniques from other processes of geographic data acquisition. To this end, we must define user agency in the context of VGI. This chapter defines agency as the capacity to exercise control over one's thought process, motivation, and action. This definition encapsulates the cognitive processes of imagining what one wants to implement, being motivated to do so, and believing in one's capacity to implement it without suffering too steep adverse effects or costs in the process (Bandura 2001). In the context of VGI, agency translates into understanding GI to the point of identifying oneself as an agent (either a producer or editor of information) and believing in one's capacity to register or to analyze GI with the available means once the motivation to do so exists. The trade-offs involved in this definition of agency in VGI pitch technical capacity (Robinson et al. 2017), on the one side, and motivation to use or create GI on the other (Verplanke et al. 2016).

The recent evolution in GI effectively demonstrates how decreased technical barriers to data production (e.g., Web 2.0 technologies) sparked a flow of interactive production of information, breaking the virtual monopoly of specialists over GI (Zhang 2019) and creating VGI (Goodchild 2007). In this process, the advent of participatory mapping tools and methods increased users' perceived capacity to create new information by themselves. This capacity increase led to more ambitious goals from users, generating new solutions that further challenged previous restrictions in GI authorship.

VGI is still arguably torn between its contributors' active or passive character (Haklay 2013; Zhang 2019), despite user agency's importance in its evolution. Passive approaches analyze the digital spatial footprint from research subjects (e.g., geotags from social media) independently from their control (Yan et al. 2020). Intermediate approaches include crowdsourcing efforts (e.g., Missing Maps, Wikimapia, and OSM) that help eliminate gaps in mapping, but their goals are not necessarily participation per se, but the data generated by them (Sui et al. 2013a). Direct subject involvement is the mark of active approaches. Participatory mapping and PPGIS (Harvey 2013; Zhang 2019), such as Slum Dwellers International (SDI) and Mapping Kibera, often feature active approaches. These aspects beg the investigation on the levels of citizen participation in VGI, how they relate to empowerment, and the lasting benefits of VGI beyond the data itself.

VGI research seldom measures citizen empowerment, although it is often implicit in VGI campaigns and studies (Corbett et al. 2016). In this sense, it is helpful to make the relationship between citizen participation, empowerment, and VGI explicit. According to Sherry Arnstein (1969), citizen participation is a prerequisite for empowerment, as it assumes active citizen engagement in decision-making and community development processes. Following this line of thought, to empower citizens through VGI, there must be methods, tools, and goals accessible to citizens, even non-specialists. Moreover, as the accessibility of VGI methods increases, there are new opportunities to achieve more plural and representative GI. From a technological perspective, though, accessibility can lead to an oversimplification of the available tools, therefore, constraining the use of the resulting GI. To avoid this contradiction, VGI should adapt its tools and processes to maximize citizen participation in the production, interpretation, and use (or reuse) of GI without compromising the quality of the data. This improvement may enhance citizen participation, and change policy and intervention priorities thanks to more diverse information and better-informed citizens.

We propose to describe the spectrum of VGI between two extremes in user participation: on the one hand, there is technical capacity and access to resources; on the other, there are the users' perceived capacity to exercise control over their GI and the motivations behind its production (agency). The following section presents a framework for comparing and evaluating VGI applications based on their ability to be replicated by ordinary citizens – and thus, to effectively foster citizen participation in the production of GI. Arnstein's "Ladder of citizen participation"(1969) is the inspiration for the framework, as it is a rung-based structure that hierarchically sorts VGI applications in a synthetic index. The latter builds on four assessment categories: (1) user agency in VGI; (2) required material resources to implement mapping; (3) necessary GIS literacy level to achieve results; (4) degree of involvement of research subjects. These four categories encompass criteria shared among most VGI applications and allow comparisons between applications in different contexts.

2 Methods

This section presents the comparison framework for citizen participation in VGI. This framework assesses citizen participation and empowerment in VGI initiatives, providing a novel, multi-dimensional and hierarchically structured comparison tool. Ultimately, the framework aims to improve VGI research and practice by making explicit the resources (e.g., material, informational, and capacity), the agents (i.e., the users, producers, and subjects of GI), and their involvement (e.g., agency and stages of direct participation) in the VGI processes. This framework introduces an innovation by bringing critical factors in GI production to light that are usually subsumed in traditional analysis, revealing the purpose, tools, participation, and empowerment in VGI practices.

Table 12.1 presents the framework and includes 16 criteria. The criteria belong to four categories that describe the tension between technical resources and GIS literacy, on the one hand, and user GI agency and the degree of involvement of research subjects (i.e., people living in the observed area), on the other. Each criterion may receive a value of zero or one, identifying the absence (zero) or presence (one) of that criterion in the case under study. Therefore, each category may receive a value from zero to four points, which adds up to a total VGI Participation Score (VPS). A high VPS (beyond 9 points, for example) would indicate a significant level

Table 12.1 Categories and criteria for the VGI Citizen Participation Score

Categories	Criteria	Categories	Criteria
User GI agency	1. Transparency 2. Editing capability 3. Two-way data flow 4. Control over data format and publication	GIS literacy	5. Specialization 6. Experience 7. Geomatics 8. GIS software
Resources	9. Software license 10. Data license 11. Mobile hardware 12. Human resources	Involvement of research subjects	13. Data collection 14. Data management 15. Data interpretation 16. Usage and impact of data

The criteria add up to a VGI participation score (VPS), which ranges between 16 (total citizen participation) and 0 (no citizen participation)

Table 12.2 Evaluation and interpretation of the VGI participation score

VGI participation score	Interpretation	Examples
13–16	Citizen empowerment	Participation and replication are possible even by the general population. Users have control over data reuse. Little to no resources are prerequisites.
9–12	Significant participation	Overall data controlled by researchers, there may be supervision or mediation by specialists. Non-specialized resources.
5–8	Limited participation	Participation is constrained to predetermined options of agency, technology, and goals. Some specialized resources are necessary.
0–4	Non-participation	Lack of GI knowledge hinders citizen participation, technology, and resources. Users have no control of the results (e.g., veiled GI collection).

of citizen participation in the VGI process (Table 12.2). Researchers analyzing the VGI practices in a particular case may assign a point for each criterion as a qualitative appraisal (e.g., expert opinion).

The qualitative assessment advances on a structured approach to evaluate the processes and practices involved in VGI. By focusing on the process rather than on the resulting data, this framework seeks to distance VGI from a technocratic discourse. Instead, the framework emphasizes the social relevance of GI in the specific context in which it is generated – i.e., to what extent the process and resulting GI contribute or harm people directly related to that context. The analytical categories in the framework highlight the conditions of the data subjects to participate in VGI processes, the degree to which the processes are proposed or designed to work jointly with the subjects (e.g., high, or low dependency on sophisticated techniques, and knowledge transfer potential). These characteristics allow researchers to understand VGI practices and data in connection to the social context that they describe.

Ultimately, the framework seeks to support VGI practitioners and researchers to address more explicitly the purposes and motivations behind data acquisition, utilization, and the degree to which they are accessible and under the control of the subjects described in the data.

VPS scores build a ladder of participation and empowerment in VGI, mirroring the example from Arnstein (1969) in creating a hierarchical evaluation of practices that involve communities and tecno-scientific content. In Table 12.2, scores between 0 and 4 fit into the class of non-participation, in which citizen participation is constrained. User agency is limited or nonexistent (e.g., veiled GI collection, absence of derivative uses), required literacy limits the effective use of the application to experts, while the necessary resources curb dissemination or replication by the public (e.g., expensive proprietary software or required coding or geodesy knowledge). Scores between 5 and 8 depict limited participation; they signal that some participation exists but is usually constrained to predetermined options and goals defined independently from data subjects and users. In this class, influence on the agency and goals of VGI still weigh away from citizens. Scores between 9 and 12 mean significant participation. In this class, users generally have high agency levels, controlling data usage and transparency. Applications may still need resources but do not require specialization (e.g., volunteer engagement, free GIS software, mobile phones as GPS data sources). In this class, research subjects have overall control of the data but are not yet at the helm of the mapping process (e.g., external parties may set the purpose of data or custody). Scores between 13 and 16 mean citizen empowerment, which supports open participation and replication of methods by any citizen interested in VGI. The top tier means nearly full user agency (e.g., users know, control, and reuse the data as they wish). There are few prerequisites in GIS knowledge, few necessary resources (e.g., user-friendly applications with very low technical literacy, little to no ground-truthing), and direct involvement of research subjects in knowledge production through VGI.

In the framework, four criteria describe user agency in GI. The first is the capacity for users to know they generate geographic data that are being collected and reused by others. High-ranking applications will provide transparency and fine-tuned control over geographic data and meta-data collection, while low ranking applications will be opaque or even misleading in presenting their data collection methods. The second criteria are the capacity for users to visualize, share and edit data in the application. Low ranking applications will limit user edits, while high-ranking applications will provide practical tools that are easy to master. Next, data flow should be accessible in both directions, meaning users may input and access information in the application, allowing derivative works. Finally, applications should provide complete data in editable formats, avoiding proprietary or simplified formats that limit derivative works to lower quality than the original input (e.g., image formats, data without geolocation).

GIS literacy stems from specialized knowledge, practical experience with GI, proficiency in geomatics, and proficiency in GIS software packages required to obtain and analyze the data. Indeed, these technical aspects may constitute substantial barriers to applying VGI methods, which often require facilitators between the technology and the public (Robinson et al. 2017). In this sense, the user of high-ranking applications could be a layperson, while the tools would only require a cursory understanding of GI (map reading) and GIS software (visualizing or adding information to non-specialized Earth observation platforms). Conversely, in low-ranking applications, the user would need specialized knowledge, and tools would require good cartographic skills and knowledge of geodesy, GIS software, and, if applicable, spatial statistics.

Resources in VGI applications refer to access to software licenses, complete and timely support data (e.g., Earth observation imagery), mobile hardware (e.g., portable GPS devices), and the level of dependence on human resources (either specialized or not) to achieve the necessary results. The costs for licensed software and hardware and volunteers' availability can be highly restrictive to implementing VGI methods (Reynard 2018). On this basis, high-ranking applications would only rely on free software and openly accessible data without the need for on-site data validation or intensive use of human resources. Conversely, low-ranking applications rely on licensed software and data, on-site data collection, and specialized hardware.

Finally, the research subjects' level of involvement in collecting, interpreting, and using geodata is determinant to distinguish different data collection methods (Verplanke et al. 2016). Indeed, the control of research subjects over data is particularly relevant when GI supports social integration through citizen empowerment (Corbett and Keller 2005). In this sense, high-ranking applications would present active research subjects' involvement in geodata collection, management, and interpretation, notably towards the subjects' goals and motivations. Conversely, low ranking applications could exclude the research subjects or implement data interpretation and use without the subjects' knowledge or control.

3 Results

Using the framework of citizen participation in VGI, we compared two VGI projects that mapped informal settlements in Latin America (Table 12.3). In the first case, researchers from the University of Hamburg (Germany) combine human and remote sensing data in a hot spot analysis framework to map informal settlements in Mexico City's fringes. In the second case, the NGO Teto uses a participatory GIS approach to map communities in São Paulo. Both projects aim to fill the gap of authoritative geographic data on informal settlements, resulting in similar outputs, albeit through different methods and with differing purposes. In the São Paulo project, volunteers produced VGI with the communities' consent using a participatory approach. This effort aimed to foster local changes to improve the living conditions in selected informal settlements. The Mexico City project brings two data sources together: VGI and remote sensing data to develop hot spot maps that explicitly aim to conflict between nature preservation and urgent housing needs.

Table 12.3 Comparison of two VGI projects in informal settlements in Latin America. Teto uses a participatory GIS approach to map communities in São Paulo (left). Researchers from the University of Hamburg (Germany, right) combine human and remote sensing to map informal settlements in Mexico City

VGI aspect	São Paulo case	Mexico City case
Volunteer workforce (data agents)	Volunteers of the NGO Teto: Mainly college students and professionals (usually studying or working in architecture, engineering, geography, and urban planning).	Locals file a complaint at the *Procuraduria Ambiental y del Ordenamiento territorial del Distrito Federal* (PAOT). VGI comes from PAOT (September 3, 2015), and the researchers further analyzed it.
Other sources of data (not volunteered)	Raster data: Georeferenced orthomosaic generated from drone images (online GeoTIFF imported in QGIS)/georeferenced VHR satellite imagery openly available online. When available, polygon information on topography, hazards, and other themes.	Raster data: Researchers obtained RapidEye satellite imagery (from the German Aerospace Center) through the German Federal Ministry of Economy and Energy funding. Vector and demographic data from Mexico's National Census 2010 (INEGI 2010), vector data of road systems from OpenStreetMap (OSM 2015)
Data proprietors	Satellite imagery distributed by Google. Drone imagery and resulting maps jointly owned by Teto and the communities where the surveys take place	RapidEye data from the German Aerospace Center. Ecological complaints from PAOT and 2010 census data from Mexico's National Census Bureau (INEGI 2010), both distributed under open access.
Input data	Raster data: Georeferenced orthomosaic generated from drone images (GeoTIFF imported in QGIS)/georeferenced VHR satellite imagery (dynamic XML or URL layer imported in QGIS).	Raster data: RapidEye satellite images, vector data: ecological complaints (PAOT), demographic data (INEGI 2010), road systems (OpenStreetMap 2015)
Site visits required	*In situ* work is required	*In situ* work is not required
Data resolution/ accuracy	Very high resolution (<1 m)	High-resolution satellite imagery (5 m, input)
Tools	QGIS (free, open software)	ArcGIS (commercial software)
Targeted audience/ application	Project designers and advocates in Teto and community members	Government, NGOs, and researchers
Purpose of VGI	To collect settlement data for Teto's development/advocacy projects	To bring to light a conflict between nature preservation and housing needs
Output data	More accurate geographic information: Filling gaps in existing (authoritative) sources, increased resolution, updated information	More accurate geographic information: Filling gaps in existing (authoritative) sources, increased resolution, updated information

3.1 Research in Mexico: Human and Remote Sensing Perspective

Mexico City's rural-urban area lies in the Federal District's southern part. The city depends on water sources outside the urban area (e.g., the Magdalena River south of Mexico City). Land management is especially sensitive in the so-called "preservation zone", where informal expansion may contaminate the water supply (Jujnovsky et al. 2012). The mapping process aimed at creating more transparency in the conflict between nature conservation and housing demand in an unequal society. When conflicts are visible, society can dialogue to develop solutions. Societal dialogue is a critical response to unequal development dynamics (Harvey 2006), especially those that present conflicts between vulnerable groups and common social goods.

In this case, the data agents are local citizens and researchers. Any resident from Mexico City may file environmental complaints voluntarily in person with the "Procuraduría Ambiental y del Ordenamiento Territorial del Distrito Federal" (PAOT), through the phone, or electronically (e.g., via email or on PAOT's website). Complaints include animal abuse, water misuse, noise, or irregular settlement in the preservation areas, among others. Each complaint generates a record in a database that includes descriptive fields and the geographic coordinates and address of the problem. The researchers included complaints filed between 2002 and 2013 in their analysis, representing the "human sensing" data (i.e., people generating geographic information) (Rodriguez Lopez et al. 2017a).

The research team then combined the "human sensed" data with high-resolution satellite imagery (5 m) from the RapidEye satellites. The German Federal Ministry of Economy and Energy covered the costs of Rapid Eye imagery for that research project (Rodriguez Lopez et al. 2017a). The researchers classified the satellite data into two land cover classes, urban and non-urban, for years 2009 and 2014. Further change-detection analysis described urban expansion (i.e., the difference in urban area between 2009 and 2014) in and around nature preservation areas. The study then quantified urban growth in the protected areas, detecting and highlighting the hot spots of this type of dynamics (i.e., areas in which there was an intense concentration of urbanization). The Getis-Ord Gi* statistic detected the hot spots and provided increased precision for research and policy about the ongoing environmental and social conflict. Finally, the analysis integrated the OpenStreetMap road system (as another VGI source) with census data to assess socioeconomic conditions and the drivers of peri-urbanization (Rodriguez Lopez et al. 2017a; Heider et al. 2018). The analysis output (Fig. 12.1) included the combined hot spots derived from human and remote sensing in a grid of polygons with a spatial resolution of 561 m. The authors published the data, results, and further methodological details under open access (Rodriguez Lopez et al. 2017b).

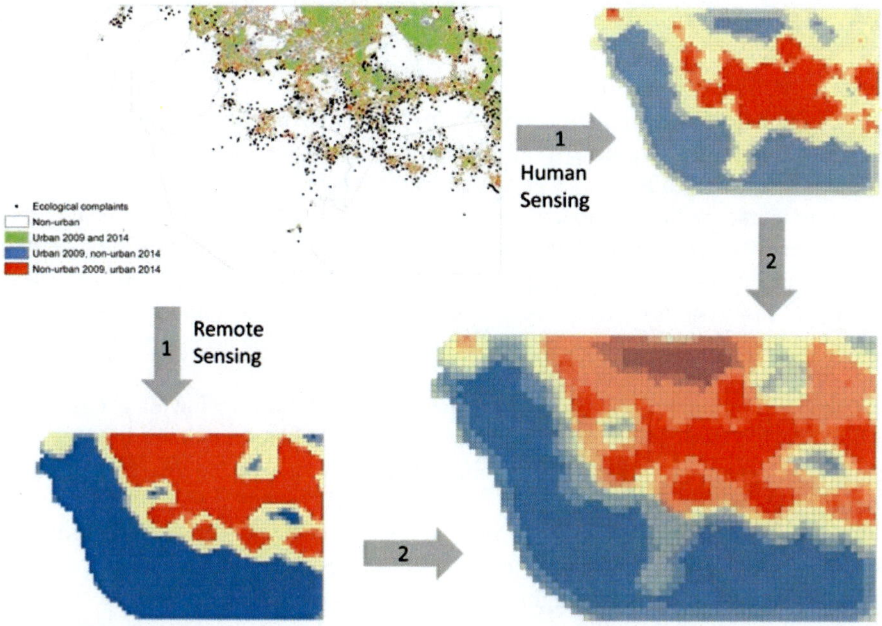

Fig. 12.1 A graphic summary of the hot spot analysis of VGI and remote sensing data in Mexico City and their combination. (**a**) area of interest; (**b**) the human sensing hot spots; (**c**) the remote sensing hot spots; and (**d**) the combined hot spots (human and remotely sensed). Red cells represent hot spots (high concentration of complaints or identified urbanization); blue cells represent cold spots (low concentration of complaints or identified urbanization). The confidence range of the hot spots analysis was 90–99%, source: Rodriguez Lopez et al. (2017a).

The goal of this research was twofold: first, to increase transparency by providing new data to the academic and policy development sectors on this dynamics. This transparency opens the debate beyond local power brokers (e.g., local legislators and public officials involved in land grabbing) and provides evidence to local advocacy groups such as housing rights or environmental NGOs. Second, the research aimed at shining a light on tradeoffs between housing rights and environmental protection policies in an unequal development setting. Within this context, the most vulnerable will suffer from the enforcement of regulations (e.g., expulsion from informal settlements in preservation areas). At the same time, the root causes remain untouched (e.g., lack of land-market regulation or inefficient housing policies), reproducing prejudices in regulatory instances and keeping encroachment-exclusion cycles in place (Zérah 2007). This research followed an ongoing investigation that included Mexican academics who constantly dialogued with local authorities (Rodriguez Lopez et al. 2015). With its results, this dialogue can better address the preservation-housing conflict and provide a more leveled playing field, exposing bias in information (e.g., complaints are more frequent in affluent areas) that stem from the inequality of the social process itself.

3.2 Research in São Paulo: Participatory GIS

Since 2014, Teto has conducted community assessment activities in different informal settlements located in peripheral areas of São Paulo. These activities take place with the community's informed consent and include surveys that combine mapping campaigns and the collection of georeferenced household data to analyze the communities' demographic, socioeconomic, and spatial characteristics. The NGO and the communities use these data to support slum-upgrading projects such as constructing emergency shelters or improving shared open spaces. This process includes a round of discussions on the possible collaborations with Teto that result in a joint agreement regarding the scope of Teto's participation and the necessity of conducting spatial and socioeconomic surveys.

Teto's community assessments rely on high-resolution spatial data collected through VGI. The mapping process results in detailed community maps locating the settlement extent and identifying significant features that include each building's footprint and use (e.g., residential or community facility) and primary road access. Data also indicates household locations, basic socioeconomic data (e.g., household members and housing conditions), and household-specific demands for infrastructure (e.g., need for more streetlights or better road access). Teto, its volunteers, and the community collaborate in these activities, and the resulting data supports projects and advocacy initiatives co-developed by the community and Teto.

Teto's volunteers are university students or recent graduates (often from architecture and urbanism) engaged in enhancing living conditions in informal settlements, and are generally much less vulnerable than the informal settlements' population. The volunteers provide technical expertise for the mapping effort, given their university training, even if they are not GI experts. Currently, this is a workaround for the lack of technical literacy in the communities, which are often some of the most vulnerable in São Paulo. The downside of this workaround is that community dwellers rarely participate in the vector data (i.e., point, line, and polygon data, commonly collected with GPS or similar devices) collection process, although their knowledge registers as GI through the interaction with the volunteers and surveys. Community dwellers also join the data validation and interpretation processes, as explained further below.

Vector mapping campaigns have a fixed timeframe, usually lasting eight weekends. Groups of four to six volunteers divide the work in each campaign, with each volunteer covering between 1.5 and 2 ha, depending on the settlement's complexity; hence, the number of available volunteers limits the process. Based on the aerial imagery, the volunteers first manually digitize each building's perimeter in QGIS (a free and open-source GIS software), considering each roof to correspond to one building (step 1 in Fig. 12.2). Teto prefers freely accessible satellite imagery, as its combination with accessible software enhances the replicability of the method. In some cases, private partners (e.g., DroneDeploy and Ponto360) provided higher-resolution drone aerial imagery. Then, the volunteers check the accuracy of the digitized built environment on-site (step 2 in Fig. 12.2). Satellite imagery may be

Fig. 12.2 Teto's mapping process, step-by-step: (1) acquiring drone or satellite imagery; (2) digitizing building footprints, with on-site verifications using Google My Maps; (3) validating GI data to obtain geographic information that may support upgrading projects. Elaborated by the authors, based on Google Earth (2018) and Pessoa Colombo (2019)

outdated or lack resolution; therefore, on-site verification is essential in informal settlements. The Google My Maps platform (which is free but not open) allows the visualization of the digitized building footprints on mobile phones, facilitating on-site verifications. Finally, the surveys collect land use, infrastructure, and demand information (step 3 in Fig. 12.2). Volunteers then georeference the tabular information from the survey into the centroids of the building outlines. This workflow requires each volunteer to use a smartphone and at least one computer per mapping campaign to digitize the final map. In addition, a reasonably good Internet connection is required.

At the end of each mapping campaign, Teto and community leaders organise focus groups that validate and interpret the collected data through a horizontal dialogue with the community (Santos Melo et al. 2021). Community leaders use printed maps to situate geographically specific demands. This way, they turn geographic data into information, which they use to plan future interventions. They use large, printed maps (e.g., in ISO A0 format) in these discussions, which allows more spontaneous annotations (Fig. 12.3). Such discussions based on printed materials are crucial to overcoming technological and material barriers to participation in GI. In this way, local knowledge enhances geographic data. This process

Fig. 12.3 Focus group co-organized by Teto and local community leaders to enhance geographic information. Source: Teto

also allows the co-development of community projects between the NGO and city inhabitants.

The cartographic outputs consist of two features generated in QGIS: a polygon feature containing the buildings' footprints and a point feature indicating the households' locations and non-residential structures. Teto usually manages those datasets, but community leaders can also manage them independently and locally when capacity (e.g., hardware, software, literacy) is available. However, most of the data restitution to communities is through printouts, in the form of reports illustrated by graphs and maps. Therefore, the outputs are high-resolution geo-datasets combining descriptive data of socioeconomic and environmental aspects of the community. Teto uses the outputs to support slum-upgrading projects, as they allow identifying the most vulnerable areas that require more urgent interventions. The type of interventions varies, but the most common are new single-family housing units (replacing shacks with new structures), improved accessibility (e.g., stairs, bridges) and community facilities. In collaboration with the community, Teto then plans and designs all interventions, including the election of beneficiaries in the case of housing projects.

4 Comparison Under the VGI Participation Framework

This section presents the VGI participation scores for the two cases under analysis. We evaluate whether these VGI initiatives attain citizen empowerment or significant participation (at the higher tiers of the VPS score) or are limited in participation or non-participatory at all (at the lower levels of the scale). For this analysis, we start by considering user GI agency and then observe the required GIS literacy, the required resources and finally, the degree of involvement of the research subjects in producing and managing the GI.

Overall, the results show the contrast between the two cases. When comparing VGI participation scores, the Mexico City case attained five out of 16 possible points, representing limited participation. Teto's mapping process in São Paulo shows significant participation with a total VPS of 10 points. Below, we present the assessment of each case under the VPS comparison framework (Table 12.4).

The case study in Mexico comprises data acquisition from locals, georeferencing by the planning authority, and hot spot analysis by an independent research team. This case scores two out of four points in the GIS Agency category. Residents in Mexico City (who may not live in the informal settlements) produce the data through complaints filed over multiple media (e.g., phone, email) and inform a geographic

Table 12.4 VGI participation score calculated within the framework of citizen participation in VGI

Criteria	Case 1: São Paulo	Case 2: Mexico
GI agency	**4**	**2**
Transparency of data usage	1	1
Possibility of data editing	1	0
Two-way data flow or exchange	1	0
Format of data communication or publication	1	1
Tech literacy	**1**	**0**
No formal specialization in GI science	1	0
No practical experience with GI	0	0
No proficiency in geomatics	0	0
No proficiency in GIS software	0	0
Required resources	**2**	**2**
No licensed GIS software	1	0
No licensed data	1	0
No mobile/external hardware (GPS or drone)	0	1
No intense human resources	0	1
Involvement of research subjects	**3**	**1**
Data collection is done by or with research subjects (RS)	0	0
Data management is done by or with RS	1	0
Data interpretation by RS	1	0
Data aims to foster local changes (physical or social)	1	1
TOTAL SCORE	**10**	**5**

location. From that point on, PAOT manages the case with no further user input. In this sense, the data producers have spatial knowledge about the fact but have no data editing or exchange possibilities. However, the researchers published the project's data (including the complaints and results) under open access (Creative Commons License, by attribution – CC BY) to enable dissemination in academia (Rodriguez Lopez et al. 2017b). Open access to the PAOT's database breaks the barriers around the information on these conflicts. The Mexico case scores no points in the tech literacy category because replicating the process, especially the hot spot analysis, requires specialized knowledge in GIS and geomatics. The case meets two criteria in the "required resources" category because the research team used licensed software and remote sensing data for hotspot mapping, limiting participation. However, neither GPS, drone imagery, nor fieldwork was required, widening participation possibilities.

The involvement of research subjects is particularly complex in the Mexico case. In our appraisal, the Mexico case scores only one point because only part of the research subjects is involved in data collection and under conflictive circumstances with other residents. The locals do not manage or interpret data, the other criteria in the framework. PAOT collected the complaint data, and it was not available to citizens as aggregate information, which in turn creates a conflict in information as PAOT might use the data to foster physical or social changes independently from the goals of all inhabitants in the area. Furthermore, the locals who file the complaints may do it motivated by protecting the preservation area (which is a common good), but against the housing need of those in the informal settlements in the region. The opening of the data potentially allows more groups to see this conflict, even if limited to an academic audience. The open data, combined with the analysis of the conflict, can foster debated social action. They are, nonetheless, independent from the research subjects and do not contribute to this score.

In summary, the hot spot mapping project in Mexico City reached limited participation with a score of 5 out of 16 due to the high level of tech literacy required for the analyses, costly resources (software and data), and the lack of research subjects' involvement. However, transparency of data usage and availability in open access publications enable a medium ranking in GI agency.

The case study in São Paulo scores four out of four points in the GIS Agency category. Both data producers and users are fully aware and have control of GI's collection, management, and publishing. The research subjects also enforce their interests and data privacy concerns, controlling the shared GI content. In terms of tech literacy, the case study scores only one point. Neither data users nor data producers need any formal specialization, but previous experience with GI dramatically facilitates the work. In this sense, Teto's volunteers act as VGI facilitators, building the bridge between the community and the use of GI tools and methods. Regarding material resources, the São Paulo case meets two criteria: the data and software are freely accessible, but the method demands on-site verification, which requires mobile hardware for geolocation and also generates transportation costs. Besides these requirements, human resources affect the geographic extent of the output.

Regarding the involvement of research subjects, São Paulo's case merits three points. While Teto's volunteers collected part of the data without the active contribution from research subjects (community dwellers), the latter oversaw the process and maintained control over data retrieval, reproduction, or deletion at any time. At the end of each mapping campaign, the community validates the data and employs it to support its projects. The digital data are stored and managed by Teto, but data are also shared with community leaders or organizations when capacity is available (e.g., personal computers).

The São Paulo project achieves significant participation thanks to community involvement. The participatory approach is visible in the maximum rating of four in GI agency and three out of four in the involvement of research subjects. However, tech literacy for professional remote sensing and the required fieldwork resources were still high within the mapping process, leading to one and two points in these categories, respectively. The total VPS is 10 out of a possible 16, highlighting gains in agency and research subjects' involvement. The participatory approach shows compromises with the technical and resource requirements for working in GI, especially when little to no preliminary data are available.

5 Discussion

This chapter asked how different VGI approaches support citizen participation and user empowerment and what are the opportunities and limitations of VGI in mapping informal settlements in Latin America beyond authoritative data sources. To address the first question, we argue that, despite its qualities, VGI also presents potential issues to informal communities, notably regarding privacy (Elwood 2010; Sharma 2019), ownership over information (Hachmann et al. 2018; Zhang 2019), and changes in political power (Corbett and Keller 2005). Due to privacy and political power concerns, this framework makes explicit the resources and agents in the VGI processes. It decouples the relationship between data producer and data subject, revealing inherent potential conflict and cooperation. Therefore, it provides critical insights in VGI beyond data quality by potentially illuminating conflicts, considering processes (rather than products) and their societal implications. It brings light to critical factors in GI production that are usually subsumed in traditional analyses, highlighting the purpose and tools and the participation and empowerment of the agents involved in VGI.

From this perspective, the framework contrasts the case studies to reveal the importance of control over information by those represented in it. In Mexico City, observation from distant and anonymous complaints separates the data producers from their subjects. In addition, although the research team published their work in an open-access journal, their findings are hardly accessible for informal settlement dwellers and more likely to remain inside academia. In São Paulo, users have veto power over information dissemination. The decoupling of this framework provides opportunities to expand previous research, in which crowdsourced methods (also called passive or contributed) and participatory (named active or volunteered) approaches are often at odds (Harvey 2013; Zhang 2019). At the same time, the technological compromises in the São Paulo case (e.g., the necessary facilitation from the NGO staff) widen the discussion on the empowerment potential from VGI. By keeping the mapping outputs aligned with the communities' interests, this approach preserves a critical aspect of agency, where external resources collaborate to produce VGI, even if community members seldom collect vector data themselves (Hachmann et al. 2018).

When considering empowerment (Corbett and Keller 2005; Cochrane and Corbett 2018), the low VPS scores of the Mexico case in the agency and the research subjects' involvement indicate a potential decrease in the community's socio-political power. The detection if this characteristic demonstrates the capacity of the framework to assess these dimensions. The decrease in power stems from a conflict of interests in which the interest of data producers (i.e., locals who complain about informal settlements) is opposed to the interest of research subjects (i.e., locals who live in informal settlements). The framework exposes this contradiction, as it makes the agents and subjects of VGI explicit. This disclosure is a noticeable advancement from previous research, which often omits the data subjects. In this sense, VGI practices that inform and provide control to the data subjects over GI about them provide enhanced empowerment. These features are present in the São Paulo case, where collaborative and participatory VGI initiatives provide local inhabitants with control over GI about their settlements. This increased control creates new political representation capacity (e.g., advocacy for land tenure rights) and supports more precise settlement improvement plans (e.g., housing, infrastructure) than other information generation methods.

The second research question examined the potential and limitations of VGI to provide information for research and policy development in informal settlements. Our results show that VGI can offer unedited GI on informal settlements at varying spatio-temporal resolutions, in line with previous research (Beukes and Mitlin 2014; Bolay et al. 2016; Hachmann et al. 2018; Lines and Makau 2018). VGI provided the location and quantity of land cover changes over a large region in Mexico. Considering the undocumented and dynamic nature of land-use in peri-urban informal settlements, volunteered sources of GI such as the PAOT are valuable complements to conventional ones. For instance, PAOT provided timely information on environmental changes in peri-urban settlements that would otherwise remain invisible to authorities. In São Paulo's case, VGI covered a much smaller extent but at a more detailed spatial resolution. This in-depth mapping allowed tracing building footprints, a piece of information that is often nonexistent for informal settlements but vital for slum upgrading projects (Hachmann et al. 2018).

Even in relatively affluent cities like Mexico and São Paulo, data on the built environment and dwellings in informal settlements are approximate at most and, at times, inconsistent. This lack of precision and completeness leads to sub-informed decision-making (Pedro et al. 2017; Pedro and Queiroz 2019), which is especially harmful to spatial interventions (Hachmann et al. 2018), risk management (Goodchild and Glennon 2010), and health policy (Elsey et al. 2016; Corburn et al. 2020). VGI can arguably foster synergistic opportunities and prevent unnecessary problems during interventions in these areas by providing locally sourced, updated, and fine-scale data. Despite the lack of focus of the framework on data quality assessment, it still provides a relevant contribution to the methods available for mapping, analyzing, and understanding informal settlements (Kuffer et al. 2016; Kuffer et al. 2018), especially from the community perspective or at the local scale (Hachmann et al. 2018; Williams et al. 2019).

Although VGI can provide data on informal settlements with high spatial and temporal resolutions, it presents limitations. From a scientific perspective, limitations in the replicability of methods and reproducibility of results challenge VGI-related research in general. As VGI initiatives employ participatory practices, the solutions tend to be context-specific, as in São Paulo. Crowdsourced methods, with lessened empowerment, provide massive, at-a-distance data collection but are easily biased and may contradict the interests of those represented in the data, as we show in Mexico City. The lack of access to volunteered data (sometimes inevitable due to ethical considerations) often hampers reproducibility. The replicability of methods is susceptible to the evolution of VGI sources and data formats (Ostermann and Granell 2017). In Mexico and São Paulo's cases, both VGI datasets contain personal data of some kind and demand editing before sharing, limiting the reproducibility of results.

Regarding replicability, both cases relied on tools and methods discussed in previous publications (Rodriguez Lopez et al. 2017a; Pessoa Colombo 2019) and are highly replicable. Nevertheless, their replicability relies on moderate-to-high levels of tech-literacy and material resources, limiting their assimilation by lay audiences from a practical perspective, as the framework exposes. This problem reflects very different approaches regarding the public's active involvement in knowledge production within the VGI spectrum (Hachmann et al. 2018; Zhang 2019), which the framework brings to light and helps discuss. This problem is central in contexts where GI is supposed to promote the empowerment of marginalized communities. This centrality is true for informal settlements, but is a general problem of a society's relationship with technology in unequal development conditions. Therefore, the active participation of citizens in the production of VGI and the transfer of knowledge and GI tools remain critical aspects in VGI research (Corbett and Keller 2005).

6 Conclusions

This chapter provided a comparison framework highlighting the "volunteered" side of VGI. This framework reveals user agency and citizen participation as critical aspects in GI acquisition, management, and dissemination. Even though much of the literature assumes an intrinsic association between VGI, participation, and empowerment, we observe far more complexity in this relationship than previously thought. The framework makes a clear distinction between passive and active participation in VGI. Specific forms of VGI may not include participation from those mapped (i.e., the research subjects) and may even be at odds with their interests, as shown in Mexico.

The framework also shows the implications of the differences in participation intensity and data contributors' composition. Differences among the authors and subjects of data may feed specific biases into the resulting GI. These biases are present in VGI and authoritative sources, albeit for divergent reasons, but often result in the under- or derogatory representation of vulnerable populations. This framework provides tools to assess the GI acquisition processes, considering these biases and the restrictions vulnerable populations face to access methods and tools to produce information. To do so, this framework differentiates VGI practices along with their agency levels, considering the data producers, on the one hand, and data subjects, on the other. This differentiation aims at increasing the precision with which research and policy understand and use VGI as a resource to achieve "people-truthing." The framework provides increased precision to this aim, indicating that VGI practices ranking high in VPS may work as *grass-roots* data validation. A critical reflection is that VGI projects geared at vulnerable populations need facilitators to overcome the existing technological barriers to participation (e.g., expertise and resources). More research could foster collaboration in the data collection stage of VGI, which currently depends on relatively sophisticated geospatial technologies.

We must also recognize the many limitations of this framework despite its potential relevance. First, this framework does not integrate traditional data quality assessment practices (e.g., completeness and accuracy), limiting its comparison to a qualitative measure. Second, other limitations arise from analyzing only two cases, which are far from exemplifying the whole spectrum of VGI. Even if these cases provide evidence for the framework's initial design, more examples will refine the methodology and possibly lead to adjustments in the score (e.g., weights for each criterion). Third, the cases do not stem from a comparative research design. A more systematic and structured set of cases could provide increased precision and critical insights. Given these shortcomings, further research should include more systematic comparisons that range across a more comprehensive set of case studies. Research would benefit from regional diversity, including variations in socio-political systems, data landscapes, and participatory traditions.

This chapter highlighted some of the significant limitations to research and policy and revealed an overall lack of timely, complete, and precise GI on informal settlements. We propose that VGI will play a central role in filling these gaps, given the importance of informal settlements for future development, the multiplicity of actors involved, and the necessity for self-reliance and determination in these communities. Therefore, further research should encompass an information environment that integrates authoritative, open, and volunteered sources of information to the top of their potential. This approach means moving VGI beyond the physical description of the environment into other dimensions of geographic information where local participation is critical, notably on land-use conflicts (as seen in Mexico City), slum-upgrade projects (as shown in the São Paulo case), and even public health.

Informal settlements face extreme social vulnerability and exposure to risks that their own socioeconomic and spatial characteristics increase. Because VGI allows obtaining updated, longitudinal information on populations, it can provide timely and precise data to support spatio-temporal analyses on health emergencies. In health research, VGI can also foster community empowerment by shifting priorities towards marginalized populations' unmet needs. In this direction, our future research efforts will focus on how the COVID-19 pandemic exposed spatial and social vulnerabilities that are yet unaddressed by VGI research. We aim to address these problems with open, authoritative, and volunteered information sources that together provide timely and fine-scale data on vulnerability, impact, and social behavior in the pandemic context. We expect future research will provide GI science with an integrated approach to identifying spatial and temporal tipping points. This contribution will help decrease uncertainty in decision-making during present and future public health emergencies when considering the specific social and spatio-temporal features of cities in the Global South.

Acknowledgements Techo acquired the data about São Paulo and kindly provided them to the authors. This paper and its affirmations do not represent Techo policy or opinions and are of sole responsibility of its authors. This research was supported by the Deutscher Akademischer Austauschdienst (DAAD) through a doctoral research grant. This publication contributes to the Cluster of Excellence Climate, Climatic Change and Society (CLICCS), funded by the Deutsche Forschungsgemeinschaft (DFG), and to the Project "Geographic Info on the COVID-19 pandemic" (COVIDGI), supported by the Volkswagen Foundation. Underlying RapidEye data has been contributed on behalf of the German Aerospace Center through funding by the German Federal Ministry of Economy and Energy.

References

Arnstein SR (1969) A ladder of citizen participation. J Am Inst Plann 35(4):216–224. https://doi.org/10.1080/01944366908977225

Bandura A (2001) Social cognitive theory: an agentic perspective. Annu Rev Psychol 52(1):1–26. https://doi.org/10.1146/annurev.psych.52.1.1

Barton H, Grant M (2006) A health map for the local human habitat. J R Soc Promot Heal 126(6):252–253. https://doi.org/10.1177/1466424006070466

Bertone A, Burghardt D (2017) A survey on visual analytics for the Spatio-temporal exploration of microblogging content. J Geovis Spat Anal 1(1–2):2. https://doi.org/10.1007/s41651-017-0002-6

Beukes A, Mitlin D (2014) Know your City: community profiling of informal settlements. IIED Briefing:1–4

Bolay J-C, Chenal J, Pedrazzini Y (eds) (2016) Learning from the slums for the development of emerging cities. Springer, Cham. https://doi.org/10.1007/978-3-319-31794-6

Camboim S, Bravo J, Sluter C (2015) An investigation into the completeness of, and the updates to, OpenStreetMap data in a heterogeneous area in Brazil. ISPRS Int J Geo-Inf 4(3):1366–1388. https://doi.org/10.3390/ijgi4031366

Cochrane L, Corbett J (2018) Participatory mapping. In: Servaes J (ed) Handbook of communication for development and social change. Springer Singapore, Singapore, pp 1–9. https://doi.org/10.1007/978-981-10-7035-8_6-1

Corbett JM, Keller CP (2005) An analytical framework to examine empowerment associated with participatory geographic information systems (PGIS). Cartographica: Int J Geogr Info Geovis 40(4):91–102. https://doi.org/10.3138/J590-6354-P38V-4269

Corbett J, Cochrane L, Gill M (2016) Powering up: revisiting participatory GIS and empowerment. Cartogr J 53(4):335–340. https://doi.org/10.1080/00087041.2016.1209624

Corburn J et al (2020) Slum health: arresting COVID-19 and improving Well-being in urban informal settlements. J Urban Health 97(3):348–357. https://doi.org/10.1007/s11524-020-00438-6

Elsey H et al (2016) Addressing inequities in urban health: do decision-makers have the data they need? Report from the urban health data special session at international conference on urban health Dhaka 2015. J Urban Health 93(3):526–537. https://doi.org/10.1007/s11524-016-0046-9

Elwood S (2010) Geographic information science: emerging research on the societal implications of the geospatial web. Prog Hum Geogr 34(3):349–357. https://doi.org/10.1177/0309132509340711

Ezeh A et al (2017) The history, geography, and sociology of slums and the health problems of people who live in slums. Lancet 389(10068):547–558. https://doi.org/10.1016/S0140-6736(16)31650-6

Friesen J et al (2020) Slums, space, and state of health—a link between settlement morphology and health data. Int J Environ Res 17(6):2022. https://doi.org/10.3390/ijerph17062022

Goodchild MF (2007) Citizens as sensors: the world of volunteered geography. GeoJournal 69(4):211–221. https://doi.org/10.1007/s10708-007-9111-y

Goodchild MF, Glennon JA (2010) Crowdsourcing geographic information for disaster response: a research frontier. Int J Digit Earth 3(3):231–241. https://doi.org/10.1080/17538941003759255

Google Earth (2018) São Paulo Satellite Imagery. Available via earthgooglecom Accessed 1 July 2018

Hachmann S, Jokar Arsanjani J, Vaz E (2018) Spatial data for slum upgrading: volunteered geographic information and the role of citizen science. Habitat Int 72:18–26. https://doi.org/10.1016/j.habitatint.2017.04.011

Haklay M (2013) Citizen science and volunteered geographic information: overview and typology of participation. In: Sui D, Elwood S, Goodchild M (eds) Crowdsourcing geographic knowledge: volunteered geographic information (VGI) in theory and practice. Springer, Dordrecht. https://doi.org/10.1007/978-94-007-4587-2

Harvey D (2006) Spaces of global capitalism. Verso, London

Harvey F (2013) To volunteer or to contribute locational information? Towards truth in labeling for crowdsourced geographic information. In: Sui D, Elwood S, Goodchild M (eds) Crowdsourcing geographic knowledge: volunteered geographic information (VGI) in theory and practice. Springer, Dordrecht. https://doi.org/10.1007/978-94-007-4587-2

Heider K, Rodriguez Lopez JM, Scheffran J (2018) The potential of volunteered geographic information to investigate peri-urbanisation in the conservation zone of Mexico City. Environ Monit Assess 190(4):219. https://doi.org/10.1007/s10661-018-6597-3

Hohl A et al (2020) Daily surveillance of COVID-19 using the prospective space-time scan statistic in the United States. Spat Spatio-temporal Epidemiol 34:100354. https://doi.org/10.1016/j.sste.2020.100354

International Telecommunications Union (2019) Measuring digital development facts and figures. International Telecommunications Union, Geneva

Jujnovsky J et al (2012) Assessment of water supply as an ecosystem service in a rural-urban watershed in southwestern Mexico City. Environ Manag 49(3):690–702

Kuffer M, Pfeffer K, Sliuzas R (2016) Slums from space-15 years of slum mapping using remote sensing. Remote Sens 8(6). https://doi.org/10.3390/rs8060455

Kuffer M et al (2018) The scope of Earth-observation to improve the consistency of the SDG slum indicator. ISPRS Inter J Geo-Info 7(11):428. https://doi.org/10.3390/ijgi7110428

Lines K, Makau J (2018) Taking the long view: 20 years of Muungano Wa Wanavijiji, the Kenyan federation of slum dwellers. Environ Urban 30(2):407–424. https://doi.org/10.1177/0956247818785327

Mahabir R et al (2018) A critical review of high and very high-resolution remote sensing approaches for detecting and mapping slums: trends, challenges and emerging opportunities. Urban Sci 2(1):8. https://doi.org/10.3390/urbansci2010008

Ostermann FO, Granell C (2017) Advancing science with VGI: reproducibility and replicability of recent studies using VGI. Trans GIS 21(2):224–237. https://doi.org/10.1111/tgis.12195

Patel S, Baptist C (2012) Editorial: documenting by the undocumented. Environ Urban 24(1):3–12. https://doi.org/10.1177/0956247812438364

Patel S, Baptist C, D'Cruz C (2012) Knowledge is power - informal communities assert their right to the city through SDI and community-led enumerations. Environ Urban 24(1):13–26. https://doi.org/10.1177/0956247812438366

Pedro AA, Queiroz AP (2019) Slum: comparing municipal and census basemaps. Habitat Int 83(October):30–40. https://doi.org/10.1016/j.habitatint.2018.11.001

Pedro AA, Kitamura CO, Terlizzi MMF (2017) Evaluation and standardisation of the favela basemap in Sao Paulo city. Int J Cartogr 3(2):151–165. https://doi.org/10.1080/23729333.2016.1251670

Pessoa Colombo V (2019) Uso de tecnologias geo-espaciais comuns para o mapeamento de assentamentos informais. In: Pessoa Colombo V, Bassani J, Torricelli GP, Alves de Araújo S (eds) Mapeamento Participativo: Tecnologia e Cidadania – Livro de resumos do encontro internacional Técnicas de Mapeamento e Cidadania (São Paulo, 2018), 1st edn. Editora da Faculdade de Arquitetura e Urbanismo da Universidade de São Paulo, São Paulo

Reynard D (2018) Five classes of geospatial data and the barriers to using them. Geogr Compass 12(4):1–13. https://doi.org/10.1111/gec3.12364

Ricker B, Schuurman N, Kessler F (2015) Implications of smartphone usage on privacy and spatial cognition: academic literature and public perceptions. GeoJ 80(5):637–652. https://doi.org/10.1007/s10708-014-9568-4

Robinson JA, Block D, Rees A (2017) Community geography: addressing barriers in public participation GIS. Cartogr J 54(1):5–13. https://doi.org/10.1080/00087041.2016.1244322

Rodriguez Lopez JM et al (2015) Remote sensing of sustainable rural-urban land use in Mexico City: a qualitative analysis for reliability and validity. INTERdisciplina 3(7):321–342. https://doi.org/10.22201/ceiich.24485705e.2015.7.52413

Rodriguez Lopez JM, Heider K, Scheffran J (2017a) Frontiers of urbanisation: identifying and explaining urbanisation hot spots in the south of Mexico City using human and remote sensing. Appl Geogr 79:1–10. https://doi.org/10.1016/j.apgeog.2016.12.001

Rodriguez Lopez JM, Heider K, Scheffran J (2017b) Human and remote sensing data to investigate the frontiers of urbanisation in the south of Mexico City. Data Br 11(February):5–11. https://doi.org/10.1016/j.dib.2016.12.049

Santos Melo Y, Pessoa Colombo V, Espitia Riveros IJ, Simionato Costa J (2021) Desenvolvimento do capital social comunitário em assentamentos vulneráveis: a experiência da organização Teto (Techo) na Colômbia e no Brasil. In: Alvear CAS, Cruz CC, Kleba JB (eds) Engenharias e outras práticas técnicas engajadas, vol 1. Editora da Universidade Estadual da Paraíba, Brazil, pp 219–250

Scheffer M et al (2001) Catastrophic shifts in ecosystems. Nature 413(6856):591–596. https://doi.org/10.1038/35098000

Scheffer M et al (2012) Anticipating critical transitions. Science 338(6105):344–348. https://doi.org/10.1126/science.1225244

Scholz S et al (2018) Volunteered geographic information for disaster risk reduction-the missing maps approach and its potential within the red cross and red crescent movement. Remote Sens 10(8). https://doi.org/10.3390/rs10081239

Sharma S (2019) Data privacy and GDPR handbook. Wiley, Hoboken. https://doi.org/10.1002/9781119594307

Souza LG (2012) Mapeamento de logradouros e gestão territorial em favelas no Rio de Janeiro. in 8o Congresso Internacional Cidade e Território Virtual. Rio de Janeiro: Universidade Federal do Rio de Janeiro, p. 210. https://doi.org/10.5821/ctv.7898

Sui D, Elwood S, Goodchild M (eds) (2013a) Crowdsourcing geographic knowledge: volunteered geographic information (VGI) in theory and practice. Springer, Dordrecht

Sui D, Elwood S, Goodchild M (2013b) Volunteered geographic information, the Exaflood, and the growing digital divide. In: Sui D, Elwood S, Goodchild M (eds) Crowdsourcing geographic knowledge: volunteered geographic information (VGI) in theory and practice. Springer, Dordrecht. https://doi.org/10.1007/978-94-007-4587-2

UN-DESA (2021) The least developed country category: 2021 country snapshots

Verplanke J et al (2016) A shared perspective for PGIS and VGI. Cartogr J 53(4):308–317. https://doi.org/10.1080/00087041.2016.1227552

Watson V (2009) The planned city sweeps the poor away: urban planning and 21st century urbanization. Prog Plann 72(3):151–193. https://doi.org/10.1016/j.progress.2009.06.002

WHO (2010) Urban HEART: urban health equity assessment and response tool, health promotion international. Kobe. Available via http://www.who.int/kobe_centre/publications/urban_heart.pdf. Accessed 7 Oct 2021

Williams DS et al (2019) Vulnerability of informal settlements in the context of rapid urbanisation and climate change. Environ Urban 31(1):157–176. https://doi.org/10.1177/0956247818819694

Yan Y et al (2020) Volunteered geographic information research in the first decade: a narrative review of selected journal articles in GIScience. Int J Geogr Info Sci 34(9):1765–1791. https://doi.org/10.1080/13658816.2020.1730848

Zérah M-H (2007) Conflict between green space preservation and housing needs: the case of the Sanjay Gandhi National Park in Mumbai. Cities 24(2):122–132. https://doi.org/10.1016/j.cities.2006.10.005

Zhang S (2019) Public participation in the Geoweb era: defining a typology for geo-participation in local governments. Cities 85(2019):38–50. https://doi.org/10.1016/j.cities.2018.12.004

Chapter 13
Challenges and Opportunities Interdisciplinary GIScience Research on Human-Environment Dynamics in Latin America

Santiago López and David Seward Salisbury

Abstract Geospatial technologies are increasingly used to address many of the pressing socio-environmental challenges that our planet is enduring due to their analytical power and popularity in both the private and public sectors across the globe. Latin America is not an exception to this trend. In the Latin American context, the thematic scope of geographic information systems (GIS) and remote sensing (RS) has been diverse, ranging from database management and basic cartographic production to advanced spatial modeling applied to decision and policy making, program planning and implementation, capacity building, and analysis of environmental change. In this final chapter, we highlight the contributions of each group of researchers showcased in this book employing the questions presented at the outset of the volume as main overarching prompts. We acknowledge that the sample of chapters compiled in this book falls short of being fully representative of all the GIScience research being done in the region. Nevertheless, the volume summarizes major themes and approaches to socio - environmental (SE) research that centers on the use of GIS and RS frameworks spanning a multitude of knowledge domains in interdisciplinary settings. We conclude with a brief reflection on the challenges and opportunities of using RS and GIS in interdisciplinary SE research across the vast, culturally, and environmentally diverse region known as Latin America.

S. López (✉)
School of Interdisciplinary Arts and Sciences, University of Washington Bothell, Bothell, WA, USA
e-mail: cslopez@uw.edu

D. S. Salisbury
Department of Geography, Environment & Sustainability, University of Richmond, Richmond, VA, USA
e-mail: dsalisbu@richmond.edu

S. López (ed.), *Socio-Environmental Research in Latin America*, The Latin American Studies Book Series, https://doi.org/10.1007/978-3-031-22680-9_13

Keywords GIScience · Human-environment interactions · Challenges · Opportunities · Interdisciplinary · Latin America

1 Introduction

Our world is increasingly being challenged by the rapid intensification of human activities and processes led by networks and actors (individual and corporate) that result in the permanent loss of ecosystems, environmental degradation, and severe changes to the Earth system. These changes, particularly those to the climate system, pose significant threats to the future of biological systems as we know them. Geospatial technologies are increasingly used to address many of these pressing challenges due to their analytical power and popularity in both the private and public sectors across the globe. Latin America is not an exception to this trend. Interdisciplinary approaches embedded in geographic information science (GIScience) geared toward the advancement of socio-environmental (SE) knowledge and environmental sustainability have expanded enormously in the region since the introduction of geographic information systems (GIS), remote sensing (RS), and the global positioning system (GPS) several decades ago (Buzai and Robinson 2010). In particular, the last 30 years have witnessed a continuous evolution of geospatial technologies and the frameworks that encompass them, from processes of technology transfer to general and specialized research applications of GIScience. The thematic scope of GIS and RS has been diverse, ranging from database management and basic cartographic production to advanced spatial modeling applied to decision and policy making, program planning and implementation, capacity building, and analysis of environmental change.

Geography and GIScience are well positioned to carry out interdisciplinary research that aims at advancing our understanding of SE dynamics (Bracken 2017) in complex regions, where people cope with conditions of relatively high political, social, and financial instability. The advantage of Geography and GIScience may in part be because these disciplines are themselves interdisciplinary in nature and can adapt to specific SE contexts, in tandem with their different epistemological and methodological approaches. This characteristic of Geography and GIScience allows a unique dialog between a diversity of actors, disciplines, theoretical frameworks, and applications. Geographers and other social and physical scientists, for example, have used GIS and cartography to map the physical, ecological, and cultural conditions of a variety of SE systems for many years, but the last 30 years saw a surge of interest in how social, political, economic, and physical forces are negotiated in and through GIScience (Elwood 2011). The diversity within the discipline of geography should be considered a strength itself, as Sheppard and Plummer (2007) emphasize that this acts as a foundation of intellectual interactions that advances both general and specialized SE knowledge. Moreover, the emphasis of geography and GIScience on the spatial dimension makes these disciplines, and their data gathering and analytical approaches, needed in all interdisciplinary research that pursues a more nuanced understanding of place, connectivity, and socio-spatial dynamics. The variety of applications showcased in this volume constitute a sample of the wide range of themes that geospatial technologies can span as well as a selection of the ways these technologies are currently implemented by researchers doing SE work in Latin America.

In this final chapter, we highlight the contributions of each group of researchers who collaborated in this book using the questions presented at the outset of the volume as main overarching prompts. We acknowledge that the sample of chapters compiled in this book falls short of being fully representative of all the GIScientific research being done in the region. Nevertheless, the volume summarizes major themes and approaches to SE research that centers on the use of GIS and RS frameworks spanning a multitude of knowledge domains. We conclude with a brief reflection on the challenges and opportunities of using RS and GIS in interdisciplinary SE environmental research across the vast, culturally and environmentally diverse region known as Latin America.

2 Overarching Questions

2.1 How Are Interdisciplinary Approaches Using RS and GIS Uniquely Facilitating a Dialog Between the Geophysical Sciences, Natural Sciences, Social Sciences, and the Humanities as Conceived by Researchers Doing Work in Latin America?

This volume showcases the application of GIS and RS approaches to advance our understanding of human-environmental relationships within specific SE Latin American contexts. The first four chapters embrace the use of geospatial technologies in combination with other non-spatial approaches to pursue unique understandings of complex and contested tropical lowland landscapes such as those found in the Amazon region. Latin American lowlands are both culturally and ecologically mega diverse and are being rapidly and continuously reshaped by SE interactions at multiple spatio-temporal scales. At the same time, the neotropics are contentious and contested landscapes where nation states, corporate interests, conservation initiatives, and illicit economies resulting from illegal activities such as bribing, tax evasion, corporate profit shifting, money laundering, etc. overlap with legitimate claims for self-determination, sovereignty, and territorial formation led by Indigenous populations or long-term residents, and recent migrants.

Aldrich's work in Southeastern Pará, Brazil, (Chap. 1), integrates the information derived from time-series RS data, historical research, and qualitative interviews to unravel the social-process drivers of deforestation in an area historically affected by social conflict and political unrest. Through an innovative mixed methods approach, the author sheds new light on land change driven by the actions of large-holders, organized landless movement members, and government officials, who take advantage of a chaotic governance structure to acquire land and the rights to clear forests for personal or corporate benefit. As Aldrich points out, a single method or disciplinary approach alone would fail to unravel this complexity nor identify land conflict as a critically important driver of deforestation, an underlying factor that has been often neglected in the land use land cover change literature. The work by Salisbury et al. (Chap. 3) describes recent GIS mapping efforts to assist border-land communities in the Peruvian and Brazilian Amazon with their mapping needs and fill cartographic information gaps that could contribute to better understanding of SE issues. Through the efforts of the Transboundary Geographic Group of the Southwestern Amazon's (GTASO in Spanish), a multitude of actors of diverse back-grounds and disciplines were brought together to generate multi-scalar information with the goal of improving our understanding of the ecological, economic, cultural, and political dynamics of a historically neglected but politically contested, culturally diverse, and ecologically rich transboundary region. In this chapter, GIS and RS served as catalysts of such spatial data gathering efforts, facilitating a meaningful dialog among communities, government officials, and environmental organizations for the purpose of better understanding and managing SE interactions in these borderlands. Similarly, the chapter by Sierra et al. (Chap. 4) relies on GIScience and GPS field methods to explain land use allocation decisions in the Waorani ancestral territory in the Northern Ecuadorian Amazon. Through the application of a spatial model of the demand for productive land, the authors assess the value of small-scale commercial cultivation in the conservation of forest resources at the community level. This chapter demonstrates how geotechnologies can facilitate a unique understanding of the socio-spatial dimensions of Indigenous land use systems by providing accurate geo-located data on land use allocation preferences mediated by socio-economic needs and development opportunities. This new understanding allows the bridging of usually disparate goals between conservation and development, providing an opportunity to re-think conservation strategies outside of the realm of the protected areas paradigm. The work by López and Maldonado (Chap. 5) addresses the spatio-environmental factors associated with land use and land cover transformations while assessing water quality changes in the Northeastern Ecuadorian Amazon. The chapter follows a land change science approach at a regional level, combining RS data and multinomial regression analysis within a GIS framework. The study also borrows elements of watershed science to elucidate patterns and processes of environmental degradation caused by extractive industries. While GIS, RS, and GPS are central to the data collection efforts, the progressive contextualization (Vayda 1983) of environmental degradation facilitated not only by the use of geotechnologies but a nuanced reading of landscape transformations, offers a unique perspective on the factors that are detrimental to the environment and social life in that part of the Amazon.

It is clear, based on these first chapters, that with over 30 years of GIScientific work in the region, the understanding of the factors involved in the reshaping of lowland Latin American regions has improved significantly, overcoming representational issues and misconceptions of Latin American lowland regions (e.g., the pristine myth) and their long-term residents (e.g., ecologically noble savage views). GIScience has become an established research field, and to some extent, the technological *lingua franca* in interdisciplinary projects that pursue a better understanding of complex lowland SE systems. Several GIS working groups in this region (e.g., GTASO) are not only filling data gaps but promoting a conversation between the critical social sciences (including human geography), the more technical branches of GIScience, and other disciplines. GIScience is also helping to order an understanding of the physical and cultural landscape that state agencies and civil society seek to manage even as illegal and illicit actors often benefit from the norm: limited data, inadequate maps, and overlapping claims. Concurrently, GIScience also provides a discursive landscape and unifying language which allows different Amazonian actors (Indigenous, state, NGOs, etc.) to have productive dialogues concerning convulsive SE dynamics. Geotechnologies will likely continue to be used in the ways described in these chapters to not only to map landscape features or measure the structural and compositional traits of tropical forest ecosystems but to analyze critical SE processes that contribute to their transformations and permanence.

The next four chapters in this volume illustrate the application of GIS and RS in interdisciplinary settings to address issues of resilience and adaptation, air quality and public health, environmental degradation, conservation, tourism, and development in the Andean region. Like the tropical lowlands, the Andean region is characterized by its cultural and ecological diversity, shaped by millennial SE interactions, with phases of stasis and accelerated change (Young 2008). This work recognizes the complexity of past and present conditions affecting nature and society relationships with implications on the sustainability of SE systems in Latin American highlands at various spatio-temporal scales.

The chapter by Postigo et al. (Chap. 6) uses GIS and RS to propose a multi-level and cross-scale framework to analyze human-environment interactions in Andean landscapes. This chapter emphasizes the use of a spatial perspective to identify the SE conditions and complex interactions linking critical elements of the SE system (e.g., physiognomic units, peasant communities, abiotic conditions). By doing so, the authors bring forward the relevance of GIScience as an interdisciplinary framework connecting issues of conservation, adaptation, and governance of Andean landscapes. From a methodological point of view, GIS and RS approaches constitute key elements of the authors' efforts to bridge the gap between the conceptual work generated by diverse groups of researchers and the empirical work of practitioners who aim to change the social and ecological conditions of Andean locales. The work by Alvarez (Chap. 7) provides a unique perspective on the use of RS to address public health and air quality issues in Latin American cities, particularly in the Andean region. As Alvarez points out, with the continuous development of RS technologies and products, new opportunities emerge to generate sophisticated characterizations of air quality and other indicators of atmospheric health at different spatio-temporal scales. New efforts are taking place that respond to data gaps and citizen needs of environmental monitoring, allowing a deeper understanding of the connections between the environmental contexts of urban dwellers and public health outcomes. Throughout this chapter, Alvarez offers a critical perspective on the use of public platforms such as Google Earth Engine and its catalog of satellite imagery, to shed light on how researchers in Latin America are tackling environmental justice and contamination in urban areas historically affected by poor air quality. The work by Salazar et al. (Chap. 8) in the Chilean Patagonia along the Exploradores Valley is a current example of an interdisciplinary, cross-scalar, and multi-temporal assessment of environmental change in the southern Andes. Through the employment of RS and GIS methods, the researchers developed a multi-temporal characterization of land use and land cover changes in an area that has received little attention in the geographic academic literature. However, as the authors acknowledge, geotechnologies alone are not sufficient to fully understand the whole range of landscape transformations and SE dynamics in the region. By employing an interdisciplinary lens combining historical environmental analysis, field methods, and geotechnologies, this chapter offers an exceptional application of interdisciplinary GIScientific research in an ecologically diverse region affected by recent immigration processes, extensive land use practices, subsistence pressures, and financial needs of recent migrants. Finally, the work by Gerique-Zipfel and Vanselow (Chap. 9) in the Argentinean Patagonia uses a mixed methods approach to provide a contextualized understanding of forest cover change and human occupation patterns in El Chaltén, a tourism-based town inside Los Glaciares National Park. This chapter analyzes the issues required to reconcile conservation and development, particularly those that emerge from both a growing tourism industry and increasing population pressure. While qualitative interviews with key actors provided data to contextualize environmental changes inside the national park, the unique regional perspective provided by RS data in tandem with field observations offered a synoptic view of forest cover change and urban growth in the area. The authors warn about *naturbanization* (i.e., transformation of natural protected areas into urban centers) (Prados Velasco 2009), a relatively new phenomenon yet to be extensively explored in the Latin American context. As the growth of the world's population continues to surge, the search for residential space and the urbanization of natural lands will persist even in relatively remote areas of Latin America.

This next set of chapter shows that with the shift of socio-economic and biogeographical conditions resulting from contemporary environmental change including land use and climate transformations, interdisciplinary GIScientific approaches are helping to not only evaluate the social and biophysical pressures affecting Andean ecosystems, but also the adaptation and mitigation responses of past and contemporary Andean societies. Through the mapping and analysis of social, political, and environmental change in space and time, researchers are gaining new insights on the effects of land cover and climate changes on human-environment dynamics in mountain environments increasingly threatened by human pressure. Moreover, GIScience is contributing to the integration of the conceptual work generated by multi-disciplinary teams of scholars and the pragmatic work of specialists who aim to preserve or restore the socio-ecological conditions of high elevation localities. GIScientific frameworks are helping to assess the strategies of survival and resilience developed by individuals and communities to deal with SE pressures. Finally, GIScience's ability to ground theory, and place activism facilitates impactful conversations in vulnerable landscapes.

The last three chapters showcase the creative use of geotechnologies to address issues of environmental justice linked to the actions (and inactions) of the gas and oil industry, community mapping and archaeological research, and community engagement in participatory land use planning throughout different geographic regions in Latin America. The work by Fry and Hilburn (Chap. 10) demonstrates the novel application of GIS in environmental justice research applied to the oil and gas industry in Mexico and highlights the challenges and opportunities of this type of work in Latin America in general. Through a mixed methods framework combining geospatial and empirical environmental justice analysis with qualitative approaches, the authors demonstrate the benefits of this type of innovative integration to not only strengthen environmental justice outcomes but also legitimize political claims and increase community participation. By promoting this type of synergy, Fry and Hilburn underscore how GIScience can facilitate an interdisciplinary dialog between governmental and non-governmental environmental sectors and corporate actors with the goal of bringing attention to the environmental concerns of the people and communities most impacted and marginalized by extractive industrial activities. Similarly, the chapter by Alvarez-Larrain and Nesbitt (Chap. 11), offers a nuanced discussion of the integration of GIScience and participatory approaches in engaged archaeological research with the goal of improving our understanding of long-term nature-society relationships. The authors show how a participatory GIS (PGIS) framework and elements of critical geography can close some of the conceptual divides that separate the intellectual goals of archaeologists and anthropologists from the needs of Indigenous and local communities, contributing in this way to the decolonization of archaeological research practices. At the same time, engaged archaeological practice has helped connect social, historical, and ecological perspectives, providing new insights into SE dynamics in contested regions of Latin America. Along similar lines, the work by Pereira et al. (Chap. 12) emphasizes the use of volunteered geographic information (VGI) to increase local capacity and participation at both the individual and community levels. These efforts allow the creation of meaningful geographic information that could eventually empower marginalized citizens who inhabit informal settlements. By drawing from two case studies in São Paulo and Mexico City, the authors promote VGI as a tool for generating precise contextualized geospatial data while empowering individuals and creating community agency through a participatory process.

The development of PGIS and VGI approaches have created new opportunities to capture the varying views and voices of local actors regarding environmental justice, ancestral heritage and territorial sovereignty, and land use planning concerns in Latin America. By drawing insights from critical GIS, anthropology, archaeology, human geography, and other social sciences, the work highlighted in these three final chapters interrogates the politics of geographic information and knowledge generation, while exploring interdisciplinary routes for data collection, the characterization of sites/places and historical episodes, and the assemblage of geotechnical knowledge. In this way, a call is made to legitimize GIScience research agendas by moving beyond the goal of knowledge generation and by bringing forward local community needs. GIScience continues to engage with diverse research practices both remote and participatory, while expanding and diversifying or hybridizing its epistemological frameworks to respond to legitimate concerns about data generation, ownership, and access in contested landscapes of continuously shifting power relations. Thus, these three chapters, and their predecessors, elucidate the power of GIScience to bridge disciplines, methods, organizational types, and worldviews (activist, interpretivist, post-structuralist, etc.) to contribute to important conversations about sustainability, science, and environmental justice across urban and rural Latin America.

2.2 How Can Multiple Narratives Be Introduced to Change the Ways in Which RS and GIS Are Practiced in This Region?

By using GIS and RS in the ways showcased in this book, we have been able to obtain deeper insights on SE dynamics mediated through human land use and agency, in addition to a better understanding of the unique role that GIS-based spatial knowledge production plays in negotiating complex political practices that shape SE relationships. We have been provided with examples of how GIS and RS interdisciplinary approaches allow planning for geolocated interventions that may change the ways in which we interact with the Earth system, including its human elements. While several chapters in this volume rely on a land chance science approach to provide a systematic study of the changing landscapes (e.g., Aldrich, López and Maldonado, Postigo et al., Salazar et al., Gerique-Zipfel and Vanselow), none of them rely solely on this one approach to unravel the complexities of SE change. Other frameworks such as political ecology, watershed science, statistical modeling, and qualitative research methodologies are introduced to complement the views of GIScience. Through this process, different perspectives, both bottom-up and top-down, are introduced to adapt RS and GIS research practices to address concerns and questions embedded in particular SE contexts. The work by Salisbury et al., Postigo et al., Fry and Hilburn, Alvarez-Larrain and Nesbitt, and Pereira et al. describe how different sets of actors are now using GIS and RS to produce detailed narratives that respond to community priorities and needs, creating agency and authority in the co-production of geographic information and knowledge. These locally grounded spatial narratives, together with those derived from frameworks rooted in theoretical research, underscore a productive hybridity in community-based GIS work. The fusion of these type of narratives also emphasizes the epistemological flexibility and fluidity of GIScience (Elwood 2006), which has also helped introduce critical social science perspectives including participatory approaches that have contributed, to some extent, to the democratization and decolonization of GIScience.

3 Challenges in the Field of GIScience in Latin America

While researchers have been able to find creative ways of using GIScientific frameworks to bridge the gaps between different knowledge systems and epistemologies or suggest hybrid versions to advance our understanding of SE dynamics (e.g., López et al. 2017), some challenges remain latent. First, while many geotechnologies are now open source and freely available, commercial authoritative GIS dominates the field and profitability from software sales is the main factor influencing GIS development globally (Goodchild 2015). Costly licenses present a real barrier for many Latin American countries, regions, and academic institutions in the Global South affected by systemic poverty and lack of access to cutting edge technologies, which relegates the use of specialized GIS to particular elites. Although patch solutions have been proposed to deal with issues of restrictions to technology transfer or creative workarounds have been implemented to bypass licensing requirements, many GIS applications and tools remain reserved for governmental use, externally funded academic research, or the commercial sector. This means that many of the barriers to geographic information generation that were present in the early 1990s when GIS and satellite RS were being introduced in Latin America, are still present today. Some conceptual frameworks have been proposed to deal with this aspect of GIS use (e.g., neogeography – a collection of practices, tools, and users generally found outside traditional authoritative GIS) (Gyrne and Pickard 2015); however, further research is needed to investigate how these new frameworks extend the democratization of GIS and RS and if increased access to geotechnologies translates into empowerment or, conversely, increased relegation.

Second, within the Latin American context, the divide between those who have access to geospatial technologies and those who do not is still present although the situation has changed noticeably since the introduction of mobile technologies a couple of decades ago. The now ubiquitous presence of mobile technologies and the continued push for their adoption together with Cloud-based services, appear to be shifting the scale to a more decentralized control of geographic information. Geotechnologies, for example, are now being adopted by lay citizens with smart phones to perform sophisticated forms of analysis like measuring distances accurately, designing exploration routes, or finding the shortest or most (fuel) efficient path between two points for delivering products and services. Virtual 3D global exploration and mapping with user-friendly GIS technologies like Google Earth or Google Maps are radically changing how we locate, navigate, and plan our journeys at local and intercontinental levels. Nevertheless, there is still a large proportion of Latin America's population (>40%) who do not have access to mobile or cloud-based systems at all (Lustig 2018) and even a smaller proportion who would not have the option to engage with the latest technologies since the rate of technological advance surpasses people's financial capacity to acquire them. Although the proportion of Latin American population on the deprived side of the digital divide may be declining, there is no indication that the inequity will fade away soon.

Third, as Cloud systems are developed, and artificial intelligence (AI) and machine learning algorithms start to occupy a main role in people's daily lives across the globe, including Latin America, GIS and RS data and software are no longer personal but collective instruments that require careful handling and consideration because of the implications for the larger society. While many users may think of the Cloud-based GIS services as anonymous or even private, in reality, they are controlled by corporations such as Google, Amazon, ESRI, Meta, Microsoft, etc. (Goodchild 2015), which are all headquartered in the Global North and have no knowledge of the SE contexts where those technologies are being used, the individuals and communities that are employing them, or the agendas behind their use. Although Cloud services and GPS-enabled mobile technologies appear to decentralize the control of geographic information, at the same time they are also shifting the scale toward increased corporate power benefiting developed nations and their geopolitical agendas. Although some advances have been made to ensure the privacy of geographic information and personal data in general, there are significant unknowns about who can access, who can use, and who owns digital geographic information when resources are stored on the Cloud, and when Cloud-based social networks are employed to share or distribute information. Therefore, concerns of surveillance and cyber-security, privacy of tasks performed by individuals and communities, neo-colonization agendas based on the application of geotechnologies and digital data, and naive adoption of analytical tools associated with the concept of Big Data are important challenges that should be addressed. The risk of using (or misusing) these services, geotechnologies, and Big Data analytics to identify patterns of SE interactions to guide or promote social and cultural behaviors or depict the movements of millions of individuals cannot be overstated. Unfortunately, Latin America has a dark history of human rights abuse embedded in a culture of secrecy, unaccountability, and restrictions to information access. A recent article by Pisanau et al. (2021) illuminates the growth of a shadowy mass surveillance industry in Latin American countries like Brazil, Argentina, and Ecuador in the past couple of decades, and exposes the government policies and practices that are undermining people's rights in urban/rural highland and lowland settings. Given the central focus of location in geographic information, there is clearly a need for GIScientists to critically revisit the role of GIS and RS in this area.

In summary, the chapters in this book encapsulated this book's multiple narratives on some of the challenges and opportunities of interdisciplinary GIScience research in an increasingly contested and contentious Latin America embedded in a global environmental change crisis. The scholarly work presented here recognizes that environmental change is a hybrid phenomenon: a mixture of human-induced and environmentally determined conditions that evolve at multiple spatiotemporal scales. With examples covering an array of applications of geotechnologies rooted in diverse scientific epistemologies, this book allows a better understanding of spatial heterogeneity and SE dynamics across Latin America, contributing to a critically important research field within human-environment geography and the sustainability of the Americas.

References

Bracken LJ (2017) Interdisciplinarity and geography. In: Richardson D, Castree N, Goodchild MF, Kobayashi A, Liu W, Marston RA (eds) The international encyclopedia of geography. Wiley, Chichester, pp 3746–3755

Buzai GD, Robinson DJ (2010) Geographical information systems (GIS) in Latin America, 1987–2010: a preliminary overview. J Lat Am Geogr 9(3):9–31

Elwood S (2006) Beyond cooptation or resistance: urban spatial politics, community organizations, and GIS-based spatial narratives. Ann Am Assoc Geogr 96(2):323–341

Elwood S (2011) Participatory approaches in GIS and society research: foundations, practices, and future directions. In: Nyerges TL, Couclelis H, McMaster R (eds) The SAGE handbook of GIS and society. Sage, London, pp 381–383

Goodchild MF (2015) Two decades on: critical GIScience since 1993. Can Geogr 59(1):3–11

Gyrne D, Pickard AJ (2015) Neogeography and the democratization of GIS: a meta synthesis of qualitative research. Inf Commun Soc 19(11):1505–1522

López S, Jung JK, López-Sandoval M (2017) A hybrid-epistemological approach to climate chan research: linking scientific and smallholder knowledge systems in the Ecuadorian Andes. Anthropocene 17:30–45

Lustig N (2018) Latin America's growing mobile market. https://www.nathanlustig.com/latin-americas-growing-mobile-market/#:~:text=Latin%20America%20is%20the%20world's,200%20million)%20are%20smartphone%20users. Accessed 8 July 2022

Pisanau G, Arroyo V, Ucciferri L, et al. (2021) Made abroad deployed at home. https://www.accessnow.org/latam-surveillance-report/. Accessed 3 July 2022

Prados Velasco (ed) (2009) Naturbanization: new identities and processes for rural-natural areas. Taylor and Francis, London

Sheppard E, Plummer P (2007) Commentary: toward engaged pluralism in geographical debate. Environ Plan A 39:2545–2548

Vayda A (1983) Progressive contextualization: methods and research in human ecology. Hum Ecol 11(3):265–281

Young K (2008) Stasis and flux in long-inhabited locales: change in rural Andean landscapes. In: Jepson W, Millington A (eds) Land change science in the tropics. Springer, Boston, pp 11–32. https://doi.org/10.1007/978-0-387-78864-7_2

Index

S. López (ed.), *Socio-Environmental Research in Latin America*, The Latin American Studies Book Series, https://doi.org/10.1007/978-3-031-22680-9